VOLUME ONE

ADVANCES IN

NEUROTOXICOLOGY

Environmental Factors in
Neurodegenerative Diseases

ADVANCES IN
NEUROTOXICOLOGY
Environmental Factors in
Neurodegenerative Diseases

Edited by

MICHAEL ASCHNER

*Albert Einstein College of Medicine, Bronx, NY,
United States*

LUCIO G. COSTA

*University of Washington, Seattle, WA, United States;
University of Parma, Parma PR, Italy*

ACADEMIC PRESS

An imprint of Elsevier

Academic Press is an imprint of Elsevier
50 Hampshire Street, 5th Floor, Cambridge, MA 02139, United States
525 B Street, Suite 1800, San Diego, CA 92101-4495, United States
The Boulevard, Langford Lane, Kidlington, Oxford OX5 1GB, United Kingdom
125 London Wall, London, EC2Y 5AS, United Kingdom

First edition 2017

ISBN: 978-0-12-812764-3
ISSN: 2468-7480

For information on all Academic Press publications
visit our website at https://www.elsevier.com/books-and-journals

Working together
to grow libraries in
developing countries

www.elsevier.com • www.bookaid.org

Publisher: Zoe Kruze
Acquisition Editor: Kirsten Shankland
Editorial Project Manager: Alina Cleju
Production Project Manager: Vignesh Tamil
Cover Designer: Alan Studholme

Typeset by SPi Global, India

CONTENTS

7. Roles of Microglia in Inflammation-Mediated Neurodegeneration: Models, Mechanisms, and Therapeutic Interventions for Parkinson's Disease   **185**

Hui-Ming Gao, Dezhen Tu, Yun Gao, Qiyao Liu, Ru Yang, Yue Liu, Tian Guan, and Jau-Shyong Hong

8. Mitochondrial Dynamics in Neurodegenerative Diseases   **211**

Jennifer Pinnell and Kim Tieu

9. Food Plant Chemicals Linked With Neurological and Neurodegenerative Disease   **247**

Peter S. Spencer and Valerie S. Palmer

CONTRIBUTORS

Michael Aschner
Albert Einstein College of Medicine, Bronx, NY, United States

Stephen C. Bondy
Environmental Toxicology Program, Center for Occupational and Environmental Health, University of California, Irvine, CA, United States

Arezoo Campbell
Western University of Health Sciences, Pomona, CA, United States

Lucio G. Costa
University of Washington, Seattle, WA, United States; University of Parma, Parma PR, Italy

João B.T. da Rocha
Centro de Ciências Naturais e Exatas, Universidade Federal de Santa Maria, Santa Maria, RS, Brazil

Donato A. Di Monte
German Center for Neurodegenerative Diseases (DZNE), Bonn, Germany

David C. Dorman
College of Veterinary Medicine, North Carolina State University, Raleigh, NC, United States

Marcelo Farina
Centro de Ciências Biológicas, Universidade Federal de Santa Catarina, Florianópolis, SC, Brazil

Hui-Ming Gao
Model Animal Research Center and MOE Key Laboratory of Model Animal for Disease Study, Nanjing University, Nanjing, Jiangsu, China; Neurobiology Laboratory, National Institute of Environmental Health Sciences/National Institutes of Health, Research Triangle Park, NC, United States

Yun Gao
Model Animal Research Center and MOE Key Laboratory of Model Animal for Disease Study, Nanjing University, Nanjing, Jiangsu, China; Neurobiology Laboratory, National Institute of Environmental Health Sciences/National Institutes of Health, Research Triangle Park, NC, United States

Samuel M. Goldman
University of California; San Francisco Veterans Affairs Health Care System, San Francisco, CA, United States

Tian Guan
Model Animal Research Center and MOE Key Laboratory of Model Animal for Disease Study, Nanjing University, Nanjing, Jiangsu, China

Jau-Shyong Hong
Neurobiology Laboratory, National Institute of Environmental Health Sciences/National Institutes of Health, Research Triangle Park, NC, United States

Sarah A. Jewell
German Center for Neurodegenerative Diseases (DZNE), Bonn, Germany

Qiyao Liu
Model Animal Research Center and MOE Key Laboratory of Model Animal for Disease Study, Nanjing University, Nanjing, Jiangsu, China

Yue Liu
Model Animal Research Center and MOE Key Laboratory of Model Animal for Disease Study, Nanjing University, Nanjing, Jiangsu, China

Ruth E. Musgrove
German Center for Neurodegenerative Diseases (DZNE), Bonn, Germany

Valerie S. Palmer
Oregon Health & Science University, Portland, OR, United States

Nancy L. Parmalee
Albert Einstein College of Medicine, Bronx, NY, United States

Jennifer Pinnell
Florida International University, Miami, FL, United States

Peter S. Spencer
Oregon Health & Science University, Portland, OR, United States

Kim Tieu
Florida International University, Miami, FL, United States

Dezhen Tu
Model Animal Research Center and MOE Key Laboratory of Model Animal for Disease Study, Nanjing University, Nanjing, Jiangsu, China; Neurobiology Laboratory, National Institute of Environmental Health Sciences/National Institutes of Health, Research Triangle Park, NC, United States

Ru Yang
Model Animal Research Center and MOE Key Laboratory of Model Animal for Disease Study, Nanjing University, Nanjing, Jiangsu, China

PREFACE

This is the first volume of *Advances in Neurotoxicology*, a new series by Elsevier entirely devoted to the publication of state-of-the-art review in all areas of neurotoxicology and authored by authorities in the field. The series, for which we expect one to two volumes per year, will have a flexible format, with most issues having a specific thematic emphasis. Drs. Aschner and Costa will serve as Series Editors and will also edit individual volumes such as this first one; however, it is expected that other colleagues will serve as Guest Editors of some volume devoted to specific neurotoxicology issues. This first volume of *Advances of Neurotoxicology* is devoted to discussions on Environmental Factors in Neurodegenerative Diseases. The various chapters, all by well-established neurotoxicologists, discuss the role of selected environmental agents (air pollution, pesticides, metals, dietary components) in the etiology of the most significant human neurodegenerative disorders and their potential underlying mechanisms. Additional chapters focus on known targets of neurotoxicants (e.g., microglia, mitochondria) that may underlie the observed associations with neurodegenerations.

We believe that this series fills a gap in the growing field of neurotoxicology. The volumes will be of interest to academic researchers, research scientists, and graduate students in universities and industry. The Series Editors will be happy to entertain proposals for specific topics to be the subject of future volumes, and for guest editing.

Michael Aschner
Lucio G. Costa

CHAPTER ONE

Traffic-Related Air Pollution and Neurodegenerative Diseases: Epidemiological and Experimental Evidence, and Potential Underlying Mechanisms

Lucio G. Costa[1]

University of Washington, Seattle, WA, United States
University of Parma, Parma PR, Italy
[1]Corresponding author: e-mail address: lgcosta@uw.edu

Contents

1. INTRODUCTION

Air pollution is a mixture of several components, including gases, organic compounds, metals, and ambient particulate matter (PM) (Table 1); the latter is believed to be the most widespread threat and has been heavily implicated in disease (Costa et al., 2014; Møller et al., 2010). PM is usually characterized by aerodynamic diameter: for example, PM_{10} is comprised of particles $<10\ \mu m$ in diameter, while $PM_{2.5}$ (fine PM) represents particles $<2.5\ \mu m$ in diameter. Also of relevance is ultrafine PM (UFPM, with diameter $<100\ nM$), which may easily reach the general circulation and distribute to various organs including the brain (Genc et al., 2012; Oberdoerster and Utell, 2002). The populations of many countries, particularly in South and East Asia, are often exposed to relatively high levels of PM ($\geq 100\ \mu g/m^3$) (Costa et al., 2017a; Gautam et al., 2016; van Donkelaar et al., 2015) (Table 2). Traffic-related air pollution is a major contributor to global air pollution and the major contributor to urban ambient $PM_{2.5}$ (Karagulian et al., 2015), and diesel exhaust (DE) is its most important component (Ghio et al., 2012). DE contains more than 40 toxic air pollutants and is a major contributor to ambient PM, particularly of $PM_{2.5}$ and UFPM (USEPA, 2002). DE exposure is indeed often utilized as a measure of traffic-related air pollution. As diesel engines provide power to a wide range of vehicles, heavy equipment, and other machinery utilized in numerous industries, millions of workers are exposed to DE; these occupational exposures to DE–PM can also be quite high, often exceeding $200–300\ \mu g/m^3$ in

Table 1 Major Components of Air Pollution

Ozone (O_3)
Carbon monoxide (CO)
Nitrogen dioxide (NO_2)[a]
Particulate matter (PM): PM_{10}, $PM_{2.5}$, UFPM
Sulfur dioxide (SO_2)
Metals (lead, manganese, nickel)
Carbon dioxide (CO_2)
Toxic air pollutants (e.g., benzene, formaldehyde)

[a]Also other nitrogen oxides (NO_x); *UFPM*, ultrafine PM.

Table 2 Air Pollution Around the World
Annual PM$_{2.5}$ Average

City	PM$_{2.5}$ Level	Year
Zabol (Iran)	217	2012
Allahabad (India)	170	2012
Riyadh (Saudi Arabia)	155	2014
Raipur (India)	144	2012
Xingtai (China)	128	2014
Baoding (China)	126	2014
Delhi (India)	122	2012
Beijing (China)	85	2013
London (United Kingdom)	15	2013

PM$_{2.5}$ Levels on March 7–8, 2017

City	PM$_{2.5}$ Range
Nehru Nagar (India)	72–999
Căotān (China)	93–447
Iğdir (Turkey)	33–880
San Salvador (S. Salvador)	168–199
Věřňovice (Czech Republic)	5–277
Lecce (Italy)	8–62
Stockholm (Sweden)	19–72
Ashton, SC (United States)	8–141
Portola, CA (United States)	9–169

Sources: Real time air quality index visual map (aqicn.org/map/world) and The Telegraph (March 6, 2017).

bus garage, construction, and dock workers, with miners experiencing the highest exposures (up to 1000 $\mu g/m^3$) (Pronk et al., 2009).

The association between air pollution, particularly PM, and morbidity and mortality caused by respiratory and cardiovascular diseases is well established (Gill et al., 2011; Manzetti and Andersen, 2016; Pope et al., 2004; Rücker et al., 2011). Oxidative stress and inflammation are the two cardinal processes by which air pollution is believed to exert its peripheral toxicity. PM has been shown to affect lung, cardiovascular, and nervous system functions by mechanisms that involve oxidative stress, and oxidative damage has been shown to be a primary mechanism of PM toxicity. For example, alterations in expression of some oxidative stress-related genes and other markers of oxidative stress have been shown in rodents following DE exposures (Anderson et al., 2012; Lodovici and Bigagli, 2011; Manzetti and Andersen, 2016; Steiner et al., 2016; Weldy et al., 2012; Yin et al., 2013). In recent years, evidence has been slowly accumulating, suggesting that air pollution may negatively affect the central nervous system (CNS) and contribute to CNS diseases, possibly by similar oxidative and inflammatory mechanisms, as increased oxidative stress and neuroinflammation in the CNS are observed as a result of exposure to air pollution (Block and Calderon-Garciduenas, 2009; Genc et al., 2012).

2. THE CNS AS A TARGET FOR AIR POLLUTION TOXICITY

2.1 Neurotoxicity in Adults: Epidemiological and Animal Studies

Almost 50 years ago, a controlled experiment in 16 volunteers provided evidence that air pollution may affect the CNS. Compared to subjects breathing clean air, those who breathed air pumped in from the roadside in London at the level of a typical car intake displayed significant deficits in auditory vigilance, addition, digit copying, and sentence comprehension (Lewis et al., 1970). However, only in recent years has evidence been accumulating from human epidemiological and animal studies, suggesting that air pollution may negatively affect the CNS and contribute to CNS diseases (Block and Calderon-Garciduenas, 2009; Block et al., 2012; Genc et al., 2012; Xu et al., 2016). $PM_{2.5}$ and UFPM are of much concern in this regard, as these particles can enter the circulation and distribute to various organs, including the brain (Genc et al., 2012; Oberdoerster and Utell, 2002; Oberdoerster et al., 2004), in addition to gaining direct access to the brain through the nasal olfactory mucosa (Garcia et al., 2015; Lucchini et al., 2012; Oberdoerster et al., 2004; Peters et al., 2006). Human epidemiological studies have shown that exposure to elevated air pollution is associated, for example, with decreased cognitive function, olfactory dysfunction, auditory deficits, depressive symptoms, and other adverse neuropsychological effects (Costa et al., 2014, 2017a; Guxens and Sunyer, 2012; Fonken et al., 2011; Freire et al., 2010; Pun et al., 2017). In highly exposed individuals, postmortem investigations have revealed increased markers of oxidative stress and neuroinflammation (Calderon-Garciduenas et al., 2012), while a controlled acute exposure to DE has been shown to induce EEG changes (Crüts et al., 2008).

Animal studies corroborate the human observations (Costa et al., 2014). For example, dogs exposed to heavy air pollution presented evidence of chronic inflammation and neurodegeneration in various brain regions (Calderon-Garciduenas et al., 2003), and mice exposed to traffic in a highway tunnel had higher levels of proinflammatory cytokines in brain (Bos et al., 2012). Controlled exposure to DE has been reported to alter motor activity, spatial learning and memory, novel object recognition ability, and emotional behavior, and to cause oxidative stress and neuroinflammation in the CNS (Gerlofs-Nijland et al., 2010; Levesque et al., 2011; Mohan Kumar et al., 2008; Win-Shwe and Fujimaki, 2011). Even a short exposure of mice to DE ($250–300$ $\mu g/m^3$ for 6 h) was found to increase lipid peroxidation and levels of proinflammatory cytokines in several brain regions, particularly in males (Cole et al., 2016).

2.2 Developmental Neurotoxicity

Epidemiological and animal studies suggest that young individuals may be particularly susceptible to air pollution-induced neurotoxicity (Calderon-Garciduenas et al., 2008a, 2012; Freire et al., 2010; Guxens and Sunyer, 2012; Guxens et al., 2014). For example, exposure to air pollution during pregnancy was found to be associated with delayed psychomotor development (Guxens et al., 2014), and to lower cognitive development in primary school children (Sunyer et al., 2015). Much attention has been recently devoted to autism, a neurodevelopmental disorder characterized by marked reduction of social and communicative skills, and by the presence of stereotyped behaviors (Levy et al., 2009), as a number of studies have found associations between exposures to traffic-related air pollution and this syndrome (see Costa et al., 2017b for a review). For example, Volk et al. (2011, 2013) found that residential proximity to freeways and gestational and early-life exposure to traffic-related air pollution were associated with autism, and similar results were obtained in several other epidemiological studies (Becerra et al., 2013; Kalkbrenner et al., 2015). Children diagnosed as autistic also have higher levels of oxidative stress (Frustaci et al., 2012), as well as higher microglia activation, neuroinflammation, and increased systemic inflammation (Depino, 2013; El-Ansary and Al-Ayadhi, 2012).

Also for the developmental neurotoxic effects of air pollution, animal studies support human findings (Costa et al., 2014, 2017a). In utero exposure to DE caused alterations in motor activity, motor coordination, and impulsive behavior in male mice (Suzuki et al., 2010; Yokota et al., 2013), while early postnatal exposure of mice to concentrated ambient PM caused long-term impairment of short-term memory and impulsivity-like behavior (Allen et al., 2014a). Additional studies have shown that developmental DE exposure of mice alters motor activity, spatial learning and memory, and novel object recognition ability and causes changes in gene expression, neuroinflammation, and oxidative damage (Hougard et al., 2008; Tsukue et al., 2009; Win-Shwe et al., 2014). Specifically with regard to autism, prenatal and early-life exposure of mice to DE caused higher levels of motor activity, elevated levels of self-grooming, and increased rearing (Thirtamara Rajamani et al., 2013), and alterations in behavioral domains (persistent/ repetitive behaviors, communication, and social interactions) typically affected in autism (Chang et al., 2017a). Postnatal exposure to concentrated ambient ultrafine particles caused persistent glial cell activation, various neurochemical changes, and ventriculomegaly (lateral ventricular dilation), which occurred preferentially in male mice (Allen et al., 2014b).

Overall, evidence accumulating from human epidemiological studies suggests that exposure to air pollution may be associated with adverse effects on the CNS. Animal studies fully support these findings and indicate that oxidative stress and neuroinflammation may represent two cardinal mechanisms underlying the observed neurotoxic effects (Costa et al., 2014, 2017a). As discussed earlier, age is emerging as an important determinant for susceptibility to air pollution neurotoxicity, and there is much interest in the role that traffic-related air pollution may play in the etiology of neurodevelopmental diseases, such as autism spectrum disorders (Costa et al., 2017b). In addition to an enhanced susceptibility of the developing brain, the aging brain may also be particularly susceptible to air pollution-induced neurotoxicity. Several epidemiological studies investigating adverse effects of air pollution on behavior, particularly cognitive behavior, have identified significant effects in the elderly. Furthermore, various studies have shown that air pollution may represent an important etiological factor in neurodegenerative diseases, and evidence of this is discussed in the following sections.

3. AIR POLLUTION AND NEURODEGENERATIVE DISEASES

Emerging evidence suggests that the aging brain may be particularly susceptible to air pollution-induced neurotoxicity, as several of the epidemiological studies identifying adverse effects of air pollution on behavior, particularly cognitive behavior, have identified significant effects in the elderly (Chen et al., 2015; Power et al., 2011; Ranft et al., 2009; Weuve et al., 2012). Aging is often associated with a wide variety of clinical and pathological conditions, which can be classified as neurodegenerative diseases. Typical examples of such diseases are Alzheimer's disease (AD), Parkinson's disease (PD), amyotrophic lateral sclerosis (ALS), and Huntington's disease (HD). The prevalence of these diseases varies quite significantly. AD is the most common, followed by PD, while HD is relatively rare (Musgrove et al., 2015).

Aging represents the main risk factor for neurodegenerative diseases. It has been estimated that in 2060, the population in the European Union will reach >510 million, of which one-third will be aged 65+, from about 17% of today (Maresova et al., 2016). The number of people with dementia (including AD, vascular dementia, dementia with diffused Lewy bodies, PD dementia, and a few others) currently about 10 million is expected to double by that time, with an economic burden of >350 billion euros (Maresova et al., 2016). Similarly, in the United States the number of individuals with AD dementia should increase from the current 5 million to about 14 million in 2050 (Hebert et al., 2013). Neurodegenerative diseases are multifaceted heterogeneous disorders, with complex clinical and pathological pictures, and mostly unknown etiologies (Musgrove et al., 2015). While several susceptibility genes have been identified, environmental factors or gene–environment interactions are believed to play the most important role in these diseases. The following sections discuss three major neurodegenerative diseases (PD, AD, and ALS), for which there is suggestive evidence of an association with air pollution. In addition, multiple sclerosis (MS), another neurodegenerative disease often found in younger individuals, is also discussed in this regard. No information is yet available on the potential association between air pollution and HD.

3.1 Air Pollution and PD

PD is a neurodegenerative disorder characterized by a slow and progressive degeneration of dopaminergic neurons in the substantia nigra, with degeneration of nerve terminals in the striatum. Once loss of dopaminergic neurons has reached about 80%, clinical signs appear which include resting tremor, rigidity, bradykinesia, and gait disturbances (Cubo and Goetz, 2014). Though PD has been traditionally considered a motor system disorder, it presents additional clinical features that may include cognitive dysfunction and dementia, olfactory dysfunction, sleep disorders, and depression, some of which may present before motor symptoms. Olfactory dysfunction is an important early symptom of neurodegenerative diseases, particularly of PD (Doty, 2012; Mesholam et al., 1998), in which damage to the olfactory bulb actually precedes neuropathology in the motor areas, such as substantia nigra and striatum (Braak et al., 2004). The prevalence of PD in the general population is about 0.3%, and age remains the main risk factor, with prevalence increasing to 1% in people above 60 and to 3%–5% in those over 85 years (Lema Tomé et al., 2013). PD shows a sex difference, and its incidence is 90% higher in males than in females (Van Den Eeden et al., 2003). Genetic forms of PD have been associated with specific mutations in a number of genes (PARK1–PARK13) and are often of early onset (before the age of 50) and occurring in family clusters. However, the great majority of PD cases is sporadic, and may be due to environmental factors or to gene–environment interactions, i.e., neurotoxic exposures in susceptible individuals. Among environmental factors believed to be associated with PD there are certain pesticides (e.g., the herbicide paraquat or the insecticide rotenone), metals (e.g., manganese), solvents (e.g., trichloroethylene) (Lo and Tanner, 2014), or air pollution (see below). Among protective factors, cigarette smoking has emerged as the strongest one, possibly for a neuroprotective effect of nicotine. As said earlier, PD is defined biochemically as a dopamine deficiency resulting from degeneration of dopamine neurons in the substantia nigra. Protein aggregations in the form of Lewy bodies in surviving neurons of adjacent areas are also a hallmark of PD (Beier and Richardson, 2015). Lewy bodies are a dense core inclusion encircled by a halo radiating fibrils composed of misfolded α-synuclein. The protein α-synuclein came to be at the center of PD research after it was reported that a single-point mutation in its gene (PARK1/4) was responsible for the first familial form of autosomal dominant PD (Polymeropoulos et al., 1997; Ulusoy and Di Monte, 2013). Levels of α-synuclein are higher in olfactory bulb and striatum, two brain regions affected in PD, and are higher in PD brain than in normal aging (Beier and Richardson, 2015; Ulusoy and Di Monte, 2013).

A few studies have specifically investigated the possible association between air pollution and PD. An earlier retrospective study by Finkelstein and Jerrett (2007) investigated the association between PD and ambient air manganese in a cohort of 110,000 subjects in the Canadian cities of Toronto and Hamilton. A small significant association was found in Hamilton but not in Toronto. The study was prompted by the use of methylcyclopentadienyl manganese tricarbonyl as an antiknock agent in gasoline. Following combustion, manganese is emitted from the tailpipe as sulfate and phosphate, and this metal was chosen in this study as a marker of traffic-related air pollution. It should be noted, however, that manganese exposure is known to cause Parkinsonism, a syndrome with many of the characteristics of PD (Costa and Aschner, 2015); thus, this study would support the contribution of manganese to the etiology of PD rather than that of air pollution. A study by Kirrane et al. (2015) examined the incidence of PD among participants in the Agricultural Health Study in two populations in North Carolina and Iowa. Positive associations between fine particulate matter ($PM_{2.5}$) [OR (odds ratio) = 1.34] and ozone (OR = 1.39) and PD were found in North Carolina, but not in Iowa, possibly because of lower levels of air pollution in the latter state (Kirrane et al., 2015). In the PAGE (Parkinson's, Genes and Environment) study—a nested case–control study within a larger National Institute of Health Diet and Health study—Liu et al. (2016b) found evidence of an association between PM_{10} (OR = 2.34) and $PM_{2.5}$ (OR = 1.79) in female never smoker (as indicated earlier, smoking has consistently been found to be a protective factor for PD). In Denmark, a case–control study with 1600 PD patients suggested that air pollution from traffic source was associated with 9% increased risk of PD for each interquartile range increase of NO_2 (2.97 $\mu g/m^3$) (Ritz et al., 2016). ORs were larger for people having lived longer in Copenhagen (OR = 1.21), and lower for individuals living in provincial towns (OR = 1.1), whereas no association was found among rural residents. In a follow-up study on the same population, individuals were genotyped for functional polymorphisms in the genes for two proinflammatory cytokines, interleukin-1β (IL-1β) (rs16944) and tumor necrosis factor-α (TNF-α) (rs1800629) to assess possible gene–environment interactions (Lee et al., 2016a). While long-term exposure to nitrogen oxides (NO_x) increased PD risk overall (OR = 1.06), the OR for individuals with high NO_x exposure and the AA IL-1β rs16944 genotype was 3.10 (Lee et al., 2016a). The rs16944 SNP is located in the promoter region of IL-1β and causes enhanced transcription; hence, individuals with higher levels of IL-1β may be more susceptible to air pollution neurotoxicity. No associations were found with regard to the TNF-α polymorphism. Another positive association between air pollution and PD was found in a study by Lee et al. (2016b) in Taiwan. Similar to the Danish study (Ritz et al., 2016), positive associations between traffic-related air pollutants [NO_x and carbon monoxide (CO)] and PD were reported (OR = 1.37 and 1.17, respectively); however, no association between PM_{10} and PD was found (Lee et al., 2016b). This latter negative finding is in agreement with the results of a study by Palacios et al. (2014) in a large (>115,000) cohort of women, part of the Nurses' Health Study, in which no statistically significant associations were found between PM_{10} or $PM_{2.5}$ and PD. In a further study examining occupational exposures as a risk factor for PD, no association between DE and PD mortality was found in a large Dutch cohort (the Netherlands Cohort Study of Diet and Cancer) (Brouwer et al., 2015). Finally, a very recent study examined the risk of PD when living near major roads in Ontario, Canada (Chen et al., 2017). In a cohort of 2.2 million adults aged 55–85 years, there were 31,577 cases of PD; however, even living less than 50 m from a major traffic road was not associated with an increased risk of PD (Chen et al., 2017). Two studies examined the rate of hospitalization for PD in a national case-crossover analysis among Medicare enrollees (Zanobetti et al., 2014), and in 50 cities in the Northeastern United States, also based on Medicare enrollees (Kioumourzoglou et al., 2016). The latter examined the effect of long-term exposure to $PM_{2.5}$, while the former focused on the effect of short-term exposure. Both studies found that exposure to $PM_{2.5}$ was associated with a significant increase in hospitalization for PD (Kioumourzoglou et al., 2016; Zanobetti et al., 2014).

Observations in postmortem brains in humans (Calderon-Garciduenas et al., 2008a,b, 2010, 2013), and in experimental animals (Levesque et al., 2011), have shown that air pollution may cause an increase of α-synuclein. Of relevance is that in humans, α-synuclein was found in brains of Mexico City children exposed to high levels of high pollution, particularly in regions associated with PD pathology such as olfactory bulb, midbrain, and medulla oblongata (Calderon-Garciduenas et al., 2013). These increases in α-synuclein, which was mostly not detected in brains of control children, were paralleled by olfactory dysfunctions in other children living in the same conditions (Calderon-Garciduenas et al., 2010). Olfaction problems have also been reported in individuals exposed to heavy air pollution (Calderon-Garciduenas et al., 2010). It has been also proposed that peripheral inflammation (known to be elicited by air pollution) may cause an increase of α-synuclein in the olfactory bulb and its misfolding and aggregation, which may set in motion events leading to PD neuropathology (Lema Tomé et al., 2013). The progressive pattern of appearance of α-synuclein aggregates suggests that the synucleinopathy may spread via the olfactory route, from the olfactory bulb to the midbrain via other olfactory areas and the limbic system (Lema Tomé et al., 2013). In a single controlled study, Levesque et al. (2011)

found that exposure of male rats to DE ($311 \ \mu g/m^3$ or higher) for 6 months increased α-synuclein levels in the midbrain.

In summary, there is limited evidence from epidemiological studies, suggesting that air pollution may be associated with PD. Observations in humans and in controlled animal studies also suggest that air pollution may increase the expression of the PD marker α-synuclein. Oxidative stress and neuroinflammation, two cardinal effects of air pollution, are also believed to play most relevant roles in PD. Finally, an in vitro study showed that DE particles could activate microglia and that microglia-derived oxidant species cause the demise of dopaminergic neurons (Block et al., 2004).

3.2 Air Pollution and Amyotrophic Lateral Sclerosis

ALS, also known as motor neuron disease or Lou Gehrig's disease, is a progressive neurodegenerative disorder of the motor neuron system, characterized by progressive weakness and wasting of striated muscle due to motor cortical and spinal neurodegeneration (Murray, 2014). Progressive paralysis leads to death within a few years, usually from respiratory failure, though survival can range from a few months to >20 years. The prevalence of the disease is about 6/100,000 and, similar to PD, there is a slight male predominance. As for other neurodegenerative diseases, oxidative stress and neuroinflammation appear to play a primary role in the pathogenesis of ALS (Komine and Yamanaka, 2015; Numan et al., 2015; Rodriguez and Mahy, 2016). Dominant mutations in the gene for copper/zinc superoxide dismutase are a frequent cause of inherited ALS (Komine and Yamanaka, 2015), but only account for about 10% of all cases, most of which are thus sporadic. Environmental factors potentially involved in the etiopathogenesis of ALS have been investigated to a very limited degree, and only the metal lead has emerged as a possible one (Belbasis et al., 2016; Factor-Livak, 2015). A single study has examined the potential association between air pollution and ALS (Malek et al., 2015). This case–control study, with 51 cases and 51 controls, examined exposure to air pollutants and ALS in six counties near Philadelphia between 2008 and 2011. An association was found for exposure to aromatic solvents (OR = 5.03), but not for metals, pesticides, or other air contaminants (Malek et al., 2015). The study was criticized for its small sample size and possible exposure misclassifications (Kullmann, 2015; Xu et al., 2016). Overall, though sporadic ALS is believed to have an environmental etiology, very little has emerged so far as possible factor(s), and this is true also for the case of air pollution. Needless to say, additional studies are needed in this regard. Since oxidative stress and neuroinflammation are two cardinal events in ALS and are also the main effects of air pollution, these further investigations appear warranted.

3.3 Air Pollution and Multiple Sclerosis

MS is a progressive autoimmune neurodegenerative disease consisting primarily in the segmental demyelination of neurons, with relative sparing of the axons, at least in the initial phases. Inflammatory processes are at the center of MS and involve in particular T cells, B cells, and macrophages (Bailey et al., 2014). Initial symptoms are usually sensory in nature and include numbness, tingling, and loss of vision (Shin, 2014). Young individuals, particularly Caucasian and women, are mostly affected. Prevalence of MS varies around the world and ranges from <5 to ~30 case per 100,000 (Howard et al., 2016). Though some genetic loci have been associated with an increased risk of developing MS (Sawcer et al., 2014), environmental factors (e.g., infections, vitamin D deficiency) are believed to play a significant role (Howard et al., 2016). Tobacco smoking has been consistently found as an important risk factor for MS (Riise et al., 2003), and exposure to environmental tobacco smoke has been similarly found to increase risk for MS (OR = 1.3; Hedström et al., 2011). A study in Tehran, Iran, found that exposure to air pollutants (PM_{10}, NO_x, and sulfur dioxide) was associated with MS (Heydarpour et al., 2014). Another study in Serbia reported that air pollution was associated with increased relapses of MS, particularly during the winter months (Vojinovic et al., 2015). Recently, Angelici et al. (2016) found that in the Northern Italy region of Lombardy, hospital admissions for MS were found to be associated with higher PM_{10} levels. Thus, overall, very limited evidence seems to suggest that air pollution may be involved either in the etiopathogenesis of MS or in its relapses.

3.4 Air Pollution and Alzheimer's Disease and Other Dementias

The term *dementia* refers to an acquired cognitive impairment involving multiple domains of functions and is used as an umbrella term for a number of distinctive diseases (Perry, 2014). Dementia is predominantly a disease of later life, and after 65 years of age, its prevalence doubles every few years. As populations are aging at an increasing rate, the number of cases in the world is predicted to reach 115 million by 2050 (Prince et al., 2013). The main type of dementia in individuals of >65 years is AD, which accounts for 55%–75%

cases. Dementia with Lewy bodies is the second more frequent type of dementia, followed by vascular dementia, frontotemporal dementia, and other forms of the disease (Perry, 2014). AD is the prototypical dementia, with disturbances starting in the domain of episodic memory, followed by deficits in language, attention, and executive functions (see below). The very common vascular dementia (second only to AD) has an heterogeneous nature and defines dementia whose causes are found in cardiovascular pathology, with clinical signs involving deficits in attention, information processing, and executive function (O'Brien and Thomas, 2015). The clinical sign of dementia with Lewy bodies includes hallucinations, Parkinsonism, and altered cognitive status (Walker et al., 2015). Frontotemporal dementia (at times referred to as Pick's disease) is an umbrella clinical term that encompasses various neurodegenerative diseases characterized by progressive deficits in behavior, language, personality, and executive function (Bang et al., 2015). Mild cognitive impairment (MCI) is a clinical syndrome, which may affect ~15% of the population >70 years (Langa and Levine, 2014), and may or may not be related to, or evolve to, dementia. Amnestic MCI is characterized primarily by memory loss and may progress to AD, while the nonamnestic subtype involves deficits in language, executive function, or visuospatial skills and may progress to other dementias. Diagnostic criteria can help the clinician defining whether MCI may be the initial stage of AD or other dementias (Langa and Levine, 2014).

AD is by far the most common cause of dementia, and one of the great health care challenges of the 21st century (Scheltens et al., 2016). Its most common symptom is memory loss for recent events; in addition, visuospatial, language, and executive impairments are also present. The gross pathology of AD is represented by diffuse cortical and hippocampal atrophy. Accumulation of abnormally folded amyloid beta (Aβ) and of tau proteins in amyloid plaques and neuronal tangles is the neuropathological hallmark of AD (Khan and Bloom, 2016; Selkoe and Hardy, 2016). The amyloid precursor protein (APP) is cleaved by secretases (α, β, γ) to generate an Aβ polypeptide (either Aβ42 or Aβ40). Aβ42 is the major form found in amyloid plaques (Pressman and Rabinovici, 2014). Neurofibrillary tangles are composed of hyperphosphorylated tau protein, which causes disassembly of microtubules. A number of genes have been associated with AD; for example, rare mutations in APP and in PSEN1 and PSEN2 (presenilins, which provide the catalytic subunit to the γ secretase which cleaves APP) confer the highest risk for AD (Scheltens et al., 2016). However, the major genetic risk factor for AD is apolipoprotein E (APOE). Three APOE isoforms exist:

APOEε2, APOEε3, and APOEε4; the latter predisposes the carrier to AD (Riedel et al., 2016). The prevalence of APOEε4 carriers in AD cases is about 60%, while APOEε4 homozygotes are about 10%–15% (Ward et al., 2012). Sex differences are present in AD with postmenopausal women constituting >60% of the affected AD population; whether this may be related to a diminished neuroprotection by estrogens is still being debated (Pike, 2017; Riedel et al., 2016).

In the past few years, epidemiological studies have begun to explore the possible contribution of environmental pollutants to AD and other dementias (Killin et al., 2016; Yegambaram et al., 2015), and air pollution has been emerging as an important factor (Killin et al., 2016; Power et al., 2016). As cognitive impairment is an initial important aspect of AD, several studies have focused on various assessments of cognitive level (e.g., performance on cognitive tests) in relationship to air pollution (Clifford et al., 2016; Cohen and Gerber, 2017; Peters et al., 2015). Most of these latter studies have been carried out in the elderly (>65 years), and with two exceptions, all have found that increasing levels of air pollution, particularly traffic-related air pollution, was associated with diminished cognitive abilities (Table 3). An earlier study by Ranft et al. (2009) reported that in a group of 399 women (68–79 years), exposure to $PM_{2.5}$ and PM_{10}, indicated by having lived for >20 years within 50 m of a busy road, was associated with increased MCI. At about the same time, Chen and Schwartz (2009) published the results of an analysis of the Third National Health and Nutrition Examination Survey (NHANES), which showed an association between exposure to ozone and reduced performance in the Neurobehavioral Evaluation System-2 (NES2) in adults aged 20–59 years. A further series of studies, all in different US populations of older individuals, confirmed and expanded these earlier findings. Power et al. (2011) found a strong association between traffic-related air pollution and decreased cognitive function in 680 older men (mean age 71 years). Similarly, residing near major roadways was found to be associated with poor performance in a series of cognitive tests in a New England population of 765 men and women >65 years (Wellenius et al., 2012). Higher levels of $PM_{2.5}$ were associated with significantly faster cognitive decline in the Nurses' Health Study Cognitive Cohort, which included 19,409 women aged 70–81 years (Weuve et al., 2012). In two separate studies, Ailshire and Crimmins (2014) and Ailshire and Clarke (2015) found that $PM_{2.5}$ was associated with decreased cognitive functions in adults 50 years or older ($n = 18,575$), and in a study of 780 black and white non-Hispanic men and women aged 55 and older, respectively.

Table 3 Epidemiological Studies on Air Pollution and Cognitive Functions

Study	Population	Air Pollutant	Outcome
Ranft et al. (2009)	399 females (68–79 years)	$PM_{2.5}$, PM_{10}	↓ Cognitive functions
Chen and Schwartz (2009)	1764 both sexes (20–59 years)	PM_{10}, O_3	↓ Cognitive functions
Zeng et al. (2010)	15,973 both sexes (>65 years)	API[a]	↓ Cognitive functions
Power et al. (2011)	680 males (>65 years)	Black carbon	↓ Cognitive functions
Wellenius et al. (2012)	765 both sexes (>65 years)	Distance from freeway	↓ Cognitive functions
Weuve et al. (2012)	19,409 females (70–81 years)	$PM_{2.5}$, PM_{10}	↓ Cognitive functions
Loop et al. (2013)	20,150 both sexes (64 ± 9 years)	$PM_{2.5}$	No effect on cognition
Ailshire and Crimmins (2014)	13,996 both sexes (>50 years)	$PM_{2.5}$	↓ Cognitive functions
Tonne et al. (2014)	2867 both sexes (66 ± 6 years)	$PM_{2.5}$, PM_{10}	↓ Cognitive functions
Gatto et al. (2014)	1496 both sexes (60 ± 8 years)	$PM_{2.5}$, O_3, NO_2	↓ Cognitive functions
Ailshire and Clarke (2015)	780 both sexes (>55 years)	$PM_{2.5}$	↓ Cognitive functions
Schikowski et al. (2015)	789 females (>55 years)	NO_2, NO_x $PM_{2.5}$, PM_{10}	↓ Cognitive functions
Tzivian et al. (2016)	4086 both sexes (50–80 years)	PM, NO_x	↓ Cognitive functions
Wilker et al. (2016)	236 both sexes (74 ± 12 years)	$PM_{2.5}$	No effect on cognition
Cacciottolo et al. (2017)	3647 females (65–79 years)	$PM_{2.5}$	↓ Cognitive functions

[a]Air pollution index.

Interestingly, further analyses of the population of the Ailshire et al. (2017) study revealed that those living in socioeconomically disadvantaged neighborhoods, where social and environmental stressors and environmental hazards are more common, may be particularly susceptible to the effects of $PM_{2.5}$ on cognitive functions (Ailshire et al., 2017). The latter observation is of interest, as another study reported that the association between $PM_{2.5}$ and depressive and anxiety symptoms was strongest in individuals with low socioeconomic status (Pun et al., 2017). In a population of civil servants from the London, UK area ($n = 2867$, mean age 66 years), Tonne et al. (2014) found that $PM_{2.5}$ was associated with diminished cognitive functions. A cross-sectional study in Los Angeles in middle-aged and older adults ($n = 1496$, mean age 60.5 years) by Gatto et al. (2014) found associations between increasing levels of ozone, $PM_{2.5}$, and NO_2, and measures of domain-specific cognitive abilities. In a population of 592 individuals (50–80 years) in the German Ruhr area, long-term exposure to air pollution ($PM_{2.5}$ and NO_x) and noise was associated with MCI, particularly of the amnestic type (Tzivian et al., 2016). This study underscores the potential relevance of noise as a cocontributor to deficits in cognition, as shown by various human and animal studies (Basner et al., 2014; Liu et al., 2016a,b); this aspect, with few exceptions, has been rarely addressed in the studies on the effects of traffic-related air pollution on cognition. An additional study supporting an association between elevated air pollution and decreased cognitive function is that by Zeng et al. (2010), as part of a study on health indicators in the Chinese Longitudinal Healthy Longevity Survey, 2002–2005. Finally, a very recent study part of the Women's Initiative Memory Study in older (>75 years) US women found an increased risk of cognitive decline and of dementia with increasing exposure to $PM_{2.5}$ (Cacciottolo et al., 2017). In contrast to these positive findings, two studies did not find associations between air pollution and cognitive impairment. One study by Loop et al. (2013) reported a lack of effect of $PM_{2.5}$ on incident cognitive impairment in an heterogeneous biracial US cohort of men and women ($n = 20,150$), while Wilker et al. (2016) in a population in Massachusetts with lower level of $PM_{2.5}$ exposure (below EPA standard, currently 35 $\mu g/m^3$; USEPA, 2016) found no association with memory impairment. Overall, the evidence to date is sufficiently consistent in indicating that exposure to air pollution is associated with cognitive decline in the elderly. However, controlled animal studies in this area are lacking, and epidemiological studies present some limitations (e.g., end-point measured, different indices of air pollution), as recently underlined (Clifford et al., 2016; Peters et al., 2015).

An additional series of related epidemiological studies examined the association between air pollution and AD and other dementias (Table 4). Three separate studies in Taiwan reported associations between air pollution and AD. Chang et al. (2014) found that in a population-based retrospective study on 29,547 patients, 1720 were diagnosed with AD, and the HRs (hazard ratio) for the top quartile were 1.54 for NO_2 and 1.61 for CO. Another population-based cohort study by Jung et al. (2015) reported that among 95,690 individuals, 1399 were diagnosed with AD, and a significant risk was found with regard to ozone and $PM_{2.5}$. A third case–control study by Wu et al. (2015) comprising 249 AD cases and 125 patients with vascular dementia found that high PM_{10} or ozone exposures were associated with either disease (OR = 4.17 and 2.00, respectively). Exposure to $PM_{2.5}$ and

Table 4 Epidemiological Studies on Air Pollution and Dementias

Study	Population	Air Pollutant	Outcome
Chang et al. (2014)	29,547 both sexes (>50 years)	CO, NO_2	↑ Risk of dementia
Zanobetti et al. (2014)	146,172 both sexes (65–75 years)	$PM_{2.5}$	↑ Risk of hospitalization
Helou and Jaecker (2014)	156 both sexes (79 ± 7 years)	Diesel exhaust	↑ Risk of AD
Jung et al. (2015)	95,690 both sexes (>65 years)	O_3, $PM_{2.5}$	↑ Risk of AD
Wu et al. (2015)	871 both sexes (~79 years)	PM_{10}	↑ Risk of AD, VaD
Koeman et al. (2015)	1552 both sexes (55–69 years)	Diesel exhaust	No ↑ in dementias
Kioumourzoglou et al. (2016)	202,614 both sexes (>65)	$PM_{2.5}$	↑ Risk of AD, dementias
Oudin et al. (2016)	1806 both sexes (55–85 years)	NO_x	↑ Risk of AD, VaD
Chen et al. (2017)	243,611 both sexes (55–85 years)	$PM_{2.5}$, NO_2	↑ Risk of AD, dementias

VaD, vascular dementia.

to NO_x was associated with increased risk of cognitive impairment and AD in a German population of elderly women (Schikowski et al., 2015). Furthermore, in Northern Sweden, Oudin et al. (2016) found that traffic-related air pollution was a risk factor for AD (HR = 1.38; n = 191) and for vascular dementia (HR = 1.47; n = 111). Finally, in a very large population-based cohort study in Ontario, Canada, involving about 2.2 million individuals aged 55–85 years, 243,611 cases of dementia were identified; the HR for dementia for people living less than 50 m from a major traffic road was 1.07–1.12, depending on the specific region, suggesting that 7%–11% of dementia cases in patients who live near major roads are attributable to traffic exposure (Chen et al., 2017). $PM_{2.5}$ and NO_2 were positively associated with dementia, but a contribution from other factors, particularly noise, could not be ruled out (Chen et al., 2017). The rate of hospitalization for AD was examined in a national case-crossover analysis among Medicare enrollees and examining short-term exposure to $PM_{2.5}$ (Zanobetti et al., 2014), and in 50 cities in the Northeastern United States upon long-term exposure to $PM_{2.5}$ (Kioumourzoglou et al., 2016). Exposure to $PM_{2.5}$ was associated with a significant increase in hospitalization for AD in both studies, and for dementia in the latter (Kioumourzoglou et al., 2016; Zanobetti et al., 2014). Two European studies examined occupational exposure to DE and risk of AD and dementia and arrived at different conclusions. Helou and Jaecker (2014) found that occupational exposure to DE and to other fuels and certain petroleum-based solvents was associated with an increased risk of AD and of mixed-type dementia among French workers (n = 156). In contrast, Koeman et al. (2015) in a group of Dutch workers (part of The Netherlands Cohort Study) reported no associations between occupational exposure to DE and solvents and nonvascular dementia.

A few studies also examined the effects of high air pollution exposure on markers of AD in brain tissue of postmortem humans or of experimental animals, or on selected brain structures by means of magnetic resonance imaging (MRI). In a series of studies in autopsy samples from children, young adults, or middle-aged individuals from the highly polluted Mexico City metropolitan area, compared to other Mexican towns, Calderon-Garciduenas and her collaborators measured increased markers for AD in various brain regions. In an initial study (Calderon-Garciduenas et al., 2004), increased Aβ42 levels were found in frontal cortex and hippocampus of individuals exposed to air pollution (n = 10; average age = 51.2 years). In addition, cyclooxygenase-2 (COX2), an indicator of neuroinflammation, which is induced by proinflammatory cytokines and is increased in AD, was also increased in these two brain regions in individuals with high air pollution exposure. In a similar study, COX2 was found to be increased in olfactory bulb and frontal cortex of young adults from Mexico City, and Aβ42 was detected in 58% of young individuals (average age 17.4 years) and in 80% of adults (>25 years) (Calderon-Garciduenas et al., 2008a,b). An additional study by the same investigators reported that in a group of children and young adult (average age = 13 years) from Mexico City, 51% had Aβ diffuse plaques compared to 0% in controls, and 40% exhibited tau hyperphosphorylation (Calderon-Garciduenas et al., 2012). Levels of Aβ42 in cerebrospinal fluid (CSF) from highly exposed Mexico City children (average age = 11.2 years; $n = 50$) where about 30% lower than in controls (Calderon-Garciduenas et al., 2016a). This is alarming, as decreases in CSF Aβ42 are considered a very early change in AD, which precede aggregation and deposition of Aβ42 in plaques, and may occur decades before the onset of symptoms (Blennow et al., 2015). Similar changes in brain markers of AD have also been reported in a limited number of animal studies. Calderon-Garciduenas et al. (2003) compared dogs exposed to high air pollution in Mexico City to control dogs and found increased expression of COX2 and of Aβ42 in several brain regions. Exposure of male rats to DE (311 or 992 $\mu g/m^3$ for 6 months) increased levels of Aβ42 and of phosphorylated tau (pS199) in the cerebral cortex (Levesque et al., 2011). In male and female mice exposed for only 3 h to nickel nanoparticles (as a model of air pollution) a significant increase of brain Aβ40 and Aβ42 levels was found (Kim et al., 2012). Male and female rats exposed to DE nanoparticles (0.3–1.0 mg/L) for 3 months had higher levels of COX2 and of Aβ42 in frontal and temporal lobes, olfactory bulb, and cerebellum (Durga et al., 2015). In another study, C57BL/6 male mice were exposed to a relatively low level of $PM_{2.5}$ (66 $\mu g/m^3$) for 3 or 9 months. The longer exposure caused an increase in Aβ40 and of BACE (beta-site APP-cleaving enzyme), and a decrease of APP. Expression of COX1 and COX2 was also increased, but phosphorylated tau was unchanged (Bhatt et al., 2015). Finally, exposure to nanoparticles by inhalation for 15 weeks caused an increase in Aβ42 in the brain of mice (Cacciottolo et al., 2017).

As part of the Framingham Offspring study, a group of 943 dementia-free men and women (average age 68 years) underwent an MRI scan to assess total cerebral brain volume, hippocampal volume, white matter hyperintensity volume and covert brain infarcts in relationship to residential proximity to high traffic roads (Wilker et al., 2015). Findings indicate that exposure to elevated $PM_{2.5}$ was associated with smaller total cerebral brain volume (a marker of age-associated brain atrophy) and with higher odds of covert brain infarcts. No association was found between $PM_{2.5}$ and white matter intensity volume, at difference with findings obtained in postmortem children and dogs from Mexico City (Calderon-Garciduenas et al., 2008b). Prefrontal white matter pathology was also confirmed in children and teens and in dogs also exposed to high air pollution in Mexico City (Calderon-Garciduenas et al., 2016b). Also, in a population from Massachusetts where median $PM_{2.5}$ were below EPA standards, a 2 $\mu g/m^3$ increase in $PM_{2.5}$ was associated with lower white matter hyperintensity volume (Wilker et al., 2016). A prospective study of 1403 community-dwelling older women (71–89 years) without dementia enrolled in the Women's Health Initiative Memory Study showed that overall brain, and white matter volumes (in frontal and temporal lobes and in corpus callosum) were inversely correlated with exposure to $PM_{2.5}$ (Chen et al., 2015). Hippocampal volume was not affected, and neither was gray matter volume (Chen et al., 2015). In a follow-up study on the same population, however, utilizing voxel-based morphometry to better analyze local brain structures, exposure to $PM_{2.5}$ was associated with smaller white matter volume (confirming previous findings and possibly reflecting effects on oligodendrocytes and/or myelin damage), as well as with smaller gray matter volumes, which may reflect synaptic neurotoxicity (Casanova et al., 2016).

As said earlier, APOEε4 represents the strongest genetic risk factor for AD (Riedel et al., 2016). A few studies have investigated whether carrier of one or two of the APOε4 alleles would be more susceptible to the neurotoxic effects of air pollution (Calderon-Garciduenas et al., 2004), and the limited evidence available appears to confirm this hypothesis. Thus, APOEε4 carriers had greater hyperphosphorylated tau and diffuse Aβ plaques than APOEε3 carriers (Calderon-Garciduenas et al., 2008a,b, 2012). Also, APOEε4 subjects exposed to high air pollution had more pronounced olfactory deficits than APOEε3 or APOEε2 carriers (Calderon-Garciduenas et al., 2010). Similarly, the effects of $PM_{2.5}$, PM_{10}, and NO_x on cognitive functions were found to be more pronounced in women carriers of the APOEε4 allele (Schikowski et al., 2015). Further studies showed that APOEε4 carriers, particularly females, were more at risk for air pollution-induced hippocampal metabolic alterations and cognitive deficits (Calderon-Garciduenas et al., 2015, 2016c). The study by Cacciottolo et al. (2017) in older women also found that the risk of cognitive decline and dementia was greater in homozygotes for the APOEε4 allele (HR = 3.95) than for those carrying the APOEε3 allele (HR = 1.65). Only one study did not find differences between APOE genotypes in susceptibility to air pollution-associated dementia (Wu et al., 2015). Cacciottolo et al. (2017) also reported on the only animal study so far, investigating the role of APOE polymorphism in susceptibility to air pollution-induced AD. They found that a 15-week exposure to nanoparticles increased amyloid plaques significantly more in mice carrying the human APOEε4 gene than in mice carrying the human APOEε3 gene. In parallel in vitro experiments Cacciottolo et al. (2017) also found that exposure of mouse neuroblastoma cells expressing Swedish mutant APP to nanoparticles (10 µg/mL for 24 h) increased levels of Aβ42 by twofold.

4. POSSIBLE MECHANISMS LINKING AIR POLLUTION TO NEURODEGENERATION

The fact that air pollution causes systemic inflammation, microglia activation, oxidative stress, and neuroinflammation provides biological plausibility and potential underlying mechanisms for the observed association between exposures and ensuing risk of neurodegenerative diseases (Kraft and Harry, 2011). As indicated earlier, oxidative stress and neuroinflammation are two of the main effects of air pollution (Costa et al., 2014, 2017a). There is ample evidence that similar processes occur in various neurodegenerative diseases and contribute to their etiopathology (Manoharan et al., 2016; Ransohoff, 2016). Microglia activation plays an important role in PD and has been strongly linked to its pathology (Lull and Block, 2010). Activated microglia has been found by positron emission tomography in the substantia nigra of living PD patients, in human postmortem PD brains, and in animal models of PD (Lull and Block, 2010). In vitro, DE particles have been shown to activate microglia, leading to the death of neurons, particularly dopaminergic neurons (Block et al., 2004; Roqué et al., 2016). Activation of microglia causes an increase in oxidative stress and in proinflammatory cytokines. Oxidative stress is believed to play a role in PD pathogenesis and has been shown to cause α-synuclein aggregation (Takahashi et al., 2007). There is also a growing recognition of the central role of neuroinflammation in the pathogenesis of PD (Hirsch et al., 2012; Qian et al., 2010), and peripheral inflammation may initiate or contribute to the neuroinflammation and dopaminergic degeneration in the CNS. Oxidative stress and neuroinflammation also play a cardinal role in AD (Heneka et al., 2015; Huang et al., 2016; Moulton and Yang, 2012; Rubio-Perez and Morillas-Ruiz, 2012). Neuroinflammation can contribute to amyloid toxicity (Minter et al., 2016), and ApoEε4 (a strong genetic risk factor for AD) is less protective against neuroinflammation (Tai et al., 2015). Neuroinflammatory processes are also involved in ALS (Komine and Yamanaka, 2015; Rodriguez and Mahy, 2016).

An additional aspect that needs to be considered is the contribution of peripheral inflammation to neuroinflammation. Though neuroinflammation and inflammation differ at the histophenotypic and transcriptomic level (Filiou et al., 2014), evidence indicate that systemic inflammation affects inflammatory processes in the CNS (Hopkins, 2007; Lema Tomé et al., 2013; Mumaw et al., 2016). As indicated earlier, peripheral inflammation is a cardinal effect of exposure to air pollution (Anderson et al., 2012). Thus, the contribution of peripheral inflammatory processes to adverse CNS effects needs to be further investigated.

In the following sections, a selected number of hypotheses are presented which may mechanistically link air pollution and neurodegeneration, particularly AD. Considered aspects are known to or believed to be relevant for neurodegeneration, and there is evidence that they may affected by exposure to air pollution.

4.1 Adult Neurogenesis

Adult neurogenesis is defined as the generation of new neurons in the adult forebrain and was first reported in rats over 50 years ago (Altman and Das, 1965). In adult rodents, neurogenesis is believed to be confined to two specific neurogenic regions, the subgranular zone (SGZ) in the dentate gyrus of the hippocampus, where new dentate granule cells are generated, and the subventricular zone (SVZ) of the lateral ventricles, where new neurons are generated, which then migrate to the olfactory bulb to become interneurons (Gage, 2000; Lazarini and Lledo, 2011; Ming and Song, 2011). While the former is believed to occur in humans (Manganas et al., 2007), whether neurogenesis occurs in SVZ and influences the olfactory bulb in humans is still debated (Ming and Song, 2011), and possibly unlikely (Bergmann et al., 2012). Thus, while in rodents inhibition of neurogenesis in the olfactory bulb and the SVZ would be expected to impair olfactory behavior and odorant discrimination (Lazarini and Lledo, 2011), these results may not be directly extrapolated to humans. In contrast, there is strong belief that stem/progenitor cells derived neurons in both adult rodents and humans contribute to critical hippocampal functions, such as learning and memory. Though some contradictory results are present in the literature, studies in which adult neurogenesis is inhibited by antimitotic agents (e.g., MAM or irradiation) collectively suggest that adult neurogenesis provides significant contribution to spatial navigation learning and spatial memory retention (Aimone et al., 2011; Deng et al., 2010; Ming and Song, 2011; Shors et al., 2001). Findings from postmortem human brains and from animal models have shown that adult neurogenesis is impaired in neurodegenerative diseases (Horgusluoglu et al., 2017), including AD (Fuster-Matanzo et al., 2013), PD (Marxreiter et al., 2013), and ALS (Li et al., 2012). Alterations in adult neurogenesis differ among different neurodegenerative diseases but are believed to be relevant to their progression. Thus, for example, impairment of adult neurogenesis in the hippocampal region may be associated with decreased cognitive function in AD (Fuster-Matanzo et al., 2013; Horgusluoglu et al., 2017).

It has been reported that brain inflammation inhibits basal neurogenesis in the hippocampal SGZ, an effect that is prevented by minocycline, an inhibitor of microglia activation (Carpentier and Palmer, 2009; Ekdahl et al., 2003). Evidence of decreased adult neurogenesis following early postnatal inflammation in rodents has also been reported (Dinel et al., 2014; Lajud and Torner, 2015). Given that a main effect of air pollution in the brain is neuroinflammation, it is conceivable that exposure may result in an impairment of adult neurogenesis. In agreement with this hypothesis, it has been reported that exposure of adult mice to DE (250–300 $\mu g/m^3$ for 6 h) decreased neurogenesis in the SGZ in animals of both sexes, and in the SVZ in males only (Coburn et al., 2015). These findings suggest that air pollution by virtue of its ability to cause neuroinflammation can disrupt the process of adult neurogenesis and may thus contribute to the etiopathology of neurodegenerative diseases.

4.2 MicroRNAs

MicroRNAs (miRNAs) are small (21- to 24-nucleotide long) noncoding RNAs that are derived from much larger primary transcripts encoded in the genome. The miRNA genes are transcribed by RNA polymerase II or III, to yield pre-miRNAs (70–100 nucleotide long), and once exported in the cytoplasm, the pre-miRNA is cleaved to generate the mature miRNA. Approximately 500–1000 miRNAs are expressed in mammalian cells, and their expression varies depending on the tissue. Upon binding to its target mRNA, the miRNA's main function is that of inducing the degradation or translational inhibition of the target mRNA. One miRNA can downregulate the expression of hundreds of proteins, and miRNAs thus represents one of the three main epigenetic regulatory mechanisms (in addition to DNA methylation and histone modification). In the nervous system, miRNAs have been shown to play important roles in brain development, morphogenesis, and synaptic plasticity, and recent evidence suggests that they are most relevant also for neurodegeneration (Hebert and De Strooper, 2009; Qiu et al., 2014; Sonntag et al., 2012). Indeed, a number of miRNAs have been found to be down- or upregulated in AD and PD (Table 5; Delay et al., 2012; Hebert and De Strooper, 2009; Mouradian, 2012; Qiu et al., 2014; Sonntag et al., 2012). Of particular interest is that miRNA regulation is important for inflammatory (and neuroinflammatory) responses (Cardoso et al., 2016; O'Connell et al., 2012; Su et al., 2016; Thounaojam et al., 2013). For example, miRNA-155 is essential for neuroinflammation in mouse AD models (Guedes et al., 2014) and in hippocampal neurogenic dysfunction (Woodbury et al., 2015). Indeed, upregulation of miRNA-155 contributes to enhanced microglia activation and expression of proinflammatory cytokines. In a mouse model of PD, miRNA-155 is also upregulated and contributes to α-synuclein-induced neuroinflammatory responses (Thome et al., 2016).

In recent years, a number of studies have investigated modifications of miRNAs induced by environmental chemicals (Sonkoly and Pivarcsi, 2011). Air pollution, in particular, has been investigated as a source of environmental exposure likely to alter expression of miRNAs (Jardim, 2011). A study by Yamammoto et al. (2013) reported that controlled exposure to DE (300 $\mu g/m^3$ for 2 h) altered expression of a number of blood miRNAs; miRNA-21, -215, -144, and -30e increased upon DE exposure. These increases were accompanied by increased markers of oxidative stress and were antagonized by the antioxidant compound N-acetylcysteine (Yamammoto et al., 2013). In contrast, a study of $PM_{2.5}$ exposure in elderly men reported that decrease of several miRNAs (-1, 126, -135a, 146a, -155, -21, -222, and -9) in blood (Fossati et al., 2014). Clearly, this area of research is still in its initial stages, and further investigations on the effects of air pollution on miRNAs, as well as on other epigenetic mechanisms, need to be carried out.

4.3 Telomere Length

An interesting potential relationship between air pollution and certain neurodegenerative disease is represented by telomere length. Telomeres are

Table 5 Examples of MicroRNA Alterations in AD and PD

MicroRNA	AD ↑	AD ↓	PD ↑	PD ↓
miR-7			×	
miR-9	×	×		
miR-15		×		
miR-23b		×		
miR-29	×	×		
miR-34b/c	×			×
miR-101		×		
miR-106		×		
miR-107		×		×
miR-124		×		
miR-125b	×	×		
miR-126	×		×	
miR-128	×	×		
miR-133b		×		×
miR-138	×			
miR-146	×	×		
miR-181		×		
miR-205				×

↑, increased expression; ↓, decreased expression.
Sources: Hebert, S.S., De Strooper, B., 2009. Alterations of the microRNA network cause neurodegenerative disease. Trends Neurosci. 32, 199–206; Delay, C., Mandemakers, W., Hebert, S.S., 2012. MicroRNAs in Alzheimer's disease. Neurobiol. Dis. 46, 285–290; Sonntag, K.C., Woo, T.W., Krichevsky, A.M., 2012. Converging miRNA functions in diverse brain disorders: a case for miR-124 and miR-126. Exp. Neurol. 235, 427–435; Mouradian, M.M., 2012. MicroRNAs in Parkinson's disease. Neurobiol. Dis. 46, 279–284.; Qiu, L., Zhang, W., Tan, E.K., Zeng, L., 2014. Deciphering the function and regulation of microRNAs in Alzheimer's disease and Parkinson's disease. ACS Chem. Neurosci. 5, 884–894.

complexes of hexameric repeats at the distal end of chromosomes where they provide stability and protection to the coding DNA (Pieters et al., 2016). Since telomere length decreases with each cell division, it can be considered as a marker of biological aging. Shorter telomere length has been found in cognitive impairment (Roberts et al., 2014), in AD and ALS

(De Felice et al., 2014; Forero et al., 2016a), but not in PD (Forero et al., 2016b). Interestingly, a few studies have shown that air pollution decreases telomere length (Bijnens et al., 2015; Hoxha et al., 2009; McCracken et al., 2010; Pieters et al., 2016). The biological mechanism by which air pollution may affect telomere length is not clear, but it may be related to oxidative stress. Indeed, due to their high content of guanine, telomeres are highly sensitive to reactive oxygen species-induced DNA damage (Grahame and Schlesinger, 2012; Pieters et al., 2016). It should also be noted that in some studies examining short-term exposure (particularly occupational exposures to DE) an increased telomere length has been observed, which may be attributed to the fact that acute inflammatory processes increase telomerase activity in B cells (Martens and Nawrot, 2016; Weng et al., 1997).

4.4 Reelin

There is increasing evidence of an involvement of reelin in AD, and air pollution-induced alterations in reelin expression and/or signaling may indeed may play a relevant role in the observed association between exposure and AD. Reelin is an extracellular glycoprotein, secreted in the marginal zone of the developing cerebral cortex by Cajal–Retzius cells, which plays a most relevant role in neuronal migration and establishment of neuronal polarity (Jossin, 2004). In the adult CNS, reelin is expressed in GABAergic interneurons in the cortex and the hippocampus, where it modulates learning and memory processes. Reelin provides functions via proteolytic activity as a serine protease, and via activation of specific receptors, which elicit signaling cascades. The canonical reelin signaling pathway consists in the binding of reelin to VLDL (very low-density lipoprotein) receptors and ApoE receptor 2, which triggers an initial tyrosine phosphorylation of the intracellular adaptor protein disabled-1 (Dab1) by Src family tyrosine kinases. Phosphorylated Dab1 then activates a kinase cascade involving PI-3 kinase, protein kinase B/Akt, and the inhibition of glycogen synthase kinase 3β (GSK3β) (Bock and May, 2016). Altered reelin expression and the ensuing alteration in reelin signaling are believed to be involved in autism spectrum disorders (Fatemi et al., 2005) and may also contribute to AD (Yu et al., 2016). There are several lines of evidence that support a relevant involvement of reelin in AD. Brain reelin levels are decreased in natural aging (Stranahan et al., 2011), in transgenic animal models of AD (Chin et al., 2007; Kocherhans et al., 2010; Mota et al., 2014a,b; Yu et al., 2014), and in brains of AD patients, particularly in the early stages (Chin et al., 2007;

Herring et al., 2012). Reelin mRNA is instead upregulated in later stages of AD (Botella-Lopez et al., 2010), which may reflect a potentially compensatory mechanism or an effect of advanced disease processes (Yu et al., 2016). However, though increased, in these later stages reelin is less effective, as β-amyloid compromises reelin signaling in late AD (Cuchillo-Ibanez et al., 2016). Through activation of the VLDL receptor and ApoE receptor 2 and ensuing phosphorylation of Dab1, reelin can inhibit Aβ generation, promote Aβ clearance, and prevent tau phosphorylation (Kocherhans et al., 2010; Pujadas et al., 2014; Yu et al., 2016). As an example, the reeler mouse $(rl^{-/-})$ shows hyperphosphorylation of tau; this is due to the fact that reelin binding to its receptors potently downregulates the activity of GSK3β, a major tau kinase (Hiesberger et al., 1999; Ohkubo et al., 2003). Hence, low levels of reelin are associated with higher tau phosphorylation (Yu et al., 2016). In addition, decreased reelin levels may contribute to the initiation and progression of AD by impairing synaptic functions, cytoskeleton stability, and proper axonal transport (Yu et al., 2016). Reelin is known to play an important role in synaptic plasticity, by stimulating long-term potentiation through modulation of NMDA (N-methyl-D-aspartate) receptors (Weeber et al., 2002), and decreased reelin levels would cause synaptic dysfunction in the hippocampus, leading to memory and cognitive deficits which may precede neuronal loss, as found in AD (Ma and Klann, 2012; Yu et al., 2016). A transcriptomics analysis of AD found that synapse-associated pathways were the most affected, and among these, reelin signaling was one of the few significantly altered (Karim et al., 2014). A genome-wide study in elderly, nondemented individuals identified three single nucleotide polymorphisms in the reelin gene (RELN) that significantly correlated with increased tau phosphorylation and concomitant appearance of neurofibrillary tangles (Kramer et al., 2010). In addition, a more recent study found that, among men, genetic polymorphisms of RELN were associated with AD (Fehen et al., 2015). Overall, decreased reelin levels and signaling appear to be key early events in AD. Hence, experiments aimed at increasing reelin signaling in the CNS as a way to counteract the development of AD have already been underway. For example, in one study, purified recombinant reelin injected into the ventricles increases synaptic function and cognitive abilities of wild-type mice (Rogers et al., 2011). Similarly, exogenous reelin prevented cognitive deficits induced by phencyclidine in mice (Ishii et al., 2015).

In order for interference with reelin homeostasis to represent a mechanism for air pollution in AD, exposure should lead to a decreased expression/signaling of reelin. Neuroinflammation (and specifically an increase in IL-6), which follows microglia activation, may be an initial event induced by air pollution, which will lead to an ensuing decrease of reelin. A possible pathway leading from neuroinflammation to decreased levels of reelin is as follows: air pollution increases the levels of IL-6, which activate IL-6 α receptors, and subsequently the signal-transducing β-receptor glycoprotein 130. This would stimulate the JAK/STAT3 pathway, leading to an increase of DNA methyltransferase 1 (DMNT1) (Garbers et al., 2015; Shaun and Thomas, 2012). Increased methylation of the reelin promoter by DNMT1 would then cause a decrease of reelin expression (Noh et al., 2005; Palacios-Garcia et al., 2015). An increase in neuroinflammation, including an increase in IL-6 expression, has been found in humans and rodents exposed to traffic-related air pollution (Block and Calderon-Garciduenas, 2009; Cole et al., 2016). In experiments investigating the effects of developmental exposure to DE, we found that the observed increase in IL-6 was paralleled by an increase in DMNT1 expression and by a significant decrease in reelin levels (Chang et al., 2017b; Costa et al., 2017b). A related study on the anesthetic sevoflurane reported that exposure increased DMNT1, caused hypermethylation of *RELN* with subsequent decrease of reelin mRNA and protein, and associated cognitive impairment (Ju et al., 2016). Interestingly, maternal immune activation (elicited by infection during pregnancy), which leads to offspring that display neuroinflammation, has been shown to decrease the levels of reelin protein and mRNA in brain of offspring (Ghiani et al., 2011; Novais et al., 2013), and to drive AD-like neuropathology as the animal ages (Krstic et al., 2012). The notion that neuroinflammation (and oxidative stress) may play an important role in modulating reelin expression is also supported by studies showing that *N*-acetylcysteine completely prevents lipopolysaccharide (LPS)-induced decreases of reelin (Novais et al., 2013). Impairment of reelin expression and signaling is known to occur in autism and has been found in mice upon developmental exposure to DE (Chang et al., 2017b; Costa et al., 2017b). Of note is that there is the suggestion that ASD and AD may indeed share some genetic and/or etiopathological aspects (Khan et al., 2016).

4.5 Glutamate

A correct functioning of glutamatergic synapses is essential for learning and memory, and disruptions of glutamate homeostasis may play a relevant role in neurological/neurodegenerative disorders (Zhang et al., 2016). The glutamatergic synapse has been defined as a "tripartite synapse" to indicate that in addition to the pre- and postsynaptic components, a key role is played by astrocytes. Glutamate, stored in presynaptic vesicles, is released upon neuronal depolarization and calcium channel opening. It then activates receptors located on neurons or astrocytes and is removed from the synaptic cleft by transporters located on astrocytes or in the presynaptic terminal. Alterations in glutamate receptors, particularly the ionotropic N-methyl-D-aspartate receptors but also metabotropic receptors, occur in AD, leading to overactivation of the glutamatergic system (Lewerenz and Maher, 2015; Mota et al., 2014a,b; Ribeiro et al., 2017; Rudy et al., 2015). In addition, glutamate transporters, particularly in astrocytes, are altered in AD, leading to dysregulation of the tightly controlled glutamate in the extracellular milieu, and contributing to an overstimulation of the glutamatergic system (Murphy-Royal et al., 2017; Rudy et al., 2015). Limited evidence from animal and in vitro studies suggests that air pollution may interact with the glutamatergic system. A study by Win-Shwe et al. (2009) reported that exposure of mice to nanoparticle-enriched DE for 4 weeks caused an increase of the NR1 (now GluN1) subunit of the NMDA receptor in the olfactory bulb. In contrast, two other studies reported that exposure of mice to nanoparticles induced a decrease of GluN1 in the hippocampus (Cacciottolo et al., 2017; Morgan et al., 2011). All three studies found that other NMDA receptor subunits were unchanged. The significance of these findings in relationship to air pollution/AD is still unclear in light of current knowledge on the impact of NMDA receptor subunit composition in neurotoxicity and in disease (Paoletti et al., 2013). Win-Shwe et al. (2009) also found that DE exposure increased the levels of free glutamate in the olfactory bulb, a finding which is in agreement with results obtained by various investigators in vitro. Liu et al. (2015) found that $PM_{2.5}$ caused the release of glutamate from microglia; excess glutamate would then overactivate NMDA receptors on neurons leading to neuronal death. The increase in glutamate appeared to be due to an increase in the activity and release from microglia of glutaminase (which converts glutamine to glutamate) (Liu et al., 2015). In an earlier study, Ye et al. (2013) reported that proinflammatory cytokines could increase glutaminase activity and glutamate levels in neurons. Furthermore, Barger et al. (2007) found that microglia activation (by LPS) caused an increased release of glutamate. They suggested that the initial oxidative stress, particularly lipid peroxidation, would cause glutathione depletion, which would be compensated for with an increased activity of the cysteine–glutamate exchanger, leading to increased extracellular glutamate. These findings indicate that neurodegenerative consequences of neuroinflammation may result from the conversion of oxidative stress to excitotoxic stress (Barger et al., 2007). Whatever the exact mechanism(s), it would appear that air pollution may increase extracellular levels of glutamate, and this may indeed contribute to neurotoxicity observed in AD and possible other neurodegenerative diseases.

5. CONCLUSION AND RESEARCH NEEDS

The aging of the global population brings along an increase in the incidence of neurodegenerative diseases. Therapeutic approaches so far have produced successes that are far from ideal. Indeed possibly only for PD, pharmacological treatments exist that at least reduce the clinical symptoms and delay, but do not prevent more significant outcomes. However, for diseases such as dementias, no pharmacological treatments so far have proven to be significant. Hence the strategy has come to focus on prevention rather than treatment. Indeed, modification of risk factors currently offers the best hope for reducing the global burden of dementia (Peters et al., 2015).

This chapter reviewed available evidence on a possible contribution of air pollution, particularly traffic-related air pollution to the etiology of neurodegenerative diseases. Though the field of air pollution and CNS effects is rather "young," sufficient evidence, provided by human epidemiological studies, animal experiments, and in vitro mechanistic investigations, suggests that (1) exposure during development, adulthood or aging, affects the CNS; (2) main biochemical effects in the CNS are microglia activation, oxidative stress, and neuroinflammation (Costa et al., 2017a,b). The studies reviewed in this chapter also suggest that air pollution may play a role in neurodegenerative disorders. Evidence is extremely limited for ALS and MS, contradictory for PD, and strongest for dementias, particularly AD. Limitations of human epidemiological studies have been pointed out and relate primarily to exposure assessment, considerations of confounding factors, and possible publication biases (Clifford et al., 2016; Xu et al., 2016). Yet, biochemical observations in postmortem human brains and in animals indicate that relevant markers for neurodegenerative diseases are increased upon exposure to air pollution. In addition, there are a number of mechanisms that provide biological plausibility for a link between air pollution and neurodegenerations.

In addition to further human epidemiological studies in different parts of the world, there is also the need of more well-controlled animal studies that can mimic real life high exposures. Animal studies would also provide the opportunity to readily test the hypotheses of genetic susceptibility to air pollution-induced neurodegeneration. For example, wild-type and α-synuclein overexpressing mice may be compared, or the effects of air pollution may be investigated in ApoEe4 mice or other models of increased AD susceptibility. First and foremost, however, evidence available so far needs to be taken in serious consideration for policy actions that would curb air pollution. This is extremely important especially in regions of China or India, which encounter high air pollution levels in heavily populated areas; the price to pay in a few decades may be enormous.

ACKNOWLEDGMENTS

Research by the author is supported by grants from NIEHS (R01ES22949, P30ES07033, P42ES04696).

REFERENCES

Ailshire, J.A., Clarke, P., 2015. Fine particulate matter air pollution and cognitive function among US older adults. J. Gerontol. B Psychol. Sci. Soc. Sci. 70, 322–328.

Ailshire, J.A., Crimmins, E.M., 2014. Fine particulate matter air pollution and cognitive function among older US adults. Am. J. Epidemiol. 180, 359–366.

Ailshire, J., Karraker, A., Clarke, P., 2017. Neighborhood social stressors, fine particulate matter air pollution, and cognitive function among older adults. Soc. Sci. Med. 172, 56–63.

Aimone, J.B., Deng, W., Gage, F.H., 2011. Resolving new memories: a critical look at the dentate gyrus, adult neurogenesis, and pattern separation. Neuron 70, 589–596.

Allen, J.L., Liu, X., Pelkowski, S., Palmer, B., Conrad, K., Oberdörster, G., Weston, D., Mayer-Proschel, M., Cory-Slechta, D., 2014a. Early postnatal exposure to ultrafine particulate matter air pollution: persistent ventriculomegaly, neurochemical disruption, and glial activation preferentially in male mice. Environ. Health Perspect. 122, 939–945.

Allen, J.L., Liu, X., Weston, D., Prince, L., Oberdorster, G., Finkelstein, J.N., Johnston, C.J., Cory-Slechta, D.A., 2014b. Developmental exposure to concentrated ambient ultrafine particulate matter air pollution in mice results in persistent and sex dependent behavioral neurotoxicity and glial activation. Toxicol. Sci. 140, 160–178.

Altman, J., Das, G.D., 1965. Autoradiographic and histological evidence of postnatal hippocampal neurogenesis in rats. J. Comp. Neurol. 124, 319–335.

Anderson, J.O., Thundiyil, J.G., Stolbach, A., 2012. Clearing the air: a review of the effects of particulate matter air pollution on human health. J. Med. Toxicol. 8, 166–175.

Angelici, L., Piola, M., Cavalleri, T., Randi, G., Cortini, F., Bergamaschi, R., Baccarelli, A.A., Bertazzi, P.A., Pesatori, A.C., Bollati, V., 2016. Effects of particulate matter exposure on multiple sclerosis hospital admissions in Lombardy region, Italy. Environ. Res. 145, 68–73.

Bailey, M., Hafler, D.A., Pelletier, D., 2014. Multiple sclerosis and acute disseminated encephalomyelitis: immunology. In: Aminoff, M.J., Daroff, R.B. (Eds.), second ed. In: The Encyclopedia of Neurological Sciences, vol. 3. Elsevier, New York, pp. 144–147.

Bang, J., Spina, S., Miller, B.L., 2015. Frontotemporal dementia. Lancet 386, 1677–1682.

Barger, S.W., Goodwin, M.E., Porter, M.M., Beggs, M.L., 2007. Glutamate release from activated microglia requires the oxidative burst and lipid peroxidation. J. Neurochem. 101, 1205–1213.

Basner, M., Babisch, W., Davis, A., Brink, M., Clark, C., Janssen, S., Stansfeld, S., 2014. Auditory and non-auditory effects of noise on health. Lancet 383, 1325–1332.

Becerra, T.A., Wilhelm, M., Olsen, J., Cockburn, M., Ritz, B., 2013. Ambient air pollution and autism in Los Angeles County, California. Environ. Health Perspect. 121, 380–386.

Beier, E.E., Richardson, J.R., 2015. Parkinson's disease: mechanisms, models, and biological plausibility. In: Aschner, M., Costa, L.G. (Eds.), Environmental Factors in Neurodevelopmental and Neurodegenerative Disorders. Academic Press/Elsevier, New York, pp. 267–288.

Belbasis, L., Bellou, V., Evangelou, E., 2016. Environmental risk factors and amyotrophic lateral sclerosis: an umbrella review and critical assessment of current evidence from systematic reviews and meta-analyses of observational studies. Neuroepidemiology 46, 96–105.

Bergmann, O., Liebl, J., Bernard, S., Alkass, K., Yeung, M.S.Y., Steier, P., Kutschera, W., Johnson, L., Landen, M., Druid, H., Spalding, K.L., Frisen, J., 2012. The age of olfactory bulb neurons in humans. Neuron 74, 634–639.

Bhatt, D.P., Puig, K.L., Gorr, M.W., Wold, L.E., Combs, C.K., 2015. A pilot study to assess effects of long-term inhalation of airborne particulate matter on early Alzheimer's-like changes in the mouse brain. PLoS One 10 (5), e0127102.

Bijnens, E., Zeegers, M.P., Gielen, M., et al., 2015. Lower placental telomere length may be attributed to maternal residential traffic exposure: a twin study. Environ. Int. 79, 1–7.

Blennow, K., Dubois, B., Fagan, A.M., Lewczuk, P., de Leon, M.J., Hampel, H., 2015. Clinical utility of cerebrospinal fluid biomarkers in the diagnosis of early Alzheimer's disease. Alzheimers Dement. 11, 58–69.

Block, M.L., Calderon-Garciduenas, L., 2009. Air pollution: mechanisms of neuroinflammation and CNS disease. Trends Neurosci. 32, 506–516.

Block, M.L., Wu, X., Pei, Z., Li, G., Wang, T., Qin, L., Wilson, B., Yang, J., Hong, J.S., Veronesi, B., 2004. Nanometer size diesel exhaust particles are selectively toxic to dopaminergic neurons: the role of microglia, phagocytosis, and NADPH oxidase. FASEB J. 18, 1618–1620.

Block, M.L., Elder, A., Auten, R.L., Bilbo, S.D., Chen, H., Chen, J.C., Cory-Slechta, D.A., Costa, D., et al., 2012. The outdoor air pollution and brain health workshop. Neurotoxicology 33, 972–984.

Bock, H.H., May, P., 2016. Canonical and non-canonical reelin signaling. Front. Cell. Neurosci. 10, 166.

Bos, I., DeBoever, P., Emmerechts, J., Buekers, J., Vanoirbeek, J., Meeusen, R., Van Poppel, M., Nemry, B., Nawrot, T., Panis, L.I., 2012. Changed gene expression in brains of mice exposed to traffic in a highway tunnel. Inhal. Toxicol. 24, 676–686.

Botella-Lopez, A., Cuchillo-Ibanez, I., Cotrufo, T., Mok, S.S., Li, Q.X., Barquero, M.S., Diersen, M., Soriano, E., et al., 2010. Beta-amyloid controls altered reelin expression and processing in Alzheimer's disease. Neurobiol. Dis. 37, 682–689.

Braak, H., Ghebremedhin, E., Rub, U., Bratzke, H., Del Tredici, K., 2004. Stages in the development of Parkinson's disease pathology. Cell Tissue Res. 318, 121–134.

Brouwer, M., Koeman, T., van den Brandt, P.A., Kromhout, H., Schouten, L.J., Peters, S., Huss, A., Vermeulen, R., 2015. Occupational exposures and Parkinson's disease mortality in a prospective Dutch cohort. Occup. Environ. Med. 72, 448–455.

Cacciottolo, M., Wang, X., Driscoll, I., Woodward, N., Saffari, A., Reyes, J., Serre, M.L., Vizuete, W., Sioutas, C., Morgan, T.E., Gatz, M., Chui, H.C., Shumaker, S.A., Resnick, S.M., Espeland, M.A., Finch, C.E., Chen, J.C., 2017. Particulate air pollutants,

APOE alleles and their contributions to cognitive impairment in older women and to the amyloidogenesis in experimental models. Transl. Psychiatry 7, e1022.

Calderon-Garciduenas, L., Maronpot, R.R., Torres-Jardon, R., Henriquez-Roldan, C., Schoonhoven, R., Acuña-Ayala, H., Villareal-Calderon, A., Nakamura, J., Fernando, R., Reed, W., Azzarelli, B., Swenberg, J.A., 2003. DNA damage in nasal and brain tissues of canines exposed to air pollutants is associated with evidence of chronic brain inflammation and neurodegeneration. Toxicol. Pathol. 31, 524–538.

Calderon-Garciduenas, L., Reed, W., Maronpot, R.R., Henriquez-Roldan, C., Delgado-Chavez, R., Calderon-Garciduenas, A., Dragustinovis, I., Franco-Lira, M., Aragon-Flores, M., Solt, A.C., Altenburg, M., Torres-Jardon, R., Swnberg, J.A., 2004. Brain inflammation and Alzheimer's-like pathology in individuals exposed to severe air pollution. Toxicol. Pathol. 32, 650–658.

Calderon-Garciduenas, L., Mora-Tiscareño, A., Ontiveros, E., Gomez-Garza, G., Barragan-Mejia, G., Broadway, J., Chapman, S., Valencia-Salazar, G., Jewells, V., Maronpot, R.R., Henriquez-Roldan, C., Perez-Guillé, B., Torres-Jardon, R., Herrit, L., Brooks, D., Osnaya-Brizuela, N., Monroy, M.E., Gonzalez-Maciel, A., Reynoso-Robles, R., Villareal-Calderon, R., Solt, A.C., Engle, R.W., 2008a. Air pollution, cognitive deficits and brain abnormalities: a pilot study with children and dogs. Brain Cogn. 68, 117–127.

Calderon-Garciduenas, L., Solt, A.C., Henriquez-Roldan, C., Torres-Jardon, R., Nuse, B., Herritt, L., Villareal-Calderon, R., Osnaya, N., Stone, I., Garcia, R., Brooks, D.M., Gonzalez-Maciel, A., Reynoso-Robles, R., Delgado-Chavez, R., Reed, W., 2008b. Long-term air pollution exposure is associated with ultrafine particulate deposition, and accumulation of amyloid β-42 and α-synuclein in children and young adults. Toxicol. Pathol. 36, 289–310.

Calderon-Garciduenas, L., Franco-Lira, M., Henriquez-Roldan, C., Osnaya, N., Gonzalez-Maciel, A., Reynoso-Robles, R., Villareal-Calderon, R., Herritt, L., Brooks, D., Keefe, S., Palacios-Moreno, J., Villareal-Calderon, R., Torres-Jardon, R., Medina-Cortina, H., Delgado-Chavez, R., Aiello-Mora, M., Maronpot, R.R., Doty, R.L., 2010. Urban air pollution: influences on olfactory function and pathology in exposed children and young adults. Exp. Toxicol. Pathol. 62, 91–102.

Calderon-Garciduenas, L., Kavanaugh, M., Block, M., D'Angiulli, A., Delgado-Chavez, R., Torres-Jardon, R., Gonzalez-Maciel, A., Reynoso-Robles, R., Osnaya, N., Villareal-Calderon, R., Guo, R., Hua, Z., Zhu, H., Perry, G., Diaz, P., 2012. Neuroinflammation, hyperphosphorylated tau, diffuse amyloid plaques, and down-regulation of the cellular prion protein in air pollution exposed children and young adults. J. Alzheimers Dis. 28, 93–107.

Calderon-Garciduenas, L., Franco-Lira, M., Mora-Tiscareno, A., Medina-Cortina, H., Torres-Jardon, R., Kavanaugh, M., 2013. Early Alzheimer's and Parkinson's disease pathology in urban children: friend versus foe responses-It is time to face the evidence. Biomed. Res. Int. 2013, 161687.

Calderon-Garciduenas, L., Mora-Tiscareno, A., Melo-Sanchez, G., Rodriguez-Diaz, J., Torres-Jardon, R., Styner, M., Mukherjee, P.S., Lin, W., Jewells, V., 2015. A critical proton MR spectroscopy marker of Alzheimer's disease early neurodegenerative change: low hippocampal NAA/Cr ratio impacts APOEε4 Mexico City children and their parents. J. Alzheimers Dis. 48, 1065–1075.

Calderon-Garciduenas, L., Avila-Ramirez, J., Calderon-Garciduenas, A., Gonzalez-Heredia, T., Acuña-Ayala, H., Chao, C.K., Thompson, C., Ruiz-Ramos, R., Cortes-Gonzalez, V., Martinez-Martinez, L., Garcia-Perez, M.A., Resi, J., Mukherjee, P.S., Torres-Jardon, R., Lachman, I., 2016a. Cerebrospinal fluid biomarkers in highly exposed PM2.5 urbanites: the risk of Alzheimer's and Parkinson's diseases in young Mexico City residents. J. Alzheimers Dis. 54, 597–613.

Calderon-Garciduenas, L., Reynoso-Robles, R., Vargas-Martinez, J., Gomez-Maqueo-Chew, A., Perez-Guillé, B., Mukherjee, P.S., Torres-Jardon, R., Perry, G., Gonzalez-Maciel, A., 2016b. Prefrontal white matter pathology in air pollution exposed Mexico City young urbanites and their potential impact on neurovascular unit dysfunction and the development of Alzheimer's disease. Environ. Res. 146, 404–417.

Calderon-Garciduenas, L., Jewells, V., Galaz-Montoya, C., van Zundert, B., Perez-Calatayud, A., Ascencio-Ferrel, E., Valencia-Salazar, G., Sandovan-Cano, M., Carlos, E., Solorio, E., Acuña-Ayala, H., Torres-Jardon, R., D-Angiulli, A., 2016c. Interactive and additive influences of gender, BMI and apolipoprotein 4 on cognition in children chronically exposed to high concentrations of $PM_{2.5}$ and ozone. APOE 4 females are at highest risk in Mexico City. Environ. Res. 150, 411–422.

Cardoso, A.L., Guedes, J.R., Pedroso de Lima, M.C., 2016. Role of microRNAs in the regulation of innate immune cells under neuroinflammatory conditions. Curr. Opin. Pharmacol. 26, 1–9.

Carpentier, P.A., Palmer, T.D., 2009. Immune influence on adult neural stem cell regulation and function. Neuron 64, 79–92.

Casanova, R., Wang, X., Reyes, J., Akita, Y., Serra, M.L., Vizuete, W., Chui, H.C., Driscoll, I., Resnick, S.M., Espeland, M.A., Chen, J.C., The WHIMS-MRI Study Group, 2016. A voxel-based morphometry study reveals local brain structural alterations associated with ambient fine particles in older women. Front. Hum. Neurosci. 10, 495.

Chang, K.H., Chang, M.Y., Muo, C.H., Wu, T.N., Chen, C.Y., Kao, C.H., 2014. Increased risk of dementia in patients exposed to nitrogen dioxide and carbon monoxide: a population-based retrospective cohort study. PLoS One 9, e103078.

Chang, Y.C., Cole, T.B., Costa, L.G., 2017a. Pre- and post-natal exposure to diesel exhaust elicits an autism spectrum disorder behavioral phenotype in mice (in preparation).

Chang, Y.C., Cole, T.B., Hickman, E., Costa, L.G., 2017b. Developmental exposure of mice to diesel exhaust is associated with dysregulation of the reelin pathway and an increase in autism-related behavior. Toxicologist. #2642.

Chen, J.C., Schwartz, J., 2009. Neurobehavioral effects of ambient air pollution on cognitive performance in US adults. Neurotoxicology 30, 231–239.

Chen, J.C., Wang, X., Wellenius, G.A., Serre, M.L., Driscoll, I., Casanova, R., McArdle, J.J., Manson, J.E., Chiu, H.C., Espeland, M.A., 2015. Ambient air pollution and neurotoxicity on brain structure: evidence from women's health initiative memory study. Ann. Neurol. 78, 466–476.

Chen, H., Kwong, J.C., Copes, R., Tu, K., Villeneuve, P.J., van Donkelaar, A., Hystad, P., Martin, R.V., Murray, B.J., Jessiman, B., Wilton, A.S., Kapp, A., Burnett, R.T., 2017. Living near major roads and incidence of dementia, Parkinson's disease, and multiple sclerosis: a population-based cohort study. Lancet 389, 718–726.

Chin, J., Massaro, C.M., Palop, J.J., Thwin, M.T., Yu, G.Q., Bien-Ly, N., Bender, A., Mucke, L., 2007. Reelin depletion in the entorhinal cortex of human amyloid precursor protein transgenic mice and humans with Alzheimer's disease. J. Neurosci. 27, 2727–2733.

Clifford, A., Lang, L., Chen, R., Anstey, K.J., Seaton, A., 2016. Exposure to air pollution and cognitive functioning across the life course-A systematic review. Environ. Res. 147, 383–398.

Coburn, J.L., Cole, T.B., Costa, L.G., 2015. Diesel exhaust exposure suppresses adult neurogenesis in mice in a sex- and brain region-dependent manner. Toxicologist, 323. #1504.

Cohen, G., Gerber, Y., 2017. Air pollution and successful aging: recent evidence and new perspectives. Curr. Environ. Health Rep. 4, 1–11.

Cole, T.B., Coburn, J., Dao, K., Roque, P., Kalia, V., Guilarte, T.R., Dziedzic, J., Costa, L.G., 2016. Sex and genetic differences in the effects of acute diesel exhaust exposure on inflammation and oxidative stress in mouse brain. Toxicology 374, 1–9.

Costa, L.G., Aschner, M.A. (Eds.), 2015. Manganese in Health and Disease. RSC, London, p. 632.

Costa, L.G., Cole, T.B., Coburn, J., Chang, Y.C., Dao, K., Roque, P., 2014. Neurotoxicants are in the air: convergence of human and in vitro studies on the effects of air pollution on the brain. BioMed. Res. Int. 2014, 8, 736385.

Costa, L.G., Chang, Y.C., Cole, T.B., 2017a. Developmental neurotoxicity of traffic-related air pollution: focus on autism. Curr. Environ. Health Rep. 4, 156–165.

Costa, L.G., Cole, T.B., Coburn, J., Chang, Y.C., Dao, K., Roque, P., 2017b. Neurotoxicity of traffic-related air pollution. Neurotoxicology 59, 133–139.

Crüts, B., Driessen, A., van Etten, L., et al., 2008. Exposure to diesel exhaust induces changes in EEG I human volunteers. Part. Fibre Toxicol. 5, 4.

Cubo, E., Goetz, C.G., 2014. Parkinson's disease. In: Aminoff, M.J., Daroff, R.B. (Eds.), second ed. In: The Encyclopedia of Neurological Sciences, vol. 3. Elsevier, New York, pp. 828–832.

Cuchillo-Ibanez, I., Mata-Balaguer, T., Balmaceda, V., Arranz, J.J., Nimpf, J., Saez-Valero, J., 2016. The β-amyloid peptide compromises reelin signaling in Alzheimer's disease. Sci. Rep. 6, 31646.

De Felice, B., Annunziata, A., Fiorentino, G., Manfellotto, F., D'Alessandro, R., Marino, R., Borra, M., Biffali, E., 2014. Telomerase expression in amyotrophic lateral sclerosis (ALS) patients. J. Hum. Genet. 59, 555–561.

Delay, C., Mandemakers, W., Hebert, S.S., 2012. MicroRNAs in Alzheimer's disease. Neurobiol. Dis. 46, 285–290.

Deng, W., Aimone, J.B., Gage, F.H., 2010. New neurons and new memories: how does adult hippocampal neurogenesis affect learning and memory? Nat. Rev. Neurosci. 11, 339–350.

Depino, A.M., 2013. Peripheral and central inflammation in autism spectrum disorders. Mol. Cell. Neurosci. 53, 69–76.

Dinel, A.L., Joffre, C., Trifilieff, P., Aubert, A., Foury, A., Le Ruyet, P., Laye, S., 2014. Inflammation early in life is a vulnerability factor for emotional behavior at adolescence and for lipopolysaccharide-induced spatial memory and neurogenesis alteration at adulthood. J. Neuroinflammation 11, 155.

Doty, R.L., 2012. Olfaction in Parkinson's disease and related disorders. Neurobiol. Dis. 46, 527–552.

Durga, M., Devasena, T., Rajasekar, A., 2015. Determination of LC_{50} and sub-chronic neurotoxicity of diesel exhaust nanoparticles. Environ. Toxicol. Pharmacol. 40, 615–625.

Ekdahl, C.T., Claasen, J.H., Bonde, S., Kokaia, Z., Lindvall, O., 2003. Inflammation is detrimental for neurogenesis in adult brain. Proc. Natl. Acad. Sci. U.S.A 100, 13632–13637.

El-Ansary, A., Al-Ayadhi, L., 2012. Neuroinflammation in autism spectrum disorders. J. Neuroinflammation 9, 265.

Factor-Livak, P., 2015. Environmental factors and amyotrophic lateral sclerosis: what do we know? In: Costa, L.G., Aschner, M.A. (Eds.), Environmental Factors in Neurodevelopmental and Neurodegenerative Disorders. Elsevier, New York, pp. 329–353.

Fatemi, S.H., Snow, A.V., Stary, J.M., Araghi-Niknam, M., Brooks, A.I., Pearce, D.A., Reutiman, T.J., Lee, S., 2005. Reelin signaling is impaired in autism. Biol. Psychiatry 57, 777–787.

Fehen, A., Juhasz, A., Pakaski, M., Kalman, J., Janka, Z., 2015. Genetic analysis of the RELN gene: gender specific association with Alzheimer's disease. Psychiatry Res. 230, 716–718.

Filiou, M.D., Arefin, A.S., Moscato, P., Graeber, M.B., 2014. Neuroinflammation differs categorically from inflammation: transcriptomes of Alzheimer's disease, Parkinson's disease, schizophrenia and inflammatory diseases compared. Neurogenetics 15, 201–212.

Finkelstein, M.M., Jerrett, M., 2007. A study of the relationship between Parkinson's disease and markers of traffic-derived and environmental manganese air pollution in two Canadian cities. Environ. Res. 104, 420–432.

Fonken, L.K., Xu, X., Weil, Z.M., Chen, G., Sun, Q., Rajagopalan, S., Nelson, R.J., 2011. Air pollution impairs cognition, provokes depressive-like behaviors and alters hippocampal cytokine expression and morphology. Mol. Psychiatry 16, 987–995.

Forero, D.A., Gonzalez-Giraldo, Y., Lopez-Quintero, C., Castro-Vega, L.J., Barreto, G.E., Perry, G., 2016a. Meta-analysis of telomere length in Alzheimer's disease. J. Gerontol. A Biol. Sci. Med. Sci. 71, 1069–1073.

Forero, D.A., Gonzalez-Giraldo, Y., Lopez-Quintero, C., Castro-Vega, L.J., Barreto, G.E., Perry, G., 2016b. Telomere length in Parkinson's disease: a meta-analysis. Exp. Gerontol. 75, 53–55.

Fossati, S., Baccarelli, A., Zanobetti, A., Hoxha, M., Voonas, P.S., Wright, R.O., Schwartz, J., 2014. Ambient particulate air pollution and microRNAs in elderly men. Epidemiology 25, 68–78.

Freire, C., Ramos, R., Puertas, R., Lopez-Espinosa, M.J., Julvez, J., Aguilera, I., Cruz, F., Fernandez, M.F., Sunyer, J., Olea, N., 2010. Association of traffic-related air pollution with cognitive development in children. J. Epidemiol. Community Health 64, 223–228.

Frustaci, A., Neri, M., Cesario, A., Adams, J.B., Domenici, E., Della Bernardina, B., Bonassi, S., 2012. Oxidative stress-related biomarkers in autism: systematic review and meta-analyses. Free Radic. Biol. Med. 52, 2128–2141.

Fuster-Matanzo, A., Llorens-Martin, M., Hernandez, F., Avila, J., 2013. Role of neuroinflammation in adult neurogenesis and Alzheimer disease: therapeutic approaches. Mediators Inflamm. 2013, 9, 260925.

Gage, F.H., 2000. Mammalian neural stem cells. Science 287, 1433–1438.

Garbers, C., Aparicio-Siegmund, S., Rose-John, S., 2015. The IL-6/gp130/STAT3 signaling axis: recent advances towards specific inhibition. Curr. Opin. Immunol. 34, 75–82.

Garcia, G.J.M., Schroeter, J.D., Kimbell, J.S., 2015. Olfactory deposition of inhaled nanoparticles in humans. Inhal. Toxicol. 27, 394–403.

Gatto, N.M., Henderson, V.W., Hodis, H.H., St John, J.A., Lurmann, F., Chen, J.C., Mack, W.J., 2014. Components of air pollution and cognitive function in middle-aged and older adults in Los Angeles. Neurotoxicology 40, 1–7.

Gautam, S., Yadav, A., Tsai, C.J., Kumar, P., 2016. A review of recent progress in observations, sources, classification and regulations of $PM_{2.5}$ in Asian environments. Environ. Sci. Pollut. Res. Int. 23, 21165–21175.

Genc, S., Zadeoglulari, Z., Fuss, S.H., Genc, K., 2012. The adverse effects of air pollution on the nervous system. J. Toxicol. 2012, 23, 782462.

Gerlofs-Nijland, M.E., van Berlo, D., Cassee, F.R., Schins, R.P.F., Wang, K., Campbell, A., 2010. Effect of prolonged exposure to diesel engine exhaust on proinflammatory markers in different regions of the rat brain. Part. Fibre Toxicol. 7, 12.

Ghiani, C.A., Mattan, N.S., Nobuta, H., Malvar, J.S., Boles, J., Ross, M.G., Waschek, J.A., Carpenter, E.M., Fisher, R.S., de Vellis, J., 2011. Early effects of lipopolysaccharide-induced inflammation of foetal brain development in rat. ASN Neuro. 3 (4), e00068.

Ghio, A.J., Smith, C.B., Madden, M.C., 2012. Diesel exhaust particles and airway inflammation. Curr. Opin. Pulm. Med. 18, 144–150.

Gill, E.A., Curl, C.L., Adar, S.D., Allen, R.W., Auchincloss, A.H., O'Neill, M.S., Park, S.K., Ven Hee, V.C., Diez Roux, A.V., Kaufman, J.D., 2011. Air pollution and cardiovascular disease in the multi-ethnic study of atherosclerosis. Prog. Cardiovasc. Dis. 53, 353–360.

Grahame, T.J., Schlesinger, R.B., 2012. Oxidative stress-induced telomeric erosion as a mechanism underlying airborne particulate matter-related cardiovascular disease. Part. Fibre Toxicol. 9, 21.

Guedes, J.R., Custodia, C.M., Silva, R.J., de Almeida, L.P., Pedroso de Lima, M.C., Cardoso, A.L., 2014. Early miR-155 upregulation contributes to neuroinflammation in Alzheimer's disease triple transgenic animal model. Hum. Mol. Genet. 23, 6286–6301.

Guxens, M., Sunyer, J., 2012. A review of epidemiological studies on neuropsychological effects of air pollution. Swiss Med. Wkly. 141, w13322.

Guxens, M., Garcia-Esteban, R., Giorgis-Allemand, L., Forns, J., Badaloni, C., Ballester, F., Cesaroni, G., Chatzi, L., et al., 2014. Air pollution during pregnancy and childhood cognitive and psychomotor development. Epidemiology 25, 636–647.

Hebert, S.S., De Strooper, B., 2009. Alterations of the microRNA network cause neurodegenerative disease. Trends Neurosci. 32, 199–206.

Hebert, L.E., Weuve, J., Scherr, P.A., Evans, D.A., 2013. Alzheimer disease in the United States (2010-2050) estimated using the 2010 census. Neurology 80, 1778–1783.

Hedström, A.K., Bäärnhielm, M., Olsson, T., Asfredsson, L., 2011. Exposure to environmental tobacco smoke is associated with increased risk for multiple sclerosis. Mult. Scler. J. 17, 788–793.

Helou, R., Jaecker, P., 2014. Occupational exposure to mineral turpentine and heavy fuels: a possible risk factor for Alzheimer's disease. Dement. Geriatr. Cogn. Disord. 4, 160–171.

Heneka, M.T., Carson, M.J., El Khoury, J., Landreth, G.E., Brosseron, F., Feinstein, D.L., Jacobs, A.H., Wyss-Coray, T., Vitorica, J., et al., 2015. Neuroinflammation in Alzheimer's disease. Lancet 14, 388–405.

Herring, A., Donath, A., Steiner, K.M., Widera, M.P., Hamzehian, S., Kanakis, D., Kölble, K., ElALi, A., et al., 2012. Reelin depletion is an early phenomenon of Alzheimer's pathology. J. Alzheimers Dis. 30, 963–979.

Heydarpour, P., Amini, H., Khoshkish, S., Seidkhani, H., Sahraian, M.A., Yunesian, M., 2014. Potential impact of air pollution on multiple sclerosis in Tehran, Iran. Neuroepidemiology 43, 233–238.

Hiesberger, T., Trommsdorff, M., Howell, B.W., Goffinet, A., Mumby, M.C., Cooper, J.A., Herz, J., 1999. Direct binding of reelin to VLDL receptor and ApoE receptor 2 induces tyrosine phosphorylation of disabled-1 and modulates tau phosphorylation. Neuron 24, 481–489.

Hirsch, E.C., Vyas, S., Hunot, S., 2012. Neuroinflammation in Parkinson's disease. Parkinsonism Relat. Disord. 18 (S1), S210–S212.

Hopkins, S.J., 2007. Central nervous system recognition of peripheral inflammation: a neural, hormonal collaboration. Acta Biomed. 78 (Suppl. 1), 231–247.

Horgusluoglu, E., Nudelman, K., Nho, K., Saykin, A.J., 2017. Adult neurogenesis and neurodegenerative diseases: a systems biology perspective. Am. J. Med. Genet. B Neuropsychiatr. Genet. 174, 93–112.

Hougard, K.S., Jensen, K.A., Nordly, P., Taxvig, C., Vogel, U., Saber, A.T., Wallin, H., 2008. Effects of prenatal exposure to diesel exhaust particles on postnatal development, behavior, genotoxicity and inflammation in mice. Part. Fibre Toxicol. 5, 3.

Howard, J., Trevick, S., Younger, D.S., 2016. Epidemiology of multiple sclerosis. Neurol. Clin. 34, 919–939.

Hoxha, M., Dioni, L., Bonzini, M., Pesatori, A.C., Fustinoni, S., Cavallo, D., et al., 2009. Association between leukocyte telomere shortening and exposure to traffic pollution: a cross-sectional study on traffic officers and indoor office workers. Environ. Health 8, 41.

Huang, W.J., Zhang, X., Chen, W.W., 2016. Role of oxidative stress in Alzheimer's disease. Biomed. Rep. 4, 519–522.

Ishii, K., Nagai, T., Hirota, Y., Noda, M., Nabeshima, T., Yamada, K., Kubo, K., Nakajima, K., 2015. Reelin has a preventive effect on phencyclidine-induced cognitive and sensory-motor gating deficits. Neurosci. Res. 96, 30–36.

Jardim, M.J., 2011. MicroRNAs: implications for air pollution research. Mutat. Res. 717, 38–45.

Jossin, Y., 2004. Neuronal migration and the role of reelin during early development of the cerebral cortex. Mol. Neurobiol. 30, 225–251.

Ju, L.S., Jia, M., Sun, J., Sun, X.R., Zhang, H., Ji, M.H., Yang, J.J., Wang, Z.Y., 2016. Hypermethylation of hippocampal synaptic plasticity-related genes is involved in neonatal sevoflurane exposure-induced cognitive impairments in rats. Neurotox. Res. 29, 243–255.

Jung, C.R., Lin, Y.T., Hwang, B.F., 2015. Ozone, particulate matter, and newly diagnosed Alzheimer's disease: a population-based cohort study in Taiwan. J. Alzheimers Dis. 44, 573–584.

Kalkbrenner, A.E., Windham, G.C., Serre, M.L., Akita, Y., Wang, X., Hoffman, K., Thayer, B.P., Daniels, J.L., 2015. Particulate matter exposure, prenatal and postnatal windows of susceptibility, and autism spectrum disorders. Epidemiology 26, 30–42.

Karagulian, F., Belis, C.A., Dora, C.F.C., Prüss-Ustün, A.M., Bonjour, S., Adair-Rohani, H., Amann, M., 2015. Contributions to cities' ambient particulate matter: a systematic review of local source contributions at global levels. Atmos. Environ. 120, 475–483.

Karim, S., Mirza, Z., Ansari, S.A., Rasool, M., Igbal, Z., Sohrab, S.S., Kamal, M.A., Abuzenadah, A.M., Al-Qahtani, M.H., 2014. Transcriptomics study of neurodegenerative disease: emphasis on synaptic dysfunction mechanism in Alzheimer's disease. CNS Neurol. Disord. Drug Targets 13, 1202–1212.

Khan, S.S., Bloom, G.S., 2016. Tau: the center of signaling nexus in Alzheimer's disease. Front. Neurosci. 10, 31.

Khan, S.A., Khan, S.A., Narendra, A.R., Mushtaq, G., Zahran, S.A., Khan, S., Kamal, M.A., 2016. Alzheimer's disease and autistic spectrum disorder: is there any association? CNS Neurol. Disord. Drug Targets 15, 390–402.

Killin, L.O.J., Starr, J.M., Shiue, I.I., Russ, T.C., 2016. Environmental risk factors for dementia: a systematic review. BMC Geriatr. 16, 175.

Kim, S.H., Knight, E.M., Saunders, E.L., Cuevas, A.K., Popovech, M., Chen, L.C., Gandy, S., 2012. Rapid doubling of Alzheimer's amyloid-$\beta40$ and 42 levels in brain of mice exposed to a nickel nanoparticle model of air pollution. F1000Res. 1, 70.

Kioumourzoglou, M.A., Schwartz, J.D., Weisskopfs, M.G., Melly, S.J., Wang, Y., Dominici, F., Zanobetti, A., 2016. Long-term $PM_{2.5}$ exposure and neurological hospital admissions in the Northeastern United States. Environ. Health Perspect. 124, 23–29.

Kirrane, E.F., Bowman, C., Davis, J.A., Hoppin, J.A., Blair, A., Chen, H., Patel, M.M., Sandler, D.P., Tanner, C.M., Vinikoor-Imler, L., Ward, M.H., Luben, T.J., Kamel, F., 2015. Associations of ozone and PM2.5 concentrations with Parkinson's disease among participants in the agricultural health study. J. Occup. Environ. Med. 57, 509–517.

Kocherhans, S., Madhusudan, A., Doehner, J., Breu, K.S., Nitsch, R.M., Fritschy, J.M., Knuesel, I., 2010. Reduced reelin expression accelerates amyloid-β plaque formation and tau pathology in transgenic Alzheimer's disease mice. J. Neurosci. 30, 9228–9240.

Koeman, T., Schouten, L.J., van den Brandt, P.A., Slottje, P., Huss, A., Peters, S., Kromhout, H., Vermeulen, R., 2015. Occupational exposures and risk of dementia-related mortality in the prospective Netherlands cohort study. Am. J. Ind. Med. 58, 625–635.

Komine, O., Yamanaka, K., 2015. Neuroinflammation in motor neuron disease. Nagoya J. Med. Sci. 77, 537–549.

Kraft, A.D., Harry, G.J., 2011. Features of microglia and neuroinflammation relevant to environmental exposure and neurotoxicity. Int. J. Environ. Res. Public Health 8, 2980–3018.

Kramer, P.L., Xu, H., Woltjer, R.L., Westaway, S.K., Clark, D., Erten-Lyons, D., Kaye, J.A., Welsh-Bohmer, K.A., et al., 2010. Alzheimer's disease pathology in cognitive healthy elderly: a genome-wide study. Neurobiol. Aging 32, 2113–2122.

Krstic, D., Madhusudan, A., Doehner, J., Vogel, P., Notter, T., Imhof, C., Manalastas, A., Hilfiker, M., Pfister, S., Schwerder, C., Riether, C., Meyer, U., 2012. Systemic immune challenges trigger and drive Alzheimer-like neuropathology in mice. J. Neuroinflammation 9, 151.

Kullmann, J.P., 2015. Letter to the editor on "Exposure to hazardous air pollutants and the risk of amyotrophic lateral sclerosis" Environ. Pollut. 207, 431.

Lajud, N., Torner, L., 2015. Early life stress and hippocampal neurogenesis in the neonate: sexual dimorphism, long term consequences and possible mediators. Front. Mol. Neurosci. 8, 3.

Langa, K.M., Levine, D.A., 2014. The diagnosis and management of mild cognitive impairment. A clinical review. JAMA 312, 2551–2561.

Lazarini, F., Lledo, P.M., 2011. Is adult neurogenesis essential for olfaction? Trends Neurosci. 34, 20–30.

Lee, P.C., Liu, L.L., Sun, Y., Chen, Y.A., Liu, C.C., Li, C.Y., Yu, H.L., Ritz, B., 2016a. Traffic-related air pollution increased the risk of Parkinson's disease in Taiwan: a nationwide study. Environ. Int. 96, 75–81.

Lee, P.C., Raaschou-Nielsen, O., Lill, C.M., Bertram, L., Sinsheimer, J.S., Hansen, J., Ritz, B., 2016b. Gene-environment interactions linking air pollution and inflammation in Parkinson's disease. Environ. Res. 151, 713–720.

Lema Tomé, C.M., Tyson, T., Rey, N.L., Grathwohl, S., Britschgi, M., Brundin, P., 2013. Inflammation and α-synuclein prion-like behavior in Parkinson's disease—is there a link? Mol. Neurobiol. 47, 561–574.

Levesque, S., Surace, M.J., McDonald, J., Block, M.L., 2011. Air pollution and the brain: subchronic diesel exhaust exposure causes neuroinflammation and elevates early markers of neurodegenerative disease. J. Neuroinflammation 8, 105.

Levy, S.E., Mandell, D.S., Schultz, R.T., 2009. Autism. Lancet 374, 1627–1638.

Lewerenz, J., Maher, P., 2015. Chronic glutamate toxicity in neurodegenerative diseases-what is the evidence? Front. Neurosci. 9, 469.

Lewis, J., Baddeley, A.D., Bonham, K.G., Lovett, D., 1970. Traffic pollution and mental efficiency. Nature 225, 95–97.

Li, R., Strykowski, R., Meyer, M., Mulcrone, P., Krakora, D., Suzuki, M., 2012. Male-specific differences in proliferation, neurogenesis, and sensitivity to oxidative stress in neural progenitor cells derived from a rat model of ALS. PLoS One 7 (11), e48581.

Liu, F., Huang, Y., Zhang, F., Chen, Q., Wu, B., Rui, W., Zhang, J.C., Ding, W., 2015. Macrophages treated with particulate matter $PM_{2.5}$ induce selective neurotoxicity through glutaminase-mediated glutamate generation. J. Neurochem. 134, 315–326.

Liu, L., Shen, P., He, T., Chang, Y., Shi, L., Tao, S., Li, X., Xun, Q., Guo, X., Yu, Z., Wang, J., 2016a. Noise induced hearing loss impairs spatial learning/memory and hippocampal neurogenesis in mice. Sci. Rep. 6, 20374.

Liu, R., Young, M.T., Chen, J.C., Kaufman, J.D., Chen, H., 2016b. Ambient air pollution exposures and risk of Parkinson disease. Environ. Health Perspect. 124, 1759–1765.

Lo, R.Y., Tanner, C.M., 2014. Parkinson's disease epidemiology. In: Aminoff, M.J., Daroff, R.B. (Eds.), second ed. In: The Encyclopedia of Neurological Sciences, vol. 3. Elsevier, New York, pp. 833–839.

Lodovici, M., Bigagli, E., 2011. Oxidative stress and air pollution exposure. J. Toxicol. 2011, 9, 487074.

Loop, M.S., Kent, S.T., Al-Hamdan, M.Z., Crosson, W.L., Estes, S.M., Ester, M.G., Quattrochi, D.A., Hemmings, S.N., Wadley, V.G., McClure, L.A., 2013. Fine particulate matter and incident cognitive impairment in the REsons for Geographic and Racial Differences in Stroke (REGARDS) cohort. PLoS One 8, e75001.

Lucchini, R.G., Dorman, D.C., Elder, A., Veronesi, B., 2012. Neurological impacts from inhalation of pollutants and the nose-brain connection. Neurotoxicology 33, 838–841.

Lull, M.E., Block, M.L., 2010. Microglial activation and chronic neurodegeneration. Neurotherapeutics 7, 354–365.

Ma, T., Klann, E., 2012. Amyloid beta: linking synaptic plasticity failure to memory disruption in Alzheimer's disease. J. Neurochem. 120 (Suppl. 1), 140–148.

Malek, A.M., Barchowsky, A., Bowser, R., Heiman-Patterson, T., Lacomis, D., Rana, S., Youk, A., Talbott, E.O., 2015. Exposure to hazardous air pollutants and the risk of amyotrophic lateral sclerosis. Environ. Pollut. 197, 181–186.

Manganas, L.N., Zhang, X., Li, Y., Hazel, R.D., Smith, S.D., Wagshul, M.E., Henn, F., Benveniste, H., Djuric, P.M., Enikolopov, G., Maletic-Savatic, M., 2007. Magnetic resonance spectroscopy identifies neural progenitor cells in the live human brain. Science 318, 980–985.

Manoharan, S., Guillemein, G.J., Abiramasundari, R.S., Essa, M.M., Akbar, M., Akbar, M.D., 2016. The role of reactive oxygen species in the pathogenesis of Alzheimer's disease, Parkinson's disease, and Huntington's disease: a mini review. Oxid. Med. Cell. Longev. 2016, 15, 8590578.

Manzetti, S., Andersen, O., 2016. Biochemical and physiological effects from exhaust emissions. A review of the relevant literature. Pathophysiology 23, 285–293.

Maresova, P., Klimova, B., Novotny, M., Kuca, K., 2016. Alzheimer's and Parkinson's diseases: expected economic impact on Europe-A call for a uniform European strategy. J. Alzheimers Dis. 54, 1123–1133.

Martens, D.S., Nawrot, T.S., 2016. Air pollution stress and the aging phenotype: the telomere connection. Curr. Environ. Health Rep. 3, 258–269.

Marxreiter, F., Regensburger, M., Winkler, J., 2013. Adult neurogenesis in Parkinson's disease. Cell. Mol. Life Sci. 70, 459–473.

McCracken, J., Baccarelli, A., Hoxha, M., Dioni, L., Melly, S., Coulll, B., et al., 2010. Annual ambient black carbon associated with shorter telomeres in elderly men: Veterans Affairs Normative Aging Study. Environ. Health Perspect. 118, 1564–1570.

Mesholam, R.I., Moberg, P.J., Mahr, R.N., Doty, R.L., 1998. Olfaction in neurodegenerative disease. A meta-analysis of olfactory functioning in Alzheimer's and Parkinson's diseases. Arch. Neurol. 55, 84–90.

Ming, G.L., Song, H., 2011. Adult neurogenesis in the mammalian brain: significant answers and significant questions. Neuron 70, 687–702.

Minter, M.R., Taylor, J.M., Crack, P.J., 2016. The contribution of neuroinflammation to amyloid toxicity in Alzheimer's disease. J. Neurochem. 136, 457–474.

Mohan Kumar, S.M.J., Campbell, A., Block, M., Veronesi, B., 2008. Particulate matter, oxidative stress and neurotoxicity. Neurotoxicology 29, 479–488.

Møller, P., Jacobsen, N.R., Folkmann, J.K., Danielsen, P.H., Mikkelsen, L., Hemmingsen, J.G., Vesterdal, L.K., Forchhammer, L., Wallin, H., Loft, S., 2010. Role of oxidative damage in toxicity of particulates. Free Radic. Res. 44, 1–46.

Morgan, T.E., Davis, D.A., Iwata, N., Tanner, J.A., Snyder, D., Ning, Z., Kam, W., Hsu, Y.T., Winkler, J.W., Chen, J.C., Petasis, N.A., Baudry, M., Sioutas, C., Finch, C.E., 2011. Glutamatergic neurons in rodent models respond to nanoscale particulate urban air pollutants in vivo and in vitro. Environ. Health Perspect. 119, 1003–1009.

Mota, S.I., Ferreira, I.L., Rego, A.C., 2014a. Dysfunctional synapse in Alzheimer's disease—a focus on NMDA receptors. Neuropharmacology 76 (Pt. A), 16–26.

Mota, S.I., Ferreira, I.L., Valero, J., Ferreiro, E., Carvalho, A.L., Oliveira, C.R., Rego, A.C., 2014b. Impaired Src signaling and post-synaptic actin polymerization in Alzheimer's disease mice hippocampus-linking NMDA receptors and the reelin pathway. Exp. Neurol. 261, 698–709.

Moulton, P.V., Yang, W., 2012. Air pollution, oxidative stress, and Alzheimer's disease. J. Environ. Public Health 2012, 472751.

Mouradian, M.M., 2012. MicroRNAs in Parkinson's disease. Neurobiol. Dis. 46, 279–284.

Mumaw, C.L., Levesque, S., McGraw, C., Robertson, S., Lucas, S., Stafflinger, J.E., Campen, M.J., Hall, P., Norenberg, J.P., Anderson, T., Lund, A.K., McDonald, J.D., Ottens, A.K., Block, M.L., 2016. Microglial priming through the lung-brain axis: the role of air pollution-induced circulating factors. FASEB J. 30, 1880–1891.

Murphy-Royal, C., Dupuis, J., Groc, L., Oliet, S.H., 2017. Astroglial glutamate transporters in the brain: regulating neurotransmitter homeostasis and synaptic transmission. J. Neurosci. Res. (in press).

Murray, B., 2014. Amyotrophic lateral sclerosis. In: Aminoff, M.J., Daroff, R.B. (Eds.), second ed. In: The Encyclopedia of Neurological Sciences, vol. 1. Elsevier, New York, pp. 165–167.

Musgrove, R.E., Jewell, S.A., Di Monte, D.A., 2015. Overview of neurodegenerative disorders and susceptibility factors in neurodegenerative processes. In: Aschner, M., Costa, L.G. (Eds.), Environmental Factors in Neurodevelopmental and Neurodegenerative Disorders. Academic Press/Elsevier, New York, pp. 197–210.

Noh, J.S., Sharma, R.P., Veldic, M., Salvacion, A.A., Jia, X., Chen, Y., Costa, E., Guidotti, A., Grayson, D.R., 2005. DNA methyltransferase 1 regulates reelin mRNA expression in mouse primary cortical cultures. Proc. Natl. Acad. Sci. U.S.A. 102, 1749–1754.

Novais, A.R.B., Guiramand, J., Cohen-Sola, C., Crouzin, N., de Jesus Ferreira, M.C., Vignes, M., Barbanel, G., Cambonie, G., 2013. N-Acetyl-cysteine prevents pyramidal cell disarray and reelin-immunoreactive neuron deficiency in CA3 after prenatal immune challenge in rats. Pediatr. Res. 73, 750–755.

Numan, M.S., Brown, J.P., Michou, L., 2015. Impact of air pollutants on oxidative stress in common autophagy-mediated aging diseases. Int. J. Environ. Res. Public Health 12, 2289–2305.

O'Brien, J.T., Thomas, A., 2015. Vascular dementia. Lancet 386, 1698–1706.

O'Connell, R.M., Rao, D.S., Baltimore, D., 2012. MicroRNAs regulation of inflammatory responses. Annu. Rev. Immunol. 30, 295–312.

Oberdoerster, G., Utell, M.J., 2002. Ultrafine particles in the urban air: to the respiratory tract—and beyond? Environ. Health Perspect. 110, A440–A441.

Oberdoerster, G., Sharp, Z., Atudorei, V., Elder, A., Gelein, R., Kreyling, W., Cox, C., 2004. Translocation of inhaled ultrafine particles to the brain. Inhal. Toxicol. 16, 437–445.

Ohkubo, N., Lee, Y.D., Morishima, A., Terashima, T., Kikkawa, S., Tohyama, M., Sakanaka, M., Tanaka, J., Maeda, N., Vitek, M.P., Mitsuda, N., 2003. Apolipoprotein E and reelin ligands modulate tau phosphorylation through apolipoprotein E receptor/disabled-1/glycogen synthase kinase-3β cascade. FASEB J. 17, 295–297.

Oudin, A., Forsberg, B., Adolfsson, A.N., lind, N., Modig, L., Nordin, M., Nordin, S., Adolfsson, R., Nilsson, L.G., 2016. Traffic-related air pollution and dementia incidence in Northern Sweden: a longitudinal study. Environ. Health Perspect. 124, 306–312.

Palacios, N., Fitzgerald, K.C., Hart, J.E., Weisskopf, M.G., Schwarzschild, M.A., Ascherio, A., Laden, F., 2014. Particulate matter and risk of Parkinson's disease in a large prospective study of women. Environ. Health 13, 80.

Palacios-Garcia, I., Lara-Vasquez, A., Montiel, J.F., Diaz-Veliz, G.F., Sepulveda, H., Utreras, E., Montecino, M., Gonzalez-Billault, C., Aboitiz, F., 2015. Prenatal stress down-regulates reelin expression by methylation of its promoter and induces adult behavioral impairment in rats. PLoS One 10 (2), e0117680.

Paoletti, P., Bellone, C., Zhou, Q., 2013. NMDA receptor subunit diversity: impact on receptor properties, synaptic plasticity and disease. Nat. Rev. Neurosci. 14, 383–400.

Perry, D.C., 2014. Dementia. In: Aminoff, M.J., Daroff, R.B. (Eds.), second ed. In: The Encyclopedia of Neurological Sciences, vol. 3. Elsevier, New York, pp. 962–969.

Peters, A., Veronesi, B., Calderon-Garciduenas, L., Gehr, P., Chen, L.C., Geiser, M., Reed, W., Rothen-Rutishauser, B., Schurch, S., Schulz, H., 2006. Translocation and potential neurological effects of fine and ultrafine particles a critical update. Part. Fibre Toxicol. 3, 13.

Peters, R., Peters, J., Booth, A., Mudway, I., 2015. Is air pollution associated with increased risk of cognitive decline? A systematic review. Age Ageing 44, 755–760.

Pieters, N., Janssen, B.G., Dewitte, H., Cox, B., Cuypers, A., Lefebvre, W., Smeets, K., Vanpoucke, C., Plusquin, M., Nawrot, T.S., 2016. Biomolecular markers within the core axis of aging and particulate air pollution exposure in the elderly: a cross-sectional study. Environ. Health Perspect. 124, 943–950.

Pike, C.J., 2017. Sex and the development of Alzheimer's disease. J. Neurosci. Res. 95, 671–680.

Polymeropoulos, M.H., Lavedan, C., Leroy, E., Ide, S.E., Dehejia, A., Dutra, A., pike, B., Root, H., et al., 1997. Mutation in the α-synuclein gene identified in families with Parkinson's disease. Science 276, 2045–2047.

Pope, C.A., Burnett, R.T., Thurston, G.D., Thun, M.J., Calle, E.E., Krewski, D., Godleski, J.J., 2004. Cardiovascular mortality and long-term exposure to particulate air pollution. Circulation 109, 71–77.

Power, M.C., Weisskopf, M.G., Alexeeff, S.E., Coull, B.A., Spiro III, A., Schwartz, J., 2011. Traffic-related air pollution and cognitive function in a cohort of older men. Environ. Health Perspect. 119, 682–687.

Power, M.C., Adar, S.D., Yanosky, J.D., Weuve, J., 2016. Exposure to air pollution as a potential contributor to cognitive function, cognitive decline, brain imaging, and dementia: a systematic review of epidemiologic research. Neurotoxicology 56, 235–253.

Pressman, P., Rabinovici, G.D., 2014. Alzheimer's disease. In: Aminoff, M.J., Daroff, R.B. (Eds.), second ed. In: The Encyclopedia of Neurological Sciences, vol. 3. Elsevier, New York, pp. 122–127.

Prince, M., Bryce, R., Albanese, E., Wimo, A., Ribeiro, W., Ferri, C.P., 2013. The global prevalence of dementia: a systematic review and meta-analysis. Alzheimers Dement. 9, 63–75.

Pronk, A., Coble, J., Stewart, P.A., 2009. Occupational exposure to diesel engine exhaust: a literature review. J. Expo. Sci. Environ. Epidemiol. 19, 443–457.

Pujadas, L., Rossi, D., Andres, R., Teixeira, C.M., Serra-Vidal, B., Parcerisa, A., Maldonado, R., Giralt, E., Carulla, N., Soriano, E., 2014. Reelin delays amyloid-beta fibril formation and rescues cognitive deficits in a model of Alzheimer's disease. Nat. Commun. 5, 3443.

Pun, V.C., Manjourides, J., Suh, H., 2017. Association of ambient air pollution with depressive and anxiety symptoms in older adults: results from the NSHAP study. Environ. Health Perspect. 125, 342–348.

Qian, L., Flood, P.M., Hong, J.S., 2010. Neuroinflammation is a key player in Parkinson's disease and a prime target for therapy. J. Neural Transm. 117, 971–979.

Qiu, L., Zhang, W., Tan, E.K., Zeng, L., 2014. Deciphering the function and regulation of microRNAs in Alzheimer's disease and Parkinson's disease. ACS Chem. Nerosci. 5, 884–894.

Ranft, U., Schikowski, T., Sugiri, D., Krutmann, J., Krämer, U., 2009. Long-term exposure to traffic-related particulate matter impairs cognitive function in the elderly. Environ. Res. 109, 1004–1011.

Ransohoff, R.M., 2016. How neuroinflammation contributes to neurodegeneration. Science 353, 777–783.

Ribeiro, F.M., Vieira, L.B., Pires, R.G., Olmo, R.P., Ferguson, S.S., 2017. Metabotropic glutamate receptors and neurodegenerative diseases. Pharmacol. Res. 115, 179–191.

Riedel, B.C., Thompson, P.M., Brinton, R.D., 2016. Age, APOE and sex: triad of risk of Alzheimer's disease. J. Steroid Biochem. Mol. Biol. 160, 134–147.

Riise, T., Nortvedt, M.W., Ascherio, A., 2003. Smoking is a risk factor for multiple sclerosis. Neurology 61, 1122–1124.

Ritz, B., Lee, P.C., Hansen, J., Lassen, C.F., Ketzel, M., Sorensen, M., Raaschou-Nielsen, O., 2016. Traffic related air pollution and Parkinson's disease in Denmark: a case-control study. Environ. Health Perspect. 124, 351–356.

Roberts, R.O., Boardman, L.A., Cha, R.H., Pankratz, V.S., Johnson, R.A., Druliner, B.R., et al., 2014. Short and long telomeres increase risk of amnestic mild cognitive impairment. Mech. Ageing Dev. 141–142, 64–69.

Rodriguez, M.J., Mahy, N., 2016. Neuron-microglia interactions in motor neuron degeneration. The inflammatory hypothesis in amyotrophic lateral sclerosis revisited. Curr. Med. Chem. 23, 4753–4772.

Rogers, J.T., Rusiana, I., Trotter, J., Zhao, L., Donaldson, E., Pak, D.T., Babus, L.W., Peters, M., et al., 2011. Reelin supplementation enhances cognitive ability, synaptic plasticity, and dendritic spine density. Learn. Mem. 18, 558–564.

Roqué, P.J., Dao, K., Costa, L.G., 2016. Microglia mediate diesel exhaust particle-induced cerebellar neuronal toxicity through neuroinflammatory mechanisms. Neurotoxicology 56, 204–214.

Rubio-Perez, J.M., Morillas-Ruiz, J.M., 2012. A review: inflammatory process in Alzheimer's disease, role of cytokines. ScientificWorldJournal 2012, 15, 756357.

Rücker, R., Schneider, A., Breitner, S., Cyrys, J., Peters, A., 2011. Health effects of particulate air pollution: a review of the epidemiological evidence. Inhal. Toxicol. 23, 555–592.

Rudy, C.C., Hunsberger, H.C., Weitzner, D.S., Reed, M.N., 2015. The role of the tripartite glutamatergic synapse in the pathophysiology of Alzheimer's disease. Aging Dis. 6, 131–148.

Sawcer, S., Franklin, R.J., Ban, M., 2014. Multiple sclerosis genetics. Lancet Neurol. 13, 700–709.

Scheltens, P., Blennow, K., Breteler, M.B.B., de Strooper, B., Salloway, S., Van der Flier, W.M., 2016. Alzheimer's disease. Lancet 388, 505–517.

Schikowski, T., Vossoughi, M., Vierkötter, A., Schulte, T., Teichert, T., Sugiri, D., Fehsel, K., Tzivian, L., Bae, I.S., Ranft, U., Hoffmann, B., Probst-Hensch, N., Herder, C., Krämer, U., Luckhaus, C., 2015. Association of air pollution with cognitive functions and its modification by APOE gene variants in elderly women. Environ. Res. 142, 10–16.

Selkoe, D.J., Hardy, J., 2016. The amyloid hypothesis of Alzheimer's disease at 25 years. EMBO Mol. Med. 8, 595–608.

Shaun, N., Thomas, B., 2012. The STAT3-DMNT1 connection. JAKSTAT 1 (4), 257–260.

Shin, R.K., 2014. Multiple sclerosis: diagnosis. In: Aminoff, M.J., Daroff, R.B. (Eds.), second ed. In: The Encyclopedia of Neurological Sciences, vol. 3. Elsevier, New York, pp. 148–152.

Shors, T.J., Miesegas, G., Beylin, A., Zhao, M., Rydel, T., Gould, E., 2001. Neurogenesis in the adult is involved in the formation of trace memories. Nature 410, 372–376.

Sonkoly, E., Pivarcsi, A., 2011. MicroRNAs in inflammation and response to injuries induced by environmental pollution. Mutat. Res. 717, 46–53.

Sonntag, K.C., Woo, T.W., Krichevsky, A.M., 2012. Converging miRNA functions in diverse brain disorders: a case for miR-124 and miR-126. Exp. Neurol. 235, 427–435.

Steiner, S., Bisg, C., Petri-Fink, A., Rothen-Rutishauser, B., 2016. Diesel exhaust: current knowledge of adverse effects and underlying cellular mechanisms. Arch. Toxicol. 90, 1541–1553.

Stranahan, A.M., Haberman, R.P., Gallagher, M., 2011. Cognitive decline is associated with reduced reelin expression in the entorhinal cortex of aged rats. Cereb. Cortex 21, 392–400.

Su, W., Aloi, M.S., Garden, G.A., 2016. MicroRNAs mediating CNS inflammation: small regulators with powerful potential. Brain Behav. Immun. 52, 1–8.

Sunyer, J., Esnaola, M., Alvarez-Pedrerol, M., Forns, J., Rivas, I., Lopez-Vicente, M., Suades-Gonzales, E., Foraster, M., Garcia-Esteban, R., et al., 2015. Association between traffic-related air pollution in schools and cognitive development in primary school children: a prospective cohort study. PLoS Med. 12, e1001792.

Suzuki, T., Oshio, S., Iwata, M., Saburi, H., Odagiri, T., Udagawa, T., Sugawara, I., Umezawa, M., Takeda, K., 2010. In utero exposure to a low concentration of diesel exhaust affects spontaneous locomotor activity and monoaminergic system in male mice. Part. Fibre Toxicol. 7, 7.

Tai, L.M., Ghura, S., Koster, K.P., Liakaite, V., Maiensen-Cline, M., Kanabar, P., Collins, N., Ben-Aissa, M., Lei, A.Z., Bahroos, N., Green, S., Hendrickson, B., Van Eldik, L.J., LaDu, M.J., 2015. APOE-modulated Aβ-induced neuroinflammation in Alzheimer's disease: current landscape, novel data, and future perspective. J. Neurochem. 133, 465–488.

Takahashi, M., Ko, L.W., Kulathingal, J., Jiang, P., Sevlever, D., Yen, S.H.C., 2007. Oxidative stress-induced phosphorylation, degradation and aggregation of α-synuclein are linked to upregulated CK2 and cathepsin D. Eur. J. Neurosci. 26, 863–874.

Thirtamara Rajamani, K., Doherty-Lyons, S., Bolden, C., Willis, D., Hoffman, C., Zelikoff, J., Chen, L.C., Gu, H., 2013. Prenatal and early life exposure to high level diesel exhaust particles leads to increased locomotor activity and repetitive behaviors in mice. Autism Res. 6, 248–257.

Thome, A.D., Harms, A.S., Volpicelli-Daley, L.A., Standaert, D.G., 2016. MicroRNA-155 regulates alpha-synuclein-induced inflammatory responses in models of Parkinson disease. J. Neurosci. 36, 2383–2390.

Thounaojam, M.C., Kaushik, D.K., Basu, A., 2013. MicroRNAs in the brain: its regulatory role I neuroinflammation. Mol. Neurobiol. 47, 1034–1044.

Tonne, C., Elbaz, A., Beevers, S., Singh-Manoux, A., 2014. Traffic-related air pollution in relation to cognitive function in older adults. Epidemiology 25, 674–681.

Tsukue, N., Watanabe, M., Kumamoto, T., Takano, H., Takeda, K., 2009. Perinatal exposure to diesel exhaust affects gene expression in mouse cerebrum. Arch. Toxicol. 83, 985–1000.

Tzivian, L., Dlugaj, M., Winkler, A., Weinmayr, G., Hennig, F., Fuks, K.B., Vossoughi, M., Schikowski, T., Weimar, C., Erbel, R., Jöckel, K.H., Moebus, S., Hoffman, B., On behalf of the Heinz Nixdorf Recall study Investigative Group, 2016. Long-term air pollution and traffic noise exposures and mild cognitive impairment in older adults: a cross-sectional analysis of the Heinz Nixdorf Recall study. Environ. Health Perspect. 124, 1361–1368.

Ulusoy, A., Di Monte, D.A., 2013. Alpha-synuclein elevation in human neurodegenerative diseases: experimental, pathogenetic, and therapeutic implications. Mol. Neurobiol. 47, 484–494.

USEPA (United States Environmental Protection Agency), 2002. Health Assessment Document for Diesel Engine Exhaust. National Center for Environmental Assessment, USEPA, Washington, DC, p. 669.

USEPA (United States Environmental Protection Agency), 2016. Integrated Review Plan for the National Ambient Air Quality Standards for Particulate Matter. 2016 USEPA, Washington, DC, p. 173.

Van Den Eeden, S.K., Tanner, C.M., Bernstein, A.L., Fross, R.D., Leimpeter, A., Bloch, D.A., Nelson, L.M., 2003. Incidence of Parkinson's disease: variation by age, gender, and race/ethnicity. Am. J. Epidemiol. 157, 1015–1022.

van Donkelaar, A., Martin, R.V., Brauer, M., Boys, B.L., 2015. Use of satellite observations for long-term exposure of global concentrations of fine particulate matter. Environ. Health Perspect. 123, 135–143.

Vojinovic, S., Savic, D., Lukic, S., Savic, L., Vojinovic, J., 2015. Disease relapses in multiple sclerosis can be influenced by air pollution and climate seasonal conditions. Vojnosanit. Pregl. 72, 44–49.

Volk, H.E., Hertz-Picciotto, I., Delwiche, L., Lurmann, F., McConnell, R., 2011. Residential proximity to freeways and autism in the CHARGE study. Environ. Health Perspect. 119, 873–877.

Volk, H.E., Lurmann, F., Penfold, B., Hertz-Picciotto, I., McConnell, R., 2013. Traffic-related air pollution, particulate matter, and autism. JAMA Psychiatry 70, 71–77.

Walker, Z., Possin, K.L., Boeve, B.F., Aarsland, D., 2015. Lewy body dementias. Lancet 386, 1683–1697.

Ward, A., Crean, S., Mercald, C.J., Collins, J.M., 2012. Prevalence of apolipoprotein E4 genotype and homozygotes (APOE e4/4) among patients diagnosed with Alzheimer's disease: a systematic review and meta-analysis. Neuroepidemiology 38, 1–17.

Weeber, E.J., Beffert, U., Jones, C., Christian, J.M., Forster, E., Sweatt, J.D., Herz, J., 2002. Reelin and APoE receptors cooperate to enhance hippocampal synaptic plasticity and learning. J. Biol. Chem. 277, 39944–39952.

Weldy, C.S., White, C.C., Wilkerson, H.W., Larson, T.V., Stewart, J.A., Gill, S.E., Parks, W.C., Kavanagh, T.J., 2012. Heterozygosity in the glutathione synthesis gene Gclm increases sensitivity to diesel exhaust particulate induced lung inflammation in mice. Inhal. Toxicol. 2012, 23: 724–735.

Wellenius, G.A., Boyle, L.D., Coull, B.A., Milberg, W.P., Gryparis, A., Schwartz, J., Mittleman, M.A., Lipsitz, L.A., 2012. Residential proximity to nearest major roadway and cognitive function in community-dwelling seniors: results from the MOBILIZE Boston study. J. Am. Geriatr. Soc. 60, 2075–2080.

Weng, N.P., Granger, L., Hodes, R.J., 1997. Telomere lengthening and telomerase activation during human B cell differentiation. Proc. Natl. Acad. Sci. U.S.A. 94, 10827–10832.

Weuve, J., Puett, R.C., Schwartz, J., Yanosky, J.D., Laden, F., Grodstein, F., 2012. Exposure to particulate air pollution and cognitive decline in older women. Arch. Intern. Med. 172, 219–227.

Wilker, E.H., Preis, S.R., Beiser, A.S., Wolf, P.A., Au, R., Kloog, I., Li, W., Schwartz, J., Koutrakis, P., DeCArli, C., Seshadri, S., Mittleman, M.A., 2015. Long-term exposure to fine particulate matter, residential proximity to major roads and measures of brain structure. Stroke 46, 1161–1166.

Wilker, E.H., Mertinez-Ramirez, S., Kloog, I., Schwarz, J., Motofsky, E., Koutrakis, P., Mittleman, M.A., Viswanathan, A., 2016. Fine particulate matter, residential proximity to major roads, and markers of small vessel disease in a memory study population. J. Alzheimers Dis. 53, 1315–1323.

Win-Shwe, T.T., Fujimaki, H., 2011. Nanoparticles and neurotoxicity. Int. J. Mol. Sci. 12, 6267–6280.

Win-Shwe, T.T., Mitsushima, D., Yamamoto, S., Fujitani, Y., Funabashi, T., Hirano, S., Fujimaki, H., 2009. Extracellular glutamate level and NMDA receptor subunit expression in mouse olfactory bulb following nanoparticle-rich diesel exhaust exposure. Inhal. Toxicol. 21, 828–836.

Win-Shwe, T.T., Fujitani, Y., Kyi-Tha-Thu, C., Furuyama, A., Michikawa, T., Tsukahara, S., Nitta, H., Hirano, S., 2014. Effects of diesel engine exhaust origin secondary organic aerosols on novel object recognition ability and maternal behavior in BALB/C mice. Int. J. Environ. Res. Public Health 11, 11286–11307.

Woodbury, M.E., Freilich, R.W., Chang, C.J., Asai, H., Ikezu, S., Boucher, J.D., Slack, F., Ikezu, T., 2015. MiR-155 is essential for inflammation-induced hippocampal neurogenic dysfunction. J. Neurosci. 35, 9764–9781.

Wu, Y.C., Lin, Y.C., Yu, H.L., Chen, J.H., Chen, T.F., Sun, Y., Wen, L.L., Yip, P.K., Chu, Y.M., Chen, Y.C., 2015. Association between air pollutants and dementia risk in the elderly. Alzheimers Dement. 1, 220–228.

Xu, X., Ha, S.U., Basnet, R., 2016. A review of epidemiological research on adverse neurological effects of exposure to ambient air pollution. Front. Public Health 4, 157.

Yamammoto, M., Singh, A., Sava, F., Pui, M., Tebbutt, S.J., Carlsten, C., 2013. MicroRNA expression in response to controlled exposure to diesel exhaust: attenuation by the antioxidant N-acetylcysteine in a randomized crossover study. Environ. Health Perspect. 121, 670–675.

Ye, L., Huang, Y., Zhao, L., Li, Y., Sun, L., Zhou, Y., Qian, G., Zhang, J.C., 2013. IL-1β and TNF-α induce neurotoxicity through glutamate production: a potential role for neuronal glutaminase. J. Neurochem. 125, 897–908.

Yegambaram, M., Manivannan, B., Beach, T.G., Halden, R.U., 2015. Role of environmental contaminants in the etiology of Alzheimer's disease: a review. Curr. Alzheimer Res. 12, 116–146.

Yin, F., Lawal, A., Ricks, J., et al., 2013. Diesel exhaust induces systemic lipid peroxidation and development of dysfunctional pro-oxidant and pro-inflammatory high density lipoproteins. Arterioscler. Thromb. Vasc. Biol. 33, 1153–1161.

Yokota, S., Moriya, N., Iwata, M., Umezawa, M., Oshio, S., Takeda, K., 2013. Exposure to diesel exhaust during fetal period affects behavior and neurotransmitters in male offspring mice. J. Toxicol. Sci. 38, 13–23.

Yu, D., Fan, W., Wu, P., Deng, J., Liu, J., Niu, Y., Li, M., Deng, J., 2014. Characterization of hippocampal Cajal-retzius cells during development in a mouse model of Alzheimer's disease (Tg2576). Neural Regen. Res. 9, 394–401.

Yu, N.N., Tan, M.S., Yu, J.T., Xie, A.M., Tan, L., 2016. The role of reelin signaling in Alzheimer's disease. Mol. Neurobiol. 53, 5692–5700.

Zanobetti, A., Dominici, A., Wang, Y., Schwartz, J.D., 2014. A national case-crossover analysis of the short-term effect of $PM_{2.5}$ on hospitalizations and mortality in subjects with diabetes and neurological disorders. Environ. Health 13, 38.

Zeng, Y., Gu, D., Purser, J., Hoenig, H., Christakis, N., 2010. Associations of environmental factors with elderly health and mortality in China. Am. J. Public Health 100, 298–305.

Zhang, Y., Li, P., Feng, J., Wu, M., 2016. Dysfunction of NMDA receptors in Alzheimer's disease. Neurol. Sci. 37, 1039–1047.

FURTHER READING

Atik, A., Stewart, T., Zhang, J., 2016. Alpha-synuclein as a biomarker for Parkinson's disease. Brain Pathol. 26, 410–418.

Heusinkveld, H.J., Wahle, T., Campbell, A., Westerink, R.H.S., Tran, L., Johnston, H., Stone, V., Casse, F.R., Schins, R.P.F., 2016. Neurodegenerative and neurological disorders by small inhaled particles. Neurotoxicology 56, 94–106.

Qiu, C., Fratiglioni, L., 2015. A major role for cardiovascular burden in age-related cognitive decline. Nat. Rev. Cardiol. 12, 267–277.

CHAPTER TWO

The Catecholaminergic Neurotransmitter System in Methylmercury-Induced Neurotoxicity

Marcelo Farina*, Michael Aschner†,1, João B.T. da Rocha‡

*Centro de Ciências Biológicas, Universidade Federal de Santa Catarina, Florianópolis, SC, Brazil
†Albert Einstein College of Medicine, Bronx, NY, United States
‡Centro de Ciências Naturais e Exatas, Universidade Federal de Santa Maria, Santa Maria, RS, Brazil
1Corresponding author: e-mail address: michael.aschner@einstein.yu.edu

Contents

1. INTRODUCTION

Catecholamines, including dopamine and norepinephrine, are biological amines that display important roles in the homeostasis of several biological functions. Within the central nervous system (CNS), dopamine and norepinephrine play major roles as neurotransmitters, taking part in the proper functioning of critical neuronal pathways linked to motor function, cognition, emotion, memory, and endocrine modulation. Impairments in catecholaminergic neurotransmission are implicated in neurologic and neuropsychiatric disorders. Moreover, catecholaminergic neurons represent important targets that mediate the deleterious effects of neurotoxic agents, such as metals and metalloids, pesticides, among others. Methylmercury (MeHg) is an organic mercury compound ubiquitously present in nature due to both natural and anthropogenic sources. MeHg is produced mainly as a consequence of the methylation of inorganic mercury in the aquatic environment (in a reaction catalyzed by reducing bacteria) and is bio-magnified within the aquatic food chain, reaching high levels in predatory fish. Human populations, exposed to MeHg, mainly due to the ingestion of predatory fish, may develop neurological symptoms. Even though the glut-amatergic neurotransmitter system has been noted as a major target medi-ating MeHg-induced neurotoxicity, several studies also affirm that the catecholaminergic system is involved in several toxic effects of MeHg. This chapter will focus on the toxic effects of MeHg toward the catecholamin-ergic system, with particular emphasis on the potential neurotoxic effects and neurological consequences resulting from MeHg exposure, as well as related mechanisms.

2. THE CATECHOLAMINERGIC NEUROTRANSMITTER SYSTEM: GENERAL ASPECTS

Catecholamines, including dopamine and norepinephrine (also called noradrenaline), are biological amines derived from the amino acid tyrosine. In order to synthesize dopamine, tyrosine is first converted to dihydroxy-phenylalanine (DOPA) in a reaction catalyzed by tyrosine hydroxylase (TH). DOPA is then decarboxylated by DOPA decarboxylase to yield dopa-mine. The generation of norepinephrine from dopamine involves the hydroxylation catalyzed by dopamine β–hydroxylase (DBH). Fig. 1 depicts the synthesis of dopamine and norepinephrine from the nonessential amino acid tyrosine.

Within the CNS, dopamine and norepinephrine act as neurotransmit-ters; the cell groups producing catecholamines are localized in discrete brain regions and project their axon terminals to a wide range of target areas that play important roles in CNS functions (Kobayashi, 2001). *Dopaminergic* cells are mainly localized in the substantia nigra (A9 cell group) and ventral teg-mental area (A10 cell group) (Lindvall et al., 1983). Neurons from the A9 group innervate the caudate–putamen to form the nigrostriatal pathway, which has a pivotal role in motor control (Gerfen, 1992). Neurons from the A10 group project their fibers to the nucleus accumbens, amygdala,

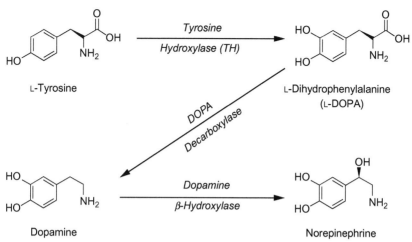

Fig. 1 Dopamine and norepinephrine biosynthesis. Tyrosine, a nonessential amino acid, is first converted to dihydroxyphenylalanine (DOPA) in a reaction catalyzed by tyrosine hydroxylase (TH). DOPA is then decarboxylated by DOPA decarboxylase to yield dopamine. The generation of norepinephrine from dopamine involves the hydroxylation catalyzed by dopamine β-hydroxylase, which is commonly used as a marker for noradrenergic neurons.

and prefrontal cortex, forming the mesocorticolimbic pathway, which is involved in emotion, motivation, and memory formation (Le Moal and Simon, 1991). Dopaminergic neurons from the A11–A14 groups are present in the mediobasal region of the hypothalamus, projecting their fibers to the median eminence and pituitary gland (Moore and Lookingland, 1995), modulating pituitary gland function.

The *norepinephrine* system originates from the locus coeruleus (A6 cell group) and the lateral tegmental area in the brain stem (A1, A2, A5, and A7 cell groups) (Moore and Card, 1984). Neurons from the A6 group innervate the cerebral cortex, amygdala, hippocampus, and thalamus, forming the dorsal norepinephrinergic pathway. The ventral norepinephrinergic pathway represents an additional way that innervates mainly the hypothalamus and septum. These neurons are implicated in both cognitive (attention, memory) and vegetative functions, such as neuroendocrine and autonomic regulation (Robbins and Everitt, 1995).

Several behavioral and pharmacological approaches have been used to understand brain functions mediated by catecholamines. Using a mutant phenotype of mice genetically impaired in dopamine or norepinephrine biosynthesis, Kobayashi and coworkers reported behavioral and physiologic roles of these two catecholamines in the CNS (Kobayashi, 2001). Using mutant mice defective in dopamine biosynthesis, in which the expression of TH in norepinephrinergic neurons was rescued with the DBH gene promoter (functional in norepinephrinergic cells), the authors observed a significant reduction of dopamine in various encephalic structures (forebrain, midbrain, hindbrain, and pituitary gland) compared with wild-type animals, whereas norepinephrine levels were normal. The dopamine-deficient mice displayed a significant reduction in spontaneous locomotor activity, exhibited cataleptic behavior, and were insensitive to the methamphetamine treatment, which normally induces the hyperactivity of locomotion (Kobayashi, 2001). These observations, derived from studies with genetically modified mice, are in line with the crucial role of dopamine in motor function in humans (Ko and Strafella, 2012).

In addition to motor dysfunction, dopamine-deficient mice displayed defects in certain emotional learning paradigms, such as the active avoidance; these defects seem to be related to dopamine depletion in nucleus accumbens (McCullough et al., 1993). In agreement with these data from experimental studies with genetically modified mice, pharmacological studies have shown the crucial role of dopamine in memory and cognition. In an in vivo study using mice treated with the D2 dopamine receptor agonist quinpirole (administered into the ventral pallidum), Lénárd et al. (2017) showed that the activation of the D2 dopamine receptors in the ventral pallidum facilitates memory consolidation as well as memory retention in inhibitory avoidance paradigm. In humans, striatal dopamine D2 receptors play a critical role in enabling the flexible updating and manipulation of information in working memory (Dodds et al., 2009). However, the role of dopamine in memory seems to depend on the type of receptor involved, given that D3 receptor blockade may enhance cognitive performance in healthy individuals and treat cognitive dysfunction in individuals with a neuropsychiatric disorder (Nakajima et al., 2013).

With respects to norepinephrine, genetic manipulation procedures were able to generate heterozygous mutant mice with reduced norepinephrine metabolism in the brain; the release level through repetitive high-K^+ stimulation was reduced in the mutant mice (56% of the wild-type level). These heterozygous mutant mice displayed deficits in three kinds of associative learning paradigms, including active avoidance, cued fear conditioning, and conditioned taste aversion (Kobayashi, 2001). Notably, treatment with desipramine, an inhibitor of norepinephrine uptake, recovered the memory deficits in the mutant mice, confirming that the central norepinephrine system plays a key role in long-term memory formation of conditioned learning. With respect to brain regions involved in long-term memory of conditioned learning, several behavioral studies have indicated that the associative learning requires the amygdala and its linking pathways, including the cerebral cortex (Everitt et al., 1991; Yamamoto et al., 1995). In line with the aforementioned evidence, Veyrac et al. (2009) showed that labetalol, a mixed β- and α1-adrenoceptor antagonist, blocked the improve in short-term olfactory memory and neurogenesis in the olfactory bulb of mice subjected to the environmental enrichment. In addition, desipramine, an inhibitor of norepinephrine uptake, has shown to improve cognition/memory in both animals (Feltmann et al., 2015) and humans (Mokhber et al., 2014).

The aforementioned evidence indicates that both catecholamines play important roles in motor control, emotional learning, and memory formation. In line with this, disruption of these neurotransmitter systems leads to motor and cognitive impairments, as discussed later.

3. CATECHOLAMINES IN NEUROLOGICAL AND NEUROPSYCHIATRIC DISORDERS

As already mentioned, catecholaminergic neurotransmission is involved in the regulation of variety of neurophysiological and behavioral processes, from relatively simple endocrine events (for instance, peripheral regulation of blood pressure) to complex mental activities (i.e., motivation, attention, learning, cognition, thinking, etc.) (Bromberg-Martin et al., 2010; Cools and D'Esposito, 2011; Huys et al., 2014; Johansen et al., 2011; Montes et al., 2015; Nieoullon, 2002). Disruption in the normal neurophysiology of catecholaminergic system has been implicated in the etiology of important neuropsychiatric disorders (Del Campo et al., 2011; Sigitova et al., 2017; Tritsch and Sabatini, 2012; Van Os and Kapur, 2009). In this context, a significant progress in our knowledge concerning the role of dopaminergic transmission in neuropsychiatry diseases was achieved by showing that the first generation of therapeutic antipsychotic agents (i.e., chlorpromazine, haloperidol), used to treat schizophrenia, had the ability of blocking dopaminergic receptors (Creese et al., 1976). Likewise, the reported relationship between dopamine depletion in the caudate nucleus of rodents and the appearance of Parkinson's diseases (PD)-related motor symptoms was instrumental in linking dopaminergic transmission with the neuropathology of this neurodegenerative condition. The major symptoms of PD, such as tremor, rigidity, difficulty to initiate movement, and postural hypotension, are associated with dopamine depletion in the substantia nigra and other interconnected brain regions. Although the degeneration of dopaminergic cells is more pronounced in PD, noradrenergic neurotransmission plays also an important role in the modulation of PD symptoms (Barbeau, 1970; Teychenne et al., 1985).

Nowadays, the importance of catecholaminergic neurotransmission in different types of motor, cognitive, and affective disorders has been validated (Nieoullon, 2002). The pivotal role of dopamine and norepinephrine in the regulation of normal behavior in experimental animals and humans and the crucial causal role of disruption of catecholaminergic neurotransmission in the etiology of neuropsychiatric disorders is now well-established. For instance, in addition to schizophrenia and PD, disturbances of catecholaminergic neurotransmission have been implicated in attention-deficit hyperactivity disorders (Van Enkhuizen et al., 2015), bipolar disorders (Sigitova et al., 2017; Van Enkhuizen et al., 2015), depression (Hamon and Blier, 2013), substance use or addiction disorders (Beaulieu and Gainetdinov, 2011; Ewing and Myers, 1985; Fitzgerald, 2013; Huys et al., 2014), eating disorders (Kaye et al., 2013; Volkow et al., 2011, 2013), Alzheimer's disease (Trillo et al., 2013), obsessive-compulsive disorder (Pauls et al., 2014; Tritsch and Sabatini, 2012), autism spectrum disorders (ASD) (Nguyen et al., 2014; Quaak et al., 2013), among others.

In the case of schizophrenia, the dopaminergic hyperactivity in the prefrontal cortex is thought to play a central role in the psychotic symptoms of this condition. Although both pre- and postsynaptic dopamine receptors located in different brain areas seem to be involved in schizophrenia (Howes et al., 2015; Nikolaus et al., 2014), particular attention has been given to the increase in the D2 dopamine receptors in the high-affinity state in schizophrenic patients (Beaulieu and Gainetdinov, 2011).

Alterations in dopaminergic neurotransmission can explain satisfactorily the symptomatology found in about 2/3 of schizophrenic patients, but the remaining cases cannot be attributed exclusively to dopaminergic hyperactivity. In fact, other neurotransmitter systems interact with the dopaminergic transmission to produce the typical psychotic (hallucination, disorganized speech, delusions, etc.) or negative symptoms (social withdraw, apathy, cognitive deficits, etc.) found in this neurological condition. For instance, changes in the glutamatergic neurotransmission have been implicated in the etiology of schizophrenia, particularly the N-methyl-D-aspartate (NMDA) receptor hypofunctioning in specific brain areas (Howes et al., 2015). Noradrenergic neurotransmission has also been suggested to contribute in the etiology of some cases of schizophrenia (Fitzgerald, 2014). Thus, as in the case of other neurological or psychiatry disorders, the disruption in the delicate interaction between different neurotransmitters seems to be more important than isolated changes in a single neurotransmitter system.

Factors that can be important to the emergence of neuropathological disorders are the exposure to stressful situations or neurotoxic agents during critical periods of brain development (Andersen, 2003; Rakers et al., 2017; Rice and Barone, 2000). Indeed, the delicate ontogeny of the nervous system can be disrupted by different stressors or neurotoxic chemicals (Andersen, 2003; Grandjean and Landrigan, 2014; Rakers et al., 2017; Rice and Barone, 2000), and the normal prenatal and postnatal development of catecholaminergic neurotransmission systems can be permanently altered by environmental stressors or neurotoxicants (Cory-Slechta et al., 2008; Spear, 2000). Accordingly, the association between exposures to environmental or industrial neurotoxicants with neurological disorders has been advocated in the literature. For instance, the increase in the incidence of autism or ASD has been attributed to mercury exposure during brain development (Austin and Shandley, 2008; Kern et al., 2011, 2016; Mutter et al., 2005; Sealey et al., 2016). However, the points of evidence supporting the link between mercury and ASD are highly questionable (Rossignol et al., 2014; Stehr-Green et al., 2003).

Exposure of experimental animals and humans to high levels of neurotoxic xenobiotics during critical periods of brain development has been consistently linked with permanent behavioral abnormalities later in life (Heyer and Meredith, 2017; Karri et al., 2016; Rice and Barone, 2000). In contrast, epidemiological data showing a potential causal relationship between exposures to low levels of neurotoxicants (for instance, MeHg, Hg, Pb, polychlorinated biphenyls (PCBs), organophosphates, pyrethroids, etc.) and the occurrence of behavioral abnormalities are scarce (Petersen et al., 2008). Despite of this, the available studies indicate that exposure to MeHg, lead and PCBs have to be avoided (Heyer and Meredith, 2017; Kraft et al., 2016; Rice and Barone, 2000).

4. THE CATECHOLAMINERGIC NEUROTRANSMITTER SYSTEM AS A POTENTIAL TARGET OF NEUROTOXICANTS AND RELATED CONSEQUENCES

Several effects resulting from exposures to neurotoxicants are linked to misbalances in the proper functioning of specific neurotransmitter systems.

Both dopaminergic and noradrenergic neurotransmission have been implicated. Of note, some neurochemical, histological, and behavioral effects resulting from exposures to neurotoxicants have been useful to understand mechanisms concerning the role of such toxicants toward a specific neurotransmitter system. In this section, we will briefly introduce two noxious agents, paraquat (PQ) and N-(2-chloroethyl)-N-ethyl-2-bromobenzylamine hydrochloride (DSP-4), which are known to cause neurotoxicity in the dopaminergic and noradrenergic neurotransmitter systems, respectively. Classical dopaminergic toxins, such as 1-methyl-4-phenyl-1,2,3,6-tetrahydropyridine (MPTP) and 6-hydroxydopamine (6-OHDA), are not discussed here because of its nonspecific toxic roles; they act on both dopaminergic and noradrenergic neurotransmitter systems (Archer and Fredriksson, 2006; Miyoshi et al., 1988; Szot et al., 2012), making difficult to discriminate between specific outcomes.

PQ (N,N'-dimethyl-4-4'-bipiridinium) is an herbicide widely used in several countries, and its toxicity has been reported since approximately 50 years ago (Clark et al., 1966). Even though initial studies with PQ have reported its effects on lung, liver, and kidney, significant damage to the brain was seen in individuals who died from PQ intoxication (Grant et al., 1980). The molecular mechanisms mediating PQ-induced dopaminergic neurotoxicity seem to involve a redox cycling that generates superoxide anion (Day et al., 1999), leading to oxidative stress and neurodegeneration.

Experimental evidence has reported that PQ possesses marked neurotoxicity and induces degeneration of the rat nigrostriatal dopaminergic system (Liou et al., 1996). In addition, elevated levels of α-synuclein have been reported in the frontal cortex and ventral midbrain, as well as α-synuclein-positive inclusions in substantia nigra neurons of mice treated with PQ (Manning-Bog et al., 2002). The occurrence of dopaminergic neurotoxicity and α-synuclein upregulation and aggregation suggests that the experimental PQ model may serve a useful tool to study PD-like disorders (i.e., nigral cell loss and synuclein pathology). Notably, PQ exposure has been significantly associated with PD (2.2-fold increase in risk for ever having used the chemical) (Pezzoli and Cereda, 2013).

It is noteworthy that experimental animals exposed to PQ have shown both motor (Kang et al., 2010; Park et al., 2005) and (Ait-Bali et al., 2016; Liu et al., 2016) cognitive impairments, consistent with the aforementioned roles of dopamine in motor control and emotional learning (presented in item 2). Even though PQ exposure has been associated with increased risk for PD (Pezzoli and Cereda, 2013), epidemiological studies on the effects of PQ toward motor and cognitive performance in humans are limited.

N-(2-chloroethyl)-N-ethyl-2-bromobenzylamine hydrochloride (DSP-4) is a neurotoxin that selectively damages the locus coeruleus noradrenergic system. The first studies with DSP-4 date from approximately 4 decades ago, when Ross and Renyl (1976) found that DSP-4 inhibited the active uptake of norepinephrine in mouse cortical cerebral slices. In the same year, Ross (2015) observed that exposure of rats to DSP-4 decreased DBH activity in the brain, suggesting the loss of noradrenergic neurons. By observing that the inhibition of adrenaline uptake did not per se cause the fall in DBH activity, this author concluded that the binding of DSP-4 to the neuronal membrane was responsible for a degenerative process. This hypothesis is reinforced by the observation that desipramine, an inhibitor of norepinephrine uptake, prevented the decrease in brain DBH activity.

Because of its relative specific effects toward the noradrenergic system (serotoninergic and dopaminergic nerves are only slightly or not at all affected), DSP-4 has been an important toxic agent to understand the roles of the norepinephrine in the CNS (for a review, see Ross and Stenfors, 2015). DSP-4 readily passes the blood–brain barrier and cyclizes to a reactive aziridinium derivative that is accumulated into the noradrenergic nerve terminals via the noradrenaline transporter. DSP-4 exposure (at the dose 50 mg/kg i.p. in rodents) causes a rapid and long-lasting loss of noradrenaline and a slower decrease in the DBH enzyme activity and immunoreactivity in the regions innervated from locus coeruleus (A6 cell group). The neurotoxic effect is counteracted by pretreatment with noradrenaline uptake inhibitors (i.e., desipramine). MAO-B inhibitors of the N-propargylamine type, such as selegiline, also counteract the DSP4-induced neurotoxicity.

Taking advantage of DSP-4 as a specific toxin for the noradrenergic neurotransmitter system, it was possible to establish critical roles of norepinephrine for several physiological functions, such as learning, memory, pain and emotion, among others. By using a protocol of DSP-4 exposure in rats, Ogren et al. (1980) showed that DSP-4-induced degeneration of the locus coeruleus noradrenergic system was paralleled by a marked impaired avoidance acquisition, indicating that the locus coeruleus noradrenergic system may play a role in aversive learning (Ogren et al., 1980). Consistent with this observation, Archer et al. (1984), studying the effects of DSP-4 on active avoidance acquisition in rats, observed that DSP-4 significantly impaired avoidance learning and that pretreatment with desipramine antagonized the active avoidance impairment. In a study with rats, Liang (1998) observed that intraamygdala infusion of DSP-4 impaired retention in the inhibitory avoidance task, suggesting the involvement of norepinephrine but not serotonin in memory storage processing in the applied experimental model.

The role of the noradrenergic system in modulating nociception was also proposed from studies using DSP-4. Zhong et al. (1985) showed that intrathecal injections of DSP-4 selectively depleted spinal noradrenaline and attenuated morphine analgesia in rats. In addition, by lesioning locus coeruleus noradrenergic neurons with DSP-4 in rats, Kudo et al. (2010) proposed that central neuropathic pain may be facilitated by DSP-4 depleting locus coeruleus noradrenergic neurons.

Based on the aforementioned studies with the toxic agent DSP-4, the noradrenergic neurotransmitter system was invoked to play a critical role in modulating learning and pain. In addition, it is noteworthy that experimental studies with DSP-4 have also pointed to important roles of the noradrenergic neurotransmitter system in modulating mesolimbic dopamine transmission (Lategan et al., 1992) and emotionality (Harro et al., 1995).

5. METHYLMERCURY: GENERAL ASPECTS AND MAJOR MECHANISMS OF NEUROTOXICITY

The mercury atom covalently bound to a carbon from the methyl group forms a reactive and soft electrophile center in the MeHg molecule. In fact, MeHg is a soft electrophile that has high affinity for soft nucleophiles (Pearson, 1963). In the biological scenario, we can found two types of physiologically relevant soft nucleophile centers, i.e., the sulfhydryl (–SH or thiol) or the selenohydryl (–SeH or selenol) groups. Thiol groups are found in a few endogenous low-molecular-mass molecules, such as cysteine and reduced and oxidized glutathione (GSH and GSSG), as well as in thousands of high-molecular-mass macromolecules (thiol-containing proteins) (Table 1; Go and Jones, 2013; Miseta and Csutora, 2000). For instance, the cysteinyl (Cys) proteome revealed that approximately 200,000 Cys residues are encoded in the human genome (Go and Jones, 2013). Evolutionary studies have indicated that cysteine was first selected to be part of the specific motif $-C-(X)_2-C-$, which is found in metal binding proteins and in oxidoreductases.

According to Miseta and Csutora (2000), in archea, the ancient motif containing the two vicinal cysteine residues was evolutionary selected to binding metals. Here, possibly those metals with physiological function (i.e., Zn^{2+}, Cu^{2+}, etc.).

Table 1 Relative Quantity of Molecules of Biological Significance Containing the Soft Nucleophile Groups Thiol (–SH) or Selenol (–SeH)

	Occurrence
High-molecular-mass thiol molecules (thiol-containing proteins)	
–Cysteinyl residues (Cys)	In thousands of proteins
High-molecular-mass selenol molecules (selenoproteins)	
–Selenocysteinyl residues (Sec)	In few number of proteins
Low-molecular-mass thiol molecules (–SH)	
Endogenous	Cysteine Glutathione (GSH)
Exogenous (synthetic)	2,3-Dimercaptopropanol (BAL) 2,3-Dimercaptosuccinic acid (DMSA) 2,3-Dimercapto sulfonic acid (DMPS)
Low-molecular-mass selenol molecules (–SeH)	
Endogenous	None
Exogenous (synthetic)	Unstable

These functional groups are the main targets of methylmercury (MeHg) and other soft electrophiles.

Glutathione (GSH and GSSG) and thioredoxin [Trx(SH)$_2$] systems have crucial roles in maintain the cellular homeostasis of redox-sensitive cysteinyl- or thiol-containing proteins (Go and Jones, 2013). In fact, Trx and GSH systems have different role as regulators of disulfide–thiol redox equilibrium, because they regulate the redox state of distinct classes of thiol–disulfide-containing proteins (Go et al., 2015; Jones, 2015). Of particular toxicological significance, MeHg can target both antioxidant systems and their disruption can have profound impact in the neurotoxicity of MeHg (Branco et al., 2017a,b; Carvalho et al., 2011; Farina et al., 2011b, 2013).

Low-molecular-mass molecules containing thiol groups have been synthesized for therapeutic purposes (Blanusa et al., 2005; Vilensky and Redman, 2003). Table 1 shows some synthetic thiol molecules used in the treatment of intoxication with mercurials, lead, and other toxic elements. In contrast, the existence of low mass selenol-containing biomolecules has not yet been clearly demonstrated and the occurrence of stable low mass selenol molecules under physiological conditions is improbable.

The occurrence of the selenol group is much more restricted than the thiol center, and the –SeH group is found only a few number of selenoproteins (Hatfield et al., 2014; Farina et al., 2011b; Table 1). For instance, the mammalian genome codifies 20–30 selenoproteins, including important oxidoreductases involved in the metabolism of reactive species (e.g., glutathione peroxidase and thioredoxin reductase isoforms; Hatfield et al., 2014).

The affinity of MeHg for the –SH is high and the formation constant of $RSH + MeHg \rightarrow RS\text{-}HgMe$ (i.e., the formation of –Hg-S– bond) is very high ($\sim$15–20; Rabenstein, 1978a). This implies that, under physiological conditions, MeHg will be not found in "a free form" or bound to other abundant ligands, such as Cl^- (chloride, e.g., in stomach) or OH^-. As a corollary, we may infer that the chemistry, biochemistry, and toxicology of MeHg will be dictated by the chemistry of thiol $\leftrightarrow$ MeHg–sulfide exchange (Figs. 2 and 3; Dórea et al., 2013; Farina et al., 2011a,b; Rabenstein, 1978a,b; Rabenstein and Fairhurst, 1975; Rabenstein and Reid, 1984; Rabenstein et al., 1974, 1982). On the other hand, MeHg can also coordinate with the selenol group of selenoproteins (Arnold et al., 1986; Rabenstein et al., 1986; and for review, see Farina et al., 2011a,b). Of particular toxicological significance, the affinity of MeHg for the selenohydryl (selenol) groups is much higher than that of for the sulfhydryl (thiol) group (Sugiura et al., 1976, 1978). However, the formation constant of $RSeH + MeHg \rightarrow RSeHgMe$ has rarely been calculated chemically (Arnold et al., 1986). Essentially, the affinity of the –SeH for MeHg is too high, which makes the determination of the constant difficult. Recent in silico studies have confirmed the superior affinity of the selenol group for electrophilic mercury forms (e.g., Hg^{2+} and MeHg) over the analog thiol molecules (Asaduzzaman and Schreckenbach, 2011; Asaduzzaman et al., 2009).

Indeed, many studies have corroborated the ability of MeHg to inhibit selenoenzymes, both in vitro and after in vivo exposure (Branco et al., 2011, 2012, 2014; Branco et al., 2017a,b; Carvalho et al., 2008, 2011; Dalla Corte et al., 2013; Farina et al., 2009; Franco et al., 2009; Meinerz et al., 2017; Wagner et al., 2010).

The higher affinity of MeHg for the –SeH than –SH groups may indicate that selenoproteins should be the primary targets of MeHg. However, the much lower concentration of –SeH groups when compared with –SH groups makes the scenario more complex and we have to weigh the superior affinity of –SeH for MeHg against the much greater concentrations of –SH groups in the biological milieu.

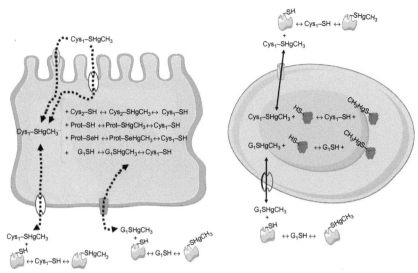

Fig. 2 Schematic representation of MeHg–thiol exchange in the body. The absorption of MeHg–Cys complex (MeHg bound to a cysteine) in the intestine can be mediated by transporters or by exchange reactions. Inside the enterocyte (*left part*), the MeHg–Cys complex can exchange either with low-molecular-mass thiols (cysteine or glutathione, GSH) or with thiol-containing proteins. The complex can also exchange with selenol-containing proteins, i.e., selenoproteins (Prot–SeH). The low-molecular-mass sulfide–methylmercury complexes (GS–HgCH$_3$ or Cys–S–HgCH$_3$) can be transported to other cells or body fluids (e.g., interstitial fluid, capillary cells, which were omitted for the sake of clarity) and then can reach the plasma and blood cells (*right part*). In the plasma, they can exchange with abundant proteins containing thiol (e.g., albumin) and can also be transported into the erythrocytes. Inside the erythrocytes, the low-molecular-mass thiols will delivery (exchange) the MeHg to hemoglobin and other thiol proteins, for instance, porphobilinogen synthase or aminolevulinate dehydratase (ALA-D, Rocha et al., 2012). The distribution of MeHg from plasma and erythrocytes to the tissues will involve the same kind of exchange reactions in the opposite direction (blood to tissue). In all the situations, the mobility of the low-molecular-mass complexes of MeHg–sulfides (MeHg–SG and MeHg–Cys) is expected to have a high facility to be redistributed than high-molecular-mass (thiol- or selenol-containing proteins).

In short, any type of molecule containing thiol or selenol groups can be a potential target of MeHg. But our knowledge about the primary targets of MeHg is still elusive. In fact, the toxicity of MeHg is rather complex and involves the hierarchical interaction with multiple targets at different levels of cellular organization (Aschner et al., 2007; Atchison, 2005; Castoldi et al., 2000; Denny and Atchison, 1996; Minnema et al., 1989; Sirois and Atchison, 2000). For instance, after being absorbed as a MeHg–cysteine complex, the MeHg derived from fish muscle proteins can interact with

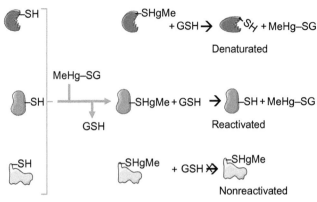

Fig. 3 Exchange reaction between low-molecular-mass methylmercury-glutathione sulfide (MeHg–SG) with different thiol-containing proteins. The first exchange (indicated by the reaction of the three proteins before the brace with GSH) results in the formation of three different protein–sulfide–methylmercury complexes. The protein–MeHg complexes are commonly inactive. The second set of exchange reactions can hypothetically reactivate the protein (represented by the second protein in the *middle* of the figure), or can release the denatured protein (i.e., the withdraw of the MeHg from the protein does not reestablish the activity, because the protein was denatured during the temporary interaction and is represented by the protein in the *top* of the figure). The reaction of MeHg with the protein thiol can cause a dramatic refolding of the protein in such a way that MeHg–S-moiety will not be accessible to the medium and will not suffer the exchange reaction (represented by the third protein in the *bottom* of the figure).

thiol- and selenol-containing molecules (Figs. 2 and 3; Dórea et al., 2013; Farina et al., 2011a,b). Since thiols are much more abundant than selenol groups, the probability that MeHg will target thiol-containing molecules is higher than targeting the selenol groups. Thus, the distribution of MeHg from nontarget tissues to target tissues (particularly the brain) will be dictated by the exchange of MeHg from one thiol to another thiol- or selenol-containing molecule (Fig. 2). However, in view of the high concentration of low mass thiol-containing molecules (i.e., GSH and cysteine), they are expected to have a central role MeHg distribution through the body and brain (Naganuma et al., 1980). In the blood, plasma and erythrocyte thiol-containing proteins (e.g., albumin and hemoglobin; Fig. 2) possibly also play an important role in the distribution of MeHg. Accordingly, hemoglobin has the ability of binding MeHg and erythrocytes can retain a great proportion of MeHg in vertebrates (Clausing et al., 1984; Doi, 1991; Doi and Tagawa, 1983; Naganuma et al., 1980). The exchange of MeHg from the thiol group of hemoglobin to low-molecular-mass thiol (e.g., GSH and cysteine) will have a central role in the distribution of MeHg (Fig. 2).

Selenoprotein P has an important role in the distribution of selenium to the brain, and it is also one of the most abundant selenoproteins in the plasma (Burk and Hill, 2005). Although elevated levels of plasma selenoprotein P levels have been recently associated with high whole blood mercury (derived from fish and whale consumption) (Ser et al., 2017), its role in the distribution of MeHg to the brain has not been investigated. It is noteworthy that selenoprotein P can bind Hg^{+2}–selenide complexes, and it has been suggested that this type of interaction can facilitate the detoxification of Hg (Suzuki et al., 1998; Yoneda and Suzuki, 1997). However, the importance of selenoprotein P in MeHg distribution has been little investigated and, in contrast to humans exposed to mercury from seafood (Ser et al., 2017), the levels of selenoprotein P were decreased in rats intoxicated with high doses of MeHg (Usuki and Fujimura, 2016).

The complex biochemistry of MeHg movement from one thiol- or selenol molecule to other molecules has not been studied in detail. Professor Dallas L. Rabenstein and collaborators were the pioneers investigating this type of MeHg–thiol exchange (Arnold et al., 1986; Rabenstein, 1978b; Rabenstein and Fairhurst, 1975; Rabenstein and Reid, 1984; Rabenstein et al., 1974, 1982, 1986). However, after them, little has been made and this subject is an important bottleneck in the molecular toxicology of MeHg. Although the exchange of MeHg from one thiol or selenol to another is well demonstrated, we cannot predict whether or not the temporary interaction of a given protein with MeHg will result in the inactivation of this protein even after MeHg leaving the thiol-containing protein (Fig. 3). Consequently, we can hypothesize that the exchange of MeHg from one class of protein to another type of thiol- or selenol-containing protein can result or not in the inactivation of the proteins that have been temporarily bound with MeHg (Fig. 3).

After reaching the brain, MeHg–thiol complex(es) are distributed to different cells types, including neural and glial cells (Aschner et al., 2007; Atchison and Hare, 1994; Colón-Rodrígue et al., 2017; Ni et al., 2012). Astrocytes play a critical role in MeHg neurotoxicity, and MeHg-induced impairment in glutamate uptake by astrocytes is tentatively one of the primary targets of MeHg (for review, see Aschner et al., 2007; Farina et al., 2011b). The inhibition of glutamate transport by MeHg causes an increase in the extracellular glutamate concentration, which can trigger excitotoxicity via activation of NMDA receptors. The activation of the glutamatergic receptors stimulates the entrance of Ca^{2+} into the neuronal cells and the activation of neurotoxic pathways. The intracellular overload of

Ca^{2+} promotes the uptake of Ca^{2+} by the mitochondria, triggering reactive oxygen species overproduction, energy failure, mitochondrial permeability transition pore, and cell demise (Atchison, 2005; Atchison and Hare, 1994; Farina et al., 2011b, 2013; Limke et al., 2003; Roos et al., 2012). The source of Ca^{2+} can be derived from the extracellular space and from intracellular stores (Atchison and Hare, 1994; Limke et al., 2003; Roos et al., 2012). Accordingly, in vitro and in vivo studies have demonstrated the protective role of Ca^{2+} channel blockers against the neurotoxicity of MeHg (Bailey et al., 2013; Castoldi et al., 2001; Gassó et al., 2001; Hare and Atchison, 1995; Ramanathan and Atchison, 2011; Sakamoto et al., 1996).

Of particular neurotoxicological significance, MeHg also stimulates the ROS in astrocytes that can overwhelm the antioxidant cellular capacity (Aschner et al., 2007). Since astrocytes have substantial reserve of antioxidants, particularly the GSH system (Peuchen et al., 1997), the overproduction of ROS/RNS can exhaust the astrocytes defenses, rendering the surroundings neurons more vulnerable to the neurotoxicity of glutamate and oxidative stress (Farina et al., 2011a,b).

MeHg can also be transported to the microglial cells and can trigger some biochemical changes similar to those provoked in astrocytes (Ni et al., 2010, 2011, 2012). However, in vitro, microglial cells responded earlier to MeHg than astrocytes, indicating a potential role of this type of glial cells in the neurotoxicity of MeHg.

Although excitotoxicity has been shown to play a crucial role in the cytotoxicity of MeHg in vitro, the points of evidence indicating the participation of glutamatergic system under in vivo conditions are still limited (Feng et al., 2014; Juárez et al., 2002, 2005). The intracerebral infusion of high concentrations of MeHg caused a substantial increase in the extracellular levels of glutamate and oxidative damage (mediated by NMDA receptor activation) in the brain of rats (Juárez et al., 2002, 2005). Indirect points of evidence support the involvement of disruption in glutamatergic neurotransmitter system in the neurotoxicity of MeHg in developing and adult rodents (Carratù et al., 2006; Deng et al., 2014; Farina et al., 2003).

In addition to glutamatergic system, MeHg can also disrupt the functionality of cholinergic, dopaminergic, adrenergic, and GABAergic neurotransmission in vertebrates (Aschner et al., 2007; Bartolome et al., 1987; Bradford et al., 2016; Gimenez-Llort et al., 2001; Hrdina et al., 1976; Limke et al., 2004b; Pereira et al., 1999; Sharma et al., 1982; Zhou et al., 1999). Indeed, MeHg can interact with thiol-containing proteins involved in the regulation of neurotransmission at different subcellular levels. For instance, MeHg can modify the activity of ionic channels, neurotransmitter receptors, neurotransmitters storage (uptake and release), and metabolism (Atchison, 2005; Atchison and Hare, 1994; Castoldi et al., 2001; Limke et al., 2004a,b; Slotkin and Bartolome, 1987). In the next sections, we specifically discuss the effects of MeHg on the catecholaminergic system.

6. METHYLMERCURY-MEDIATED CATECHOLAMINERGIC TOXICITY

6.1 Dopamine

The effects of MeHg toward the dopaminergic neurotransmitter system have been extensively reported. In 1982, Bartolome et al. (1982) reported the effects of neonatal MeHg exposure on the biochemical development of CNS catecholamine synapses by evaluating synaptosomal and synaptic vesicular uptakes of [^{3}H]catecholamines, TH activity, and levels and turnover rates of catecholamines. Most of the evaluated parameters were measured during the early postnatal (PN) period (from birth to PN day 40). The authors observed that neonatal MeHg exposure produced initial inhibition of [^{3}H] dopamine synaptosomal uptake, followed by marked elevations of uptake from PN day 20 day onward. The observed changes in synaptosomal uptake were preceded by increases in the turnover rate of dopamine. However, the authors observed normal developmental patterns for other presynaptic terminal parameters (vesicular uptake, TH activity, and dopamine content). At that time, it was concluded that the alterations in uptake and turnover for dopamine indicate that the synaptic dynamics of developing central dopaminergic neurons are indeed affected by MeHg exposure.

Several years later, two groups reported increased dopamine release after MeHg exposure. Kalisch and Racz (1996) investigated the effects of in vitro MeHg exposure on endogenous dopamine efflux from mouse striatal slices. MeHg produced a concentration-dependent increase in the efflux of dopamine from mouse striatal slices. Notably, potassium-stimulated efflux of dopamine was enhanced by MeHg in both the presence and absence of Ca^{2+} in the medium, suggesting that under depolarizing conditions, dopamine efflux induced by MeHg has a Ca^{2+}-independent component. The authors suggested that alterations in dopamine neurotransmission in the striatum might contribute to the symptoms of MeHg toxicity. In agreement with Kalisch and Racz (1996), Faro et al. (2000) investigating the effects of intrastriatal administration of MeHg on the dopaminergic system of rat striatum in conscious and freely moving animals (using microdialysis coupled to liquid chromatography) observed that intrastriatal administration of MeHg (40 µM–4 mM) led to concentration-dependent increases in dopamine release from rat striatal tissue associated with significant decreases in extracellular levels of its main metabolites dihydroxyphenylacetic acid (DOPAC) and homovanillic acid (HVA). The authors explained these effects as a result of stimulated DA release and/or decreased DA intraneuronal degradation. The results from these two research groups reinforce the idea that increased dopamine release is an important effect of MeHg toward the dopaminergic neurotransmitter system. Further studies from Faro et al. (2002a) provided additional mechanistic information concerning MeHg-induced dopamine release in a microdialysis study. Evaluating the effects of striatal MeHg administration in dopamine release, the results demonstrated that MeHg increases the spontaneous DA release from rat striatum in a transporter-dependent manner, which is independent on vesicular stores and external Ca^{2+}. MeHg also decreased KCl-evoked DA release.

Notably, the possibility of indirect effects in MeHg-induced dopamine release was proposed in 2002. Faro et al. (2002b) observed that intrastriatal infusion of MeHg in rats increased the extracellular dopamine levels with respect to basal levels. In contrast, this effect was significantly decreased in 400 µM MK-801 pretreated animals. Moreover, MeHg-induced increase in the extracellular dopamine levels was significantly decreased in 100 µM L-NAME or 7-NI (nitric oxide synthase inhibitors) pretreated animals, suggesting that MeHg acts, at least in part, through an overstimulation of NMDA receptors with possible NO production to induce DA release, and that administration of NMDA receptor antagonists and NOS inhibitors protects against MeHg-induced DA release from rat striatum. This study by Faro et al. (2002b) raised the possibility of indirect effects in MeHg-induced dopamine release through activation of other neurotransmitter systems, such as the glutamatergic one.

In agreement with these studies (Faro et al., 2000, 2002a,b; Kalisch and Racz, 1996), an in vivo study by O'Kusky et al. (1988) showed that the treatment of rats with subcutaneous injections of MeHg during early postnatal development caused a significant increase in the concentrations of dopamine (28%–29%) with a significant decrease in the concentration of 3,4-dihydroxyphenylacetic acid (DOPAC, 20%–27%) at PN days 22–24. The authors pointed to altered metabolism of dopamine in the developing CNS during the pathogenesis of MeHg-induced movement and postural disorder.

The effects of MeHg on dopamine transporters (DATs) have also been reported. By investigating whether MeHg exposure leads to changes in dopamine levels and DAT function in synaptosomes from early postnatal rats, Dreiem et al. (2009) reported that MeHg exposure led to DAT inhibition and increased levels of released DA compared to control animals. Notably, the effects were much greater in synaptosomes prepared from PN day 7 rats than in synaptosomes from PND 14 or PND 21 animals. To our knowledge, this is the only study showing the effects of MeHg toward DAT function.

The direct deleterious effects of MeHg toward dopaminergic cells under in vitro conditions have been also reposted. Götz et al. (2002) used an in vitro approach to investigate the alterations induced by MeHg in primary dopaminergic cells isolated from the ventral mesencephalon of CD-1 embryonic mice. The morphometric analysis of DA neurons exposed to $1 \mu M$ MeHg demonstrated a striking decrease in the number of neurites, indicative of cytoskeletal alteration. In addition, dopaminergic neurons displayed cell shrinkage and a significant increase in the number of nuclei with chromatin condensation. Based on these results, the authors concluded that MeHg is highly toxic to primary dopaminergic neurons. Götz et al. (2002) and Shao et al. (2015) investigated the effects of MeHg exposure on gene and protein profiles in a dopaminergic MN9D cell line. By performing proteomic analysis and evaluating differential protein expression, the authors suggested that MeHg and MPP^+ (a classical dopaminergic toxin) share many similar signaling pathways leading to the pathogenesis of PD.

MeHg also seems to affect the activity of monoamine oxidase (MAO), an enzyme involved in the metabolism of dopamine to DOPAC and/or HVA. Beyrouty et al. (2006) studied if oral exposure of adult female rats to MeHg before and during pregnancy would affect MAO activity in various brain regions of the offspring. The authors demonstrated that exposure to MeHg in rats before and/or during gestation resulted in a reduction of MAO activity in the developing embryo and brain stem of the female offspring with accompanying changes in auditory startle response. Based on these results, it is reasonable to posit a decreased metabolism of dopamine to DOPAC and HVA after MeHg exposure.

Dopamine receptors have also been proposed as potential targets involved in MeHg toxicity. Coccini et al. (2011) treated rat dams with oral administrations of MeHg during the gestational days 7 to PN day 21. After treatments, the density (B_{max}) and affinity (K_d) of dopamine D1-like (D1-Rs) and D2-like receptors (D2-Rs) were evaluated by saturation binding studies. Dopamine (DA) D1- and D2-like receptor B_{max} and K_d were assessed in brain cortical and striatal membranes by saturation binding experiments using increasing concentrations of the specific ligands. The authors observed that the cerebral dopaminergic D1-like and D2-like receptors are differently impaired by developmental exposure to MeHg according to the brain area considered and/or to animal gender and time of growth, with some early changes persisting in time (D2-like receptors in males) or disappearing with time (D1-like receptors in males). To the best of our knowledge, this is the only study showing the effects of MeHg toward dopamine receptors function.

Studies in *Caenorhabditis elegans* (*C. elegans*) have also contributed to understand molecular mechanisms mediating MeHg effects toward the dopaminergic system. Martinez-Finley et al. (2013), using *C. elegans* as experimental model, tested the hypothesis that early-life exposure to MeHg and knockout (KO) of *pdr-1* (mammalian: parkin/PARK2, mutations in this gene are a risk factor for PD) exacerbates MeHg toxicity and damage to the dopaminergic system. The authors observed that *pdr-1* KO worms were more sensitive to MeHg than wild-type worms, but MeHg did not exacerbate behavioral changes related to the absence of *pdr-1*. They concluded that the combination of early-life exposure to MeHg and *pdr-1* KO had significant effects on oxidative stress and aging, also suggesting that early-life MeHg exposure is a risk factor for loss of dopaminergic function later in life in wild-type worms, but the combination of *pdr-1* KO and early-life MeHg does not further exacerbate the already reduced dopaminergic function produced by *pdr-1* KO alone. In another study with *C. elegans*, VanDuyn and Nass (2014) showed MRP-7 (multidrug resistance protein 7) loss-of-function mutations increase susceptibility of dopaminergic neurons to MeHg, attesting to gene × environment interactions.

Based to the studies mentioned in herein in this section, it is reasonable to propose that the dopaminergic neurotransmitter system is indeed a target involved in MeHg-induced neurotoxicity, including developmental neurotoxicity. But what do the previously mentioned molecular effects represent in terms of behavioral outcomes? Reed and Newland (2009) studied the effects gestational exposure to MeHg on rat behavior during adulthood, as well as the involvement of dopaminergic system. The authors performed a protocol in which female rats were exposed in utero to MeHg, via maternal drinking water. As adults, the MeHg-exposed offspring was trained to lever press under a fixed interval schedule of reinforcement. Experiments with acute dose–effect curves were performed with dopamine (cocaine) and norepinephrine (desipramine) reuptake inhibitors, as well as a direct D1 or D2 agonist and antagonists. For high-rate behavior, MeHg-exposed rats were 2–3 times more sensitive to the rate-reducing effects of high doses of cocaine, but no differential effects of MeHg were seen with desipramine, suggesting MeHg's effect is specific to dopaminergic receptors. In addition, no differential effects were seen with the specific D1 and D2 agonists. Notably, it seems that the observed effects were formed during gestational exposure and persisted into adulthood, suggesting that gestational MeHg exposure produces irreversible sensitivity to dopamine in rats.

6.2 Epinephrine

The effects of MeHg toward the adrenergic system have been less investigated than the dopaminergic system. Early in vitro studies reported that MeHg caused inhibition of norepinephrine uptake and stimulated its release from preloaded synaptosomes (Komulainen and Tuomisto, 1981; Rajanna and Hobson, 1985) and hippocampal slices (Gassó et al., 2000). The effects of perinatal in vivo exposure to MeHg have consistently indicated changes in brain norepinephrine levels in rats. However, the magnitude of brain norepinephrine decrease or increase varied depending on the schedule of exposure (period, dose, brain area evaluated, etc.; Bartolome et al., 1982, 1984a; Hrdina et al., 1976; Lindström et al., 1991; Taylor and DiStefano, 1976; Tsuzuki, 1982). For instance, MeHg exposure of suckling rats from birth to weaning caused dose and time-dependent increase in brain norepinephrine levels and turnover. Of neurotoxicological significance, the levels of norepinephrine remained elevated even 20 days after the end of MeHg exposure (Bartolome et al., 1982). In contrast, the gestational exposure to MeHg did not cause changes in norepinephrine levels or turnover, but caused an increase in norepinephrine uptake by synaptosomes of suckling rats (Bartolome et al., 1984a).

The ontogeny of alpha- and beta-adrenergic receptors ($\alpha1$, $\alpha2$, and β) was modified by pre- and postnatal exposure to MeHg. The regions more affected were cerebellum > cerebral cortex > midbrain and brain stem and, according to authors, the effects depended on the maturational profile of each brain region, and the regions that developed early were less affected than the ones that matured latter (Bartolome et al., 1987). In another study, the exposure of rats to a low dose of MeHg from conception–gestation to 50 days of life was associated with an increase in norepinephrine levels in the cerebellum, but not in frontal cortex, striatum, hypothalamus, and hippocampus (Lindström et al., 1991).

In a comprehensive study, O'Kusky and coworkers examined the neurochemical alterations in rat brain noradrenergic system at three different stages of postnatal MeHg exposure (O'Kusky et al., 1988). MeHg exposure was started at the fifth postnatal day until day 15 (stage I, where MeHg-treated pups did not show any overt sign of behavioral intoxication and gained weight but at lower rate than controls), day 18–21 (stage II, where significant changes in body weight were observed), and day 22–24 (stage III, which corresponded to the period of the onset of neurobehavioral impairments). The authors reported that norepinephrine levels were increased in spinal cord and caudate putamen, when compared with controls and body weight gain-matched controls (which were separated from their dams to growth at the same rate as did MeHg-exposed rats). The increase in norepinephrine levels started before the installation of neurobehavioral impairments (started in the stage II and persisted until stage III), indicating that changes in norepinephrine levels occurred before the onset of neurobehavioral manifestations (O'Kusky et al., 1988). In cerebral cortex, the variations in norepinephrine levels were very complex and body weight gain-matched controls had higher levels of norepinephrine than controls, whereas MeHg-treated rats had the lowest levels of norepinephrine in the stage I. At stage II, the levels of norepinephrine were lower in controls and MeHg, when compared with body weight-matched controls. During the onset of neurological symptoms, there were no differences between the groups. Authors concluded that movement and postural disorders in suckling rats were associated with selective alterations in central catecholaminergic (dopaminergic and noradrenergic) and serotoninergic neurotransmitter systems.

The exposure of catfish (*Clarias batrachus*) to MeHg for long periods (90 and 180 days) caused a significant increase in brain norepinephrine levels. To some extent, this is agreement with results obtained with rats, where the levels of catecholamines (dopamine and norepinephrine) can be increased after different schedules of exposure to MeHg (Kirubagaran and Joy, 1990). Moreover, this indicates a conserved neurotoxic effect of MeHg in different vertebrates possibly reflecting the targeting of a similar group of conserved proteins involved in catecholaminergic neurotransmission.

One important aspect of MeHg neurotoxicity that has been little explored is its deleterious effect toward the sympathetic noradrenergic neurotransmission in heart and kidney. Bartolome and collaborators have demonstrated that

MeHg exposure during the early postnatal period accelerated the maturation of cardiac sympathetic transmission, which was associated with increased norepinephrine levels and turnover. The authors have also demonstrated that MeHg caused heart and kidney overgrowth, which could be linked to sympathetic innervations overactivity (Bartolome et al., 1984a,b). This topic should be further analyzed particularly in view of the points of evidence that MeHg exposure can cause hypertension in experimental animals (Grotto et al., 2009) or increased diastolic and systolic blood pressure in humans exposed to MeHg during prenatal life (Sørensen et al., 1999; Thurston et al., 2007). The results from human exposed during the prenatal life indicate that MeHg can modify development of cardiac homeostasis (Sørensen et al., 1999) as observed in developing rats (Bartolome et al., 1984a,b).

In fact, the disruption of normal cardiac function by inorganic mercury and the participation of sympathetic catecholaminergic system are well-documented (Beck et al., 2004; Henningsson et al., 1993; Michaeli-Yossef et al., 2007; Torres et al., 2000). However, the potential role of MeHg via fish consumption and catecholamines (particularly changes in noradrenergic sympathetic neurotransmission) as a risk for the development of cardiovascular pathologies is still debatable (Guallar et al., 2002; Roman et al., 2011; Salonen et al., 1995; Yoshizawa et al., 2002).

7. CONCLUDING REMARKS

MeHg is a toxicant that targets a great variety of proteins involved in different cellular processes. Of particular neurotoxicological significance, MeHg targets several components of different neurotransmitter systems. MeHg exposure during perinatal phases of brain development can modify the fine dynamics of brain synapses (synaptogenesis, establishment of cell–cell contact, cell migration, etc.) by interacting with neurotransmitter receptors, ion channels, transport proteins, enzymes, etc. In short, MeHg can interfere with the synthesis, degradation, storage, release, reuptake of different neurotransmitters, including dopamine and norepinephrine. The ultimate outcomes of disrupting such complex biochemical processes will be the behavioral teratology. More recently, the literature data have also indicated the dangers of interfering with normal brain development during the adolescence period. Indeed, the exposures to MeHg during adolescence have been shown to disrupt the dopaminergic neuropharmacology in rodents (Boomhower and Newland, 2016, 2017). If the experimental data derived from animal models were to be extrapolated to human, it might explain the growing incidence of behavioral and cognitive abnormalities in the human population. However, since now we can be exposed to thousands of potentially neurotoxic agents (Grandjean and Landrigan, 2014), the understanding on how much a single potent neurotoxic agent, such as MeHg, effectively contributes to the disruption of behavior in developing organism will be a difficult task. In fact, our knowledge about the primary targets of MeHg, as well as of other neurotoxicants, is still incipient. The development of in silico methodologies to search for proteins containing moieties of the type $-X-C-X-$, $X-U-X$, $-C-(X)_2-C-$, or $-C-(X)_2-U-$ (where $C =$ cysteinyl residues, $U =$ selenocysteinyl residues, and $X =$ other residues than *cys* or *sec*) will be important to define potential targets of MeHg. Furthermore, the in vitro and in vivo identification of which of these proteins are actually targeted by MeHg will help in defining primary molecular targets of MeHg. As reported here, MeHg changes several neurochemical processes related to catecholaminergic neurotransmission, but it remains to be determined whether these are direct (primary) or indirect (secondary or tertiary) effects of MeHg neurotoxicity.

ACKNOWLEDGMENTS

The author would like to thank the colleagues/coauthors who have contributed to several studies referenced in this chapter. These studies were funded in part by grants from the National Institute of Environmental Health Sciences (Grant numbers NIEHS R01ES07331, NIEHS R01ES10563, and NIEHS R01ES020852), as well as the Brazilian Agencies CNPq, CAPES, FAPERGS, and FINEP.

REFERENCES

Ait-Bali, Y., Ba-M'hamed, S., Bennis, M., 2016. Prenatal Paraquat exposure induces neurobehavioral and cognitive changes in mice offspring. Environ. Toxicol. Pharmacol. 48, 53–62.

Andersen, S.L., 2003. Trajectories of brain development: point of vulnerability or window of opportunity? Neurosci. Biobehav. Rev. 27, 3–18.

Archer, T., Fredriksson, A., 2006. Influence of noradrenaline denervation on MPTP-induced deficits in mice. J. Neural. Transm 113, 1119–1129.

Archer, T., Jonsson, G., Ross, S.B., 1984. A parametric study of the effects of the noradrenaline neurotoxin DSP4 on avoidance acquisition and noradrenaline neurones in the CNS of the rat. Br. J. Pharmacol. 82, 249–257.

Arnold, A.P., Tan, K.-S., Rabenstein, D.L., 1986. Nuclear magnetic resonance studies of the solution chemistry of metal complexes. 23. Complexation of methylmercury by selenohydryl-containing amino acids and related molecules. Inorg. Chem. 25, 2433–2437.

Asaduzzaman, A.M., Schreckenbach, G., 2011. Degradation mechanism of methyl mercury selenoamino acid complexes: a computational study. Inorg. Chem. 50, 2366–2372.

Asaduzzaman, A.M., Khan, M.A., Schreckenbach, G., Wang, F., 2009. Computational studies of structural, electronic, spectroscopic, and thermodynamic properties of methylmercury-amino acid complexes and their Se analogues. Inorg. Chem. 49, 870–878.

Aschner, M., Syversen, T., Souza, D.O., Rocha, J.B., Farina, M., 2007. Involvement of glutamate and reactive oxygen species in methylmercury neurotoxicity. Braz. J. Med. Biol. Res. 40, 285–291.

Atchison, W.D., 2005. Is chemical neurotransmission altered specifically during methylmercury-induced cerebellar dysfunction? Trends Pharmacol. Sci. 26, 549–557.

Atchison, W.D., Hare, M.F., 1994. Mechanisms of methylmercury-induced neurotoxicity. FASEB J. 8, 622–629.

Austin, D.W., Shandley, K., 2008. An investigation of porphyrinuria in Australian children with autism. J. Toxicol. Environ. Health A 71, 1349–1351.

Bailey, J.M., Hutsell, B.A., Newland, M.C., 2013. Dietary nimodipine delays the onset of methylmercury neurotoxicity in mice. Neurotoxicology 37, 108–117.

Barbeau, A., 1970. Dopamine and disease. Can. Med. Assoc. J. 103, 824–832.

Bartolome, J., Trepanier, P., Chait, E.A., Seidler, F.J., Deskin, R., Slotkin, T.A., 1982. Neonatal methylmercury poisoning in the rat: effects on development of central catecholamine neurotransmitter systems. Toxicol. Appl. Pharmacol. 65, 92–99.

Bartolome, J., Trepanier, P.A., Chait, E.A., Barnes, G.A., Lerea, L., Whitmore, W.L., Weigel, S.J., Slotkin, T.A., 1984a. Neonatal methylmercury poisoning in the rat: effects on development of peripheral sympathetic nervous system. Neuronal participation in methylmercury-induced cardiac and renal overgrowth. Neurotoxicology 5, 45–54.

Bartolome, J., Whitmore, W.L., Seidler, F.J., Slotkin, T.A., 1984b. Exposure to methylmercury in utero: effects on biochemical development of catecholamine neurotransmitter systems. Life Sci. 35, 657–670.

Bartolome, J.V., Kavlock, R.J., Cowdery, T., Orband-Miller, L., Slotkin, T.A., 1987. Development of adrenergic receptor binding sites in brain regions of the neonatal rat: effects of prenatal or postnatal exposure to methylmercury. Neurotoxicology 8, 1–14.

Beaulieu, J.M., Gainetdinov, R.R., 2011. The physiology, signaling, and pharmacology of dopamine receptors. Pharmacol. Rev. 63, 182–217.

Beck, C., Krafchik, B., Traubici, J., Jacobson, S., 2004. Mercury intoxication: it still exists. Pediatr. Dermatol. 21, 254–259.

Beyrouty, P., Stamler, C.J., Liu, J.N., Loua, K.M., Kubow, S., Chan, H.M., 2006. Effects of prenatal methylmercury exposure on brain monoamine oxidase activity and neurobehaviour of rats. Neurotoxicol. Teratol. 28, 251–259.

Blanusa, M., Varnai, V.M., Piasek, M., Kostial, K., 2005. Chelators as antidotes of metal toxicity: therapeutic and experimental aspects. Curr. Med. Chem. 12, 2771–2794.

Boomhower, S.R., Newland, M.C., 2016. Adolescent methylmercury exposure affects choice and delay discounting in mice. Neurotoxicology 57, 136–144.

Boomhower, S.R., Newland, M.C., 2017. Effects of adolescent exposure to methylmercury and d-amphetamine on reversal learning and an extra dimensional shift in male mice. Exp. Clin. Psychopharmacol. 25, 64–73.

Bradford, A.B., Mancini, J.D., Atchison, W.D., 2016. Methylmercury-dependent increases in fluo4 fluorescence in neonatal rat cerebellar slices depend on granule cell migrational stage and GABAA receptor modulation. J. Pharmacol. Exp. Ther. 356, 2–12.

Branco, V., Canário, J., Holmgren, A., Carvalho, C., 2011. Inhibition of the thioredoxin system in the brain and liver of zebra-seabreams exposed to waterborne methylmercury. Toxicol. Appl. Pharmacol. 251, 95–103.

Branco, V., Ramos, P., Canário, J., Lu, J., Holmgren, A., Carvalho, C., 2012. Biomarkers of adverse response to mercury: histopathology versus thioredoxin reductase activity. J. Biomed. Biotechnol. 2012. 359879.

Branco, V., Godinho-Santos, A., Gonçalves, J., Lu, J., Holmgren, A., Carvalho, C., 2014. Mitochondrial thioredoxin reductase inhibition, selenium status, and Nrf-2 activation are determinant factors modulating the toxicity of mercury compounds. Free Radic. Biol. Med. 73, 95–105.

Branco, V., Caito, S., Farina, M., Rocha, J.B., Aschner, M., Carvalho, C., 2017a. Biomarkers of mercury toxicity: past, present, and future trends. J. Toxicol. Environ. Health B Crit. Rev. 20, 119–154.

Branco, V., Coppo, L., Solá, S., Lu, J., Rodrigues, C.M., Holmgren, A., Carvalho, C., 2017b. Impaired cross-talk between the thioredoxin and glutathione systems is related to ASK-1 mediated apoptosis in neuronal cells exposed to mercury. Redox Biol. 13, 278–287. Available online 01 June 2017, ISSN 2213-2317. http://dx.doi.org/10.1016/j.redox.2017.05.024.

Bromberg-Martin, E.S., Matsumoto, M., Hikosaka, O., 2010. Dopamine in motivational control: rewarding, aversive, and alerting. Neuron 68, 815–834.

Burk, R.F., Hill, K.E., 2005. Selenoprotein P: an extracellular protein with unique physical characteristics and a role in selenium homeostasis. Annu. Rev. Nutr. 25, 215–235.

Carratù, M.R., Borracci, P., Coluccia, A., Giustino, A., Renna, G., Tomasini, M.C., Raisi, E., Antonelli, T., Cuomo, V., Mazzoni, E., Ferraro, L., 2006. Acute exposure to methylmercury at two developmental windows: focus on neurobehavioral and neurochemical effects in rat offspring. Neuroscience 141, 1619–1629.

Carvalho, C.M.L., Chew, E.-H., Hashemy, S.I., Lu, J., Holmgren, A., 2008. Inhibition of the human thioredoxin system: a molecular mechanism of mercury toxicity. J. Biol. Chem. 283, 11913–11923.

Carvalho, C.M.L., Lu, J., Zhang, X., Arnér, E.S.J., Holmgren, A., 2011. Effects of selenite and chelating agents on mammalian thioredoxin reductase inhibited by mercury: implications for treatment of mercury poisoning. FASEB J. 25, 370–381.

Castoldi, A., Barni, S., Turin, I., Gandini, C., Manzo, L., 2000. Early acute necrosis, delayed apoptosis and cytoskeletal breakdown in cultured cerebellar granule neurons exposed to methylmercury. J. Neurosci. Res. 59, 775–787.

Castoldi, A.F., Coccini, T., Ceccatelli, S., Manzo, L., 2001. Neurotoxicity and molecular effects of methylmercury. Brain Res. Bull. 55, 197–203.

Clark, D.G., McElligott, T.F., Hurst, E.W., 1966. The toxicity of paraquat. Br. J. Ind. Med. 23, 126–132.

Clausing, P., Riedel, B., Gericke, S., Grün, G., Müller, L., 1984. Differences in the distribution of methyl mercury in erythrocytes, plasma, and brain of Japanese quails and rats after a single oral dose. Arch. Toxicol. 56, 132–135.

Coccini, T., Roda, E., Castoldi, A.F., Poli, D., Goldoni, M., Vettori, M.V., Mutti, A., Manzo, L., 2011. Developmental exposure to methylmercury and 2,2′, 4,4′, 5,5′-hexachloro-biphenyl (PCB153) affects cerebral dopamine D1-like and D2-like receptors of weanling and pubertal rats. Arch. Toxicol. 85, 1281–1294.

Colón-Rodrígue, A., Hannon, H., Atchison, W.D., 2017. Effects of methylmercury on spinal cord afferents and efferents—A review. Neurotoxicology 60, 308–320.

Cools, R., D'Esposito, M., 2011. Inverted-U-shaped dopamine actions on human working memory and cognitive control. Biol. Psychiatry 69, e113–e125.

Cory-Slechta, D.A., Virgolini, M.B., Rossi-George, A., Thiruchelvam, M., Lisek, R., Weston, D., 2008. Lifetime consequences of combined maternal lead and stress. Basic Clin. Pharmacol. Toxicol. 102, 218–227.

Creese, I., Burt, D.R., Snyder, S.H., 1976. Dopamine receptor binding predicts clinical and pharmacological potencies of antischizophrenic drugs. Science 192, 481–483.

Dalla Corte, C.L., Wagner, C., Sudati, J.H., Comparsi, B., Leite, G.O., Busanello, A., Soares, F.A.A., Aschner, M., Rocha, J.B.T., 2013. Effects of diphenyl diselenide on methylmercury toxicity in rats. Biomed. Res. Int. 2013. 983821.

Day, B.J., Patel, M., Calavetta, L., Chang, L.Y., Stamler, J.S., 1999. A mechanism of paraquat toxicity involving nitric oxide synthase. Proc. Natl. Acad. Sci. U.S.A. 96, 12760–12765.

Del Campo, N., Chamberlain, S.R., Sahakian, B.J., Robbins, T.W., 2011. The roles of dopamine and noradrenaline in the pathophysiology and treatment of attention-deficit/hyperactivity disorder. Biol. Psychiatry 69, e145–e157.

Deng, Y., Xu, Z., Xu, B., Liu, W., Wei, Y., Li, Y., Feng, S., Yang, T., 2014. Exploring cross-talk between oxidative damage and excitotoxicity and the effects of riluzole in the rat cortex after exposure to methylmercury. Neurotox. Res. 26, 40–51.

Denny, M.F., Atchison, W.D., 1996. Mercurial-induced alterations in neuronal divalent cation homeostasis. Neurotoxicology 17, 47–61.

Dodds, C.M., Clark, L., Dove, A., Regenthal, R., Baumann, F., Bullmore, E., Robbins, T.W., Müller, U., 2009. The dopamine D2 receptor antagonist sulpiride modulates striatal BOLD signal during the manipulation of information in working memory. Psychopharmacology (Berl) 207, 35–45.

Doi, R., 1991. Individual difference of methylmercury metabolism in animals and its significance in methylmercury toxicity. In: Advances in Mercury Toxicology (Section 2). Springer, US, pp. 77–98.

Doi, R., Tagawa, M., 1983. A study on the biochemical and biological behavior of methylmercury. Toxicol. Appl. Pharmacol. 69, 407–416.

Dórea, J.G., Farina, M., Rocha, J.B., 2013. Toxicity of ethylmercury (and thimerosal): a comparison with methylmercury. J. Appl. Toxicol. 33, 700–711.

Dreiem, A., Shan, M., Okoniewski, R.J., Sanchez-Morrissey, S., Seegal, R.F., 2009. Methylmercury inhibits dopaminergic function in rat pup synaptosomes in an age-dependent manner. Neurotoxicol. Teratol. 31, 312–317.

Everitt, B.J., Morris, K.A., O'Brien, A., Robbins, T.W., 1991. The basolateral amygdala-ventral striatal system and conditioned place preference: further evidence of limbic-striatal interactions underlying reward-related processes. Neuroscience 42, 1–18.

Ewing, J.A., Myers, R.D., 1985. Norepinephrine, alcohol, and alcoholism. In: Raymond Lake, C., Ziegler, M.G. (Eds.), The Catecholamines in Psychiatric and Neurologic Disorders. Butterworth-Heinemann, pp. 137–152.

Farina, M., Dahm, K.C.S., Schwalm, F.D., Brusque, A.M., Frizzo, M.E.S., Zeni, G., Souza, D.O., Rocha, J.B.T., 2003. Methylmercury increases glutamate release from brain synaptosomes and glutamate uptake by cortical slices from suckling rat pups: modulatory effect of ebselen. Toxicol. Sci. 73, 135–140.

Farina, M., Campos, F., Vendrell, I., Berenguer, J., Barzi, M., Pons, S., Suñol, C., 2009. Probucol increases glutathione peroxidase-1 activity and displays long-lasting protection against methylmercury toxicity in cerebellar granule cells. Toxicol. Sci. 112, 416–426.

Farina, M., Aschner, M., Rocha, J.B., 2011a. Oxidative stress in MeHg-induced neurotoxicity. Toxicol. Appl. Pharmacol. 256, 405–417.

Farina, M., Rocha, J.B., Aschner, M., 2011b. Mechanisms of methylmercury-induced neurotoxicity: evidence from experimental studies. Life Sci. 89 (15–16), 555–563. http://dx.doi.org/10.1016/j.lfs.2011.05.019. Epub 2011 Jun 13. Review.

Farina, M., Avila, D.S., Da Rocha, J.B.T., Aschner, M., 2013. Metals, oxidative stress and neurodegeneration: a focus on iron, manganese and mercury. Neurochem. Int. 62, 575–594.

Faro, L.R., do Nascimento, J.L., San José, J.M., Alfonso, M., Durán, R., 2000. Intrastriatal administration of methylmercury increases in vivo dopamine release. Neurochem. Res. 25, 225–229.

Faro, L.R., do Nascimento, J.L., Alfonso, M., Durán, R., 2002a. Mechanism of action of methylmercury on in vivo striatal dopamine release. Possible involvement of dopamine transporter. Neurochem. Int. 40, 455–465.

Faro, L.R., do Nascimento, J.L., Alfonso, M., Durán, R., 2002b. Protection of methylmercury effects on the in vivo dopamine release by NMDA receptor antagonists and nitric oxide synthase inhibitors. Neuropharmacology 42, 612–618.

Feltmann, K., Konradsson-Geuken, Å., De Bundel, D., Lindskog, M., Schilström, B., 2015. Antidepressant drugs specifically inhibiting noradrenaline reuptake enhance recognition memory in rats. Behav. Neurosci. 129, 701–708.

Feng, S., Xu, Z., Liu, W., Li, Y., Deng, Y., Xu, B., 2014. Preventive effects of dextromethorphan on methylmercury-induced glutamate dyshomeostasis and oxidative damage in rat cerebral cortex. Biol. Trace Elem. Res. 159, 332–345.

Fitzgerald, P.J., 2013. Elevated norepinephrine may be a unifying etiological factor in the abuse of a broad range of substances: alcohol, nicotine, marijuana, heroin, cocaine, and caffeine. Subst. Abuse. 7, 171–183.

Fitzgerald, P.J., 2014. Is elevated norepinephrine an etiological factor in some cases of schizophrenia? Psychiatry Res. 215, 497–504.

Franco, J.L., Posser, T., Dunkley, P.R., Dickson, P.W., Mattos, J.J., Martins, R., Bainy, A.C., Marques, M.R., Dafre, A.L., Farina, M., 2009. Methylmercury neurotoxicity is associated with inhibition of the antioxidant enzyme glutathione peroxidase. Free Radic. Biol. Med. 47, 449–457.

Gassó, S., Suñol, C., Sanfeliu, C., Rodríguez-Farré, E., Cristòfol, R.M., 2000. Pharmacological characterization of the effects of methylmercury and mercuric chloride on spontaneous noradrenaline release from rat hippocampal slices. Life Sci. 67, 1219–1231.

Gassó, S., Cristofol, R.M., Selema, G., Rosa, R., Rodríguez-Farré, E., Sanfeliu, C., 2001. Antioxidant compounds and Ca^{2+} pathway blockers differentially protect against methylmercury and mercuric chloride neurotoxicity. J. Neurosci. Res. 66, 135–145.

Gerfen, C.R., 1992. The neostriatal mosaic: multiple levels of compartmental organization. J. Neural Transm. Suppl. 36, 43–59.

Gimenez-Llort, L., Ahlbom, E., Dare, E., Vahter, M., Ögren, S.O., Ceccatelli, S., 2001. Prenatal exposure to methylmercury changes dopamine-modulated motor activity during early ontogeny: age and gender-dependent effects. Environ. Toxicol. Pharmacol. 9, 61–70.

Go, Y.M., Jones, D.P., 2013. The redox proteome. J. Biol. Chem. 288, 26512–26520.

Go, Y.M., Chandler, J.D., Jones, D.P., 2015. The cysteine proteome. Free Radic. Biol. Med. 84, 227–245.

Götz, M.E., Koutsilieri, E., Riederer, P., Ceccatelli, S., Daré, E., 2002. Methylmercury induces neurite degeneration in primary culture of mouse dopaminergic mesencephalic cells. J. Neural Transm. 109, 597–605.

Grandjean, P., Landrigan, P.J., 2014. Neurobehavioural effects of developmental toxicity. Lancet Neurol. 13, 330–338.

Grant, H., Lantos, P.L., Parkinson, C., 1980. Cerebral damage in paraquat poisoning. Histopathology 4, 185–195.

Grotto, D., de Castro, M.M., Barcelos, G.R., Garcia, S.C., Barbosa Jr., F., 2009. Low level and sub-chronic exposure to methylmercury induces hypertension in rats: nitric oxide depletion and oxidative damage as possible mechanisms. Arch. Toxicol. 83, 653–662.

Guallar, E., Sanz-Gallardo, M.I., van't Veer, P., Bode, P., Aro, A., Gómez-Aracena, J., Kark, J.D., Riemersma, R.A., Martín-Moreno, J.M., Kok, F.J., Heavy Metals and Myocardial Infarction Study Group, 2002. Mercury, fish oils, and the risk of myocardial infarction. N. Engl. J. Med. 347, 1747–1754.

Hamon, M., Blier, P., 2013. Monoamine neurocircuitry in depression and strategies for new treatments. Prog. Neuropsychopharmacol. Biol. Psychiatry 45, 54–63.

Hare, M.F., Atchison, W.D., 1995. Methylmercury mobilizes Ca++ from intracellular stores sensitive to inositol 1,4,5-trisphosphate in NG108-15 cells. J. Pharmacol. Exp. Ther. 272 (3), 1016–1023.

Harro, J., Oreland, L., Vasar, E., Bradwejn, J., 1995. Impaired exploratory behaviour after DSP-4 treatment in rats: implications for the increased anxiety after noradrenergic denervation. Eur. Neuropsychopharmacol. 5, 447–455.

Hatfield, D.L., Tsuji, P.A., Carlson, B.A., Gladyshev, V.N., 2014. Selenium and selenocysteine: roles in cancer, health, and development. Trends Biochem. Sci. 39, 112–120.

Henningsson, C., Hoffmann, S., McGonigle, L., Winter, J.S., 1993. Acute mercury poisoning (acrodynia) mimicking pheochromocytoma in an adolescent. J. Pediatr. 122, 252–253.

Heyer, D.B., Meredith, R.M., 2017. Environmental toxicology: sensitive periods of development and neurodevelopmental disorders. Neurotoxicology 58, 23–41.

Howes, O., McCutcheon, R., Stone, J., 2015. Glutamate and dopamine in schizophrenia: an update for the 21st century. J. Psychopharmacol. 29, 97–115.

Hrdina, P.D., Peters, D.A., Singhal, R.L., 1976. Effects of chronic exposure to cadmium, lead and mercury of brain biogenic amines in the rat. Res. Commun. Chem. Pathol. Pharmacol. 15, 483–493.

Huys, Q.J., Tobler, P.N., Hasler, G., Flagel, S.B., 2014. The role of learning-related dopamine signals in addiction vulnerability. Prog. Brain Res. 211, 31–77.

Johansen, J.P., Cain, C.K., Ostroff, L.E., Ledoux, J.E., 2011. Molecular mechanisms of fear learning and memory. Cell 147, 509–524.

Jones, D.P., 2015. Redox theory of aging. Redox Biol. 5, 71–79.

Juárez, B.I., Martinez, M.L., Montante, M., Dufour, L., Garcia, E., Jimenez-Capdeville, M.E., 2002. Methylmercury glutamate extracellular levels in frontal cortex of awake rats. Neurotoxicol. Teratol. 24, 767–771.

Juárez, B.I., Portillo-Salazar, H., González-Amaro, R., Mandeville, P., Aguirre, J.R., Jiménez, M.E., 2005. Participation of N-methyl-D-aspartate receptors on methylmercury-induced DNA damage in rat frontal cortex. Toxicology 207, 223–229.

Kalisch, B.E., Racz, W.J., 1996. The effects of methylmercury on endogenous dopamine efflux from mouse striatal slices. Toxicol. Lett. 89, 43–49.

Kang, M.J., Gil, S.J., Lee, J.E., Koh, H.C., 2010. Selective vulnerability of the striatal subregions of C57BL/6 mice to paraquat. Toxicol. Lett. 195, 127–134.

Karri, V., Schuhmacher, M., Kumar, V., 2016. Heavy metals (Pb, Cd, As and MeHg) as risk factors for cognitive dysfunction: a general review of metal mixture mechanism in brain. Environ. Toxicol. Pharmacol. 48, 203–213.

Kaye, W.H., Wierenga, C.E., Bailer, U.F., Simmons, A.N., Bischoff-Grethe, A., 2013. Nothing tastes as good as skinny feels: the neurobiology of anorexia nervosa. Trends Neurosci. 36, 110–120.

Kern, J.K., Geier, D.A., Adams, J.B., Mehta, J.A., Grannemann, B.D., Geier, M.R., 2011. Toxicity biomarkers in autism spectrum disorder: a blinded study of urinary porphyrins. Pediatr. Int. 53, 147–153.

Kern, J.K., Geier, D.A., Sykes, L.K., Haley, B.E., Geier, M.R., 2016. The relationship between mercury and autism: a comprehensive review and discussion. J. Trace Elem. Med. Biol. 37, 8–24.

Kirubagaran, R., Joy, K.P., 1990. Changes in brain monoamine levels and monoamine oxidase activity in the catfish, Clarias batrachus, during chronic treatments with mercurial. Bull. Environ. Contam. Toxicol. 45, 88–93.

Ko, J.H., Strafella, A.P., 2012. Dopaminergic neurotransmission in the human brain: new lessons from perturbation and imaging. Neuroscientist 18 (2), 149–168. http://dx.doi.org/10.1177/1073858411401413. Epub 2011 May 2. Review.

Kobayashi, K., 2001. Role of catecholamine signaling in brain and nervous system functions: new insights from mouse molecular genetic study. J. Investig. Dermatol. Symp. Proc. 6, 115–121.

Komulainen, H., Tuomisto, J., 1981. Interference of methyl mercury with monoamine uptake and release in rat brain synaptosomes. Basic Clin. Pharmacol. Toxicol. 48, 214–222.

Kraft, A.D., Aschner, M., Cory-Slechta, D.A., Bilbo, S.D., Caudle, W.M., Makris, S.L., 2016. Unmasking silent neurotoxicity following developmental exposure to environmental toxicants. Neurotoxicol. Teratol. 55, 38–44.

Kudo, T., Kushikata, T., Kudo, M., Kudo, T., Hirota, K., 2010. A central neuropathic pain model by DSP-4 induced lesion of noradrenergic neurons: preliminary report. Neurosci. Lett. 481, 102–104.

Lategan, A.J., Marien, M.R., Colpaert, F.C., 1992. Suppression of nigrostriatal and mesolimbic dopamine release in vivo following noradrenaline depletion by DSP-4: a microdialysis study. Life Sci. 50, 995–999.

Le Moal, M., Simon, H., 1991. Mesocorticolimbic dopaminergic network: functional and regulatory roles. Physiol. Rev. 71, 155–234.

Lénárd, L., Ollmann, T., László, K., Kovács, A., Gálosi, R., Kállai, V., Attila, T., Kertes, E., Zagoracz, O., Karádi, Z., Péczely, L., 2017. Role of D2 dopamine receptors of the ventral pallidum in inhibitory avoidance learning. Behav. Brain Res. 321, 99–105.

Liang, K.C., 1998. Pretraining infusion of DSP-4 into the amygdala impaired retention in the inhibitory avoidance task: involvement of norepinephrine but not serotonin in memory facilitation. Chin. J. Physiol. 41, 223–233.

Limke, T.L., Otero-Montañez, J.K., Atchison, W.D., 2003. Evidence for interactions between intracellular calcium stores during methylmercury-induced intracellular calcium dysregulation in rat cerebellar granule neurons. J. Pharmacol. Exp. Ther. 304, 949–958.

Limke, T.L., Bearss, J.J., Atchison, W.D., 2004a. Acute exposure to methylmercury causes Ca^{2+} dysregulation and neuronal death in rat cerebellar granule cells through an M3 muscarinic receptor-linked pathway. Toxicol. Sci. 80, 60–68.

Limke, T.L., Heidemann, S.R., Atchison, W.D., 2004b. Disruption of intraneuronal divalent cation regulation by methylmercury: are specific targets involved in altered neuronal development and cytotoxicity in methylmercury poisoning? Neurotoxicology 25, 741–760.

Lindström, H., Luthman, J., Oskarsson, A., Sundberg, J., Olson, L., 1991. Effects of long-term treatment with methyl mercury on the developing rat brain. Environ. Res. 56, 158–169.

Lindvall, O., Björklund, A., Skagerberg, G., 1983. Dopamine-containing neurons in the spinal cord: anatomy and some functional aspects. Ann. Neurol. 14 (3), 255–260.

Liou, H.H., Chen, R.C., Tsai, Y.F., Chen, W.P., Chang, Y.C., Tsai, M.C., 1996. Effects of paraquat on the substantia nigra of the wistar rats: neurochemical, histological, and behavioral studies. Toxicol. Appl. Pharmacol. 137, 34–41.

Liu, W., Xu, Z., Yang, T., Xu, B., Deng, Y., Feng, S., 2016. Memantine, a low-affinity NMDA receptor antagonist, protects against methylmercury-induced cytotoxicity of rat primary cultured cortical neurons, involvement of Ca^{2+} D dyshomeostasis antagonism, and indirect antioxidation effects. Mol. Neurobiol. 18, 1–7.

Manning-Bog, A.B., McCormack, A.L., Li, J., Uversky, V.N., Fink, A.L., Di Monte, D.A., 2002. The herbicide paraquat causes up-regulation and aggregation of alpha-synuclein in mice: paraquat and alpha-synuclein. J. Biol. Chem. 277, 1641–1644.

Martinez-Finley, E.J., Chakraborty, S., Slaughter, J.C., Aschner, M., 2013. Early-life exposure to methylmercury in wildtype and pdr-1/parkin knockout C. elegans. Neurochem. Res. 38, 1543–1552.

McCullough, L.D., Sokolowski, J.D., Salamone, J.D., 1993. A neurochemical and behavioral investigation of the involvement of nucleus accumbens dopamine in instrumental avoidance. Neuroscience 52 (4), 919–925.

Meinerz, D.F., Branco, V., Aschner, M., Carvalho, C., Rocha, J.B.T., 2017. Diphenyl diselenide protects against methylmercury-induced inhibition of thioredoxin reductase and glutathione peroxidase in human neuroblastoma cells: a comparison with ebselen. J. Appl. Toxicol. 37, 1073–1081. http://dx.doi.org/10.1002/jat.3458. First published: 6 April 2017.

Michaeli-Yossef, Y., Berkovitch, M., Goldman, M., 2007. Mercury intoxication in a 2-year-old girl: a diagnostic challenge for the physician. Pediatr. Nephrol. 22, 903–906.

Minnema, D.J., Cooper, G.P., Greenland, R.D., 1989. Effects of methylmercury on neurotransmitter release from rat brain synaptosomes. Toxicol. Appl. Pharmacol. 99, 510–521.

Miseta, A., Csutora, P., 2000. Relationship between the occurrence of cysteine in proteins and the complexity of organisms. Mol. Biol. Evol. 17, 1232–1239.

Miyoshi, R., Kito, S., Ishida, H., Katayama, S., 1988. Alterations of the central noradrenergic system in MPTP-induced monkey parkinsonism. Res. Commun. Chem. Pathol. Pharmacol. 62, 93–102.

Mokhber, N., Abdollahian, E., Soltanifar, A., Samadi, R., Saghebi, A., Haghighi, M.B., Azarpazhooh, A., 2014. Comparison of sertraline, venlafaxine and desipramine effects on depression, cognition and the daily living activities in Alzheimer patients. Pharmacopsychiatry 47, 131–140.

Montes, D.R., Stopper, C.M., Floresco, S.B., 2015. Noradrenergic modulation of risk/reward decision making. Psychopharmacology 232, 2681–2696.

Moore, R.Y., Card, J.P., 1984. Noradrenaline-containing neuron systems. In: BjoÈrklund, A., HoÈkfelt, T. (Eds.), Handbook of Chemical Neuroanatomy, vol. 2. Elsevier, Amsterdam, pp. 123–156.

Moore, K.E., Lookingland, K.J., 1995. Dopaminergic neuronal systems in the hypothalamus. In: Bloom, F.E., Kupfer, D.J. (Eds.), Psychopharmacology. Raven Press, New York, pp. 245–256.

Mutter, J., Naumann, J., Schneider, R., Walach, H., Haley, B., 2005. Mercury and autism: accelerating evidence. Neuroendocrinol. Lett. 26, 439–446.

Naganuma, A., Koyama, Y., Imura, N., 1980. Behavior of methylmercury in mammalian erythrocytes. Tox. Appl. Pharmacol. 54, 405–410.

Nakajima, S., Gerretsen, P., Takeuchi, H., Caravaggio, F., Chow, T., Le Foll, B., Mulsant, B., Pollock, B., Graff-Guerrero, A., 2013. The potential role of dopamine D3 receptor neurotransmission in cognition. Eur. Neuropsychopharmacol. 23, 799–813.

Nguyen, M., Roth, A., Kyzar, E.J., Poudel, M.K., Wong, K., Stewart, A.M., Kalueff, A.V., 2014. Decoding the contribution of dopaminergic genes and pathways to autism spectrum disorder (ASD). Neurochem. Int. 66, 15–26.

Ni, M., Li, X., Yin, Z., Jiang, H., Sidoryk-Węgrzynowicz, M., Milatovic, D., Aschner, M., 2010. Methylmercury induces acute oxidative stress, altering Nrf2 protein level in primary microglial cells. Toxicol. Sci. 116, 590–603.

Ni, M., Li, X., Yin, Z., Sidoryk-Wegrzynowicz, M., Jiang, H., Farina, M., Rocha, J.B., Syversen, T., Aschner, M., 2011. Comparative study on the response of rat primary astrocytes and microglia to methylmercury toxicity. Glia 59, 810–820.

Ni, M., Li, X., Rocha, J.B., Farina, M., Aschner, M., 2012. Glia and methylmercury neurotoxicity. J. Toxicol. Environ. Health A 75, 1091–1101.

Nieoullon, A., 2002. Dopamine and the regulation of cognition and attention. Prog. Neurobiol. 67, 53–83.

Nikolaus, S., Hautzel, H., Müller, H.W., 2014. Neurochemical dysfunction in treated and nontreated schizophrenia—a retrospective analysis of in vivo imaging studies. Rev. Neurosci. 25, 25–96.

Ogren, S.O., Archer, T., Ross, S.B., 1980. Evidence for a role of the locus coeruleus noradrenaline system in learning. Neurosci. Lett. 20, 351–356.

O'Kusky, J.R., Boyes, B.E., McGeer, E.G., 1988. Methylmercury-induced movement and postural disorders in developing rat: regional analysis of brain catecholamines and indoleamines. Brain Res. 439, 138–146.

Park, J., Kim, S.Y., Cha, G.H., Lee, S.B., Kim, S., Chung, J., 2005. Drosophila DJ-1 mutants show oxidative stress-sensitive locomotive dysfunction. Gene 361, 133–139.

Pauls, D.L., Abramovitch, A., Rauch, S.L., Geller, D.A., 2014. Obsessive-compulsive disorder: an integrative genetic and neurobiological perspective. Nat. Rev. Neurosci. 15, 410–424.

Pearson, R.G., 1963. Hard and soft acids and bases. J. Am. Chem. Soc. 85, 3533–3539.

Pereira, M.E., Morsch, V.M., Christofari, R.S., Rocha, J.B., 1999. Methyl mercury exposure during post-natal brain growth alters behavioral response to SCH 23390 in young rats. Bull. Environ. Contam. Toxicol. 63, 256–262.

Petersen, M.S., Halling, J., Bech, S., Wermuth, L., Weihe, P., Nielsen, F., Jørgensen, P.J., Budtz-Jørgensen, E., Grandjean, P., 2008. Impact of dietary exposure to food contaminants on the risk of Parkinson's disease. Neurotoxicology 29 (4), 584–590.

Peuchen, S., Bolaños, J.P., Heales, S.J., Almeida, A., Duchen, M.R., Clark, J.B., 1997. Inter-relationships between astrocyte function, oxidative stress and antioxidant status within the central nervous system. Prog. Neurobiol. 52, 261–281.

Pezzoli, G., Cereda, E., 2013. Exposure to pesticides or solvents and risk of Parkinson disease. Neurology 80, 2035–2041.

Quaak, I., Brouns, M.R., Van de Bor, M., 2013. The dynamics of autism spectrum disorders: how neurotoxic compounds and neurotransmitters interact. Int. J. Environ. Res. Public Health 10, 3384–3408.

Rabenstein, D.L., 1978a. The chemistry of methylmercury toxicology. J. Chem. Educ. 54, 292–296.

Rabenstein, D.L., 1978b. The aqueous solution chemistry of methylmercury and its complexes. Acc. Chem. Res. 11, 100–107.

Rabenstein, D.L., Fairhurst, M.T., 1975. Nuclear magnetic resonance studies of the solution chemistry of metal complexes. XI. The binding of methylmercury by sulfhydryl-containing amino acids and by glutathione. J. Am. Chem. Soc. 97, 2086–2092.

Rabenstein, D.L., Reid, R.S., 1984. Nuclear magnetic resonance studies of the solution chemistry of metal complexes. 20. Ligand-exchange kinetics of methylmercury(II)-thiol complexes. Inorg. Chem. 23, 1246–1250.

Rabenstein, D.L., Ozubko, R., Libich, S., Evans, C.A., Fairhurst, M.T., Suvanprakorn, C., 1974. Nuclear magnetic resonance studies of the solution chemistry of metal complexes. X. Determination of the formation constants of the methylmercury complexes of selected amines and amino carboxylic acids. J. Coord. Chem. 3, 263–271.

Rabenstein, D.L., Isab, A.A., Reid, R.S., 1982. A proton nuclear magnetic resonance study of the binding of methylmercury in human erythrocytes. Biochim. Biophys. Acta 720, 53–64.

Rabenstein, D.L., Reid, R.S., Yamashita, G., Tan, K.S., Arnold, A.P., 1986. Determination of thiols and selenols by titration with methylmercury with end point detection by nuclear magnetic resonance spectrometry. Anal. Chem. 58, 1266–1269.

Rajanna, B., Hobson, M., 1985. Influence of mercury on uptake of [3H]dopamine and [3H]norepinephrine by rat brain synaptosomes. Toxicol. Lett. 27, 7–14.

Rakers, F., Rupprecht, S., Dreiling, M., Bergmeier, C., Witte, O.W., Schwab, M., 2017. Transfer of maternal psychosocial stress to the fetus. Neurosci. Biobehav. Rev. Available online 22 February 2017. ISSN 0149-7634. http://dx.doi.org/10.1016/j.neubiorev.2017.02.019.

Ramanathan, G., Atchison, W.D., 2011. Ca2 + entry pathways in mouse spinal motor neurons in culture following in vitro exposure to methylmercury. Neurotoxicology 32, 742–750.

Reed, M.N., Newland, M.C., 2009. Gestational methylmercury exposure selectively increases the sensitivity of operant behavior to cocaine. Behav. Neurosci. 123, 408–417.

Rice, D., Barone Jr. S., 2000. Critical periods of vulnerability for the developing nervous system: evidence from humans and animal models. Environ. Health Perspect. 108 (Suppl 3), 511–533. Review.

Robbins, T.W., Everitt, B.J., 1995. Central norepinephrine neurons and behavior. In: Bloom, F.E., Kupfer, D.J. (Eds.), Psychopharmacology: The Fourth Generation of Progress. Raven Press, New York, pp. 363–372.

Rocha, J.B., Saraiva, R.A., Garcia, S.C., Gravina, F.S., Nogueira, C.W., 2012. Aminolevulinate dehydratase (δ-ALA-D) as marker protein of intoxication with metals and other pro-oxidant situations. Toxicol. Res. 1, 85–102.

Roman, H.A., Walsh, T.L., Coull, B.A., Dewailly, É., Guallar, E., Hattis, D., Mariën, K., Schwartz, J., Stern, A.H., Virtanen, J.K., Rice, G., 2011. Evaluation of the cardiovascular effects of methylmercury exposures: current evidence supports development of a dose-response function for regulatory benefits analysis. Environ. Health Perspect. 119, 607–614.

Roos, D., Seeger, R., Puntel, R., Vargas Barbosa, N., 2012. Role of calcium and mitochondria in MeHg-mediated cytotoxicity. J. Biomed. Biotechnol. 2012. 248764.

Ross, S.B., 2015. Long-term effects of N-2-chlorethyl-N-ethyl-2-bromobenzylamine hydrochloride on noradrenergic neurones in the rat brain and heart. Br. J. Pharmacol. 58, 521–527.

Ross, S.B., Renyl, A.L., 1976. On the long-lasting inhibitory effect of N-(2-chloroethyl)-N-ethyl-2-bromobenzylamine (DSP 4) on the active uptake of noradrenaline. J. Pharm. Pharmacol. 28, 458–459.

Ross, S.B., Stenfors, C., 2015. DSP4, a selective neurotoxin for the locus coeruleus noradrenergic system. A review of its mode of action. Neurotox. Res. 27, 15–30.

Rossignol, D.A., Genuis, S.J., Frye, R.E., 2014. Environmental toxicants and autism spectrum disorders: a systematic review. Transl. Psychiatry 4, e360. http://dx.doi.org/10.1038/tp.2014.4. Review.

Sakamoto, M., Ikegami, N., Nakano, A., 1996. Protective effects of Ca^{2+} channel blockers against methyl mercury toxicity. Basic Clin. Pharmacol. Toxicol. 78, 193–199.

Salonen, J.T., Seppanen, K., Nyyssonen, K., Korpela, H., Kauhanen, J., Kantola, M., Tuomilehto, J., Esterbauer, H., Tatzber, F., Salonen, R., 1995. Intake of mercury from fish, lipid peroxidation, and the risk of myocardial infarction and coronary, cardiovascular, and any death in eastern Finnish men. Circulation 91, 645–655.

Sealey, L.A., Hughes, B.W., Sriskanda, A.N., Guest, J.R., Gibson, A.D., Johnson-Williams, L., Pace, D.G., Bagasra, O., 2016. Environmental factors in the development of autism spectrum disorders. Environ. Int. 88, 288–298.

Ser, P.H., Omi, S., Shimizu-Furusawa, H., Yasutake, A., Sakamoto, M., Hachiya, N., Konishi, S., Nakamura, M., Watanabe, C., 2017. Differences in the responses of three plasma selenium-containing proteins in relation to methylmercury-exposure through consumption of fish/whales. Toxicol. Lett. 267, 53–58.

Shao, Y., Figeys, D., Ning, Z., Mailloux, R., Chan, H.M., 2015. Methylmercury can induce Parkinson's-like neurotoxicity similar to 1-methyl-4-phenylpyridinium: a genomic and proteomic analysis on MN9D dopaminergic neuron cells. J. Toxicol. Sci. 40, 817–828.

Sharma, R.P., Aldous, C.N., Farr, C.H., 1982. Methylmercury induced alterations in brain amine syntheses in rats. Toxicol. Lett. 13, 195–201.

Sigitova, E., Fišar, Z., Hroudová, J., Cikánková, T., Raboch, J., 2017. Biological hypotheses and biomarkers of bipolar disorder. Psychiatry Clin. Neurosci. 71, 77–103.

Sirois, J.E., Atchison, W.D., 2000. Methylmercury affects multiple subtypes of calcium channels in rat cerebellar granule cells. Toxicol. Appl. Pharmacol. 167, 1–11.

Slotkin, T.A., Bartolome, J., 1987. Biochemical mechanisms of developmental neurotoxicity of methylmercury. Neurotoxicology 8, 65–84.

Sørensen, N., Murata, K., Budtz-Jørgensen, E., Weihe, P., Grandjean, P., 1999. Prenatal methylmercury exposure as a cardiovascular risk factor at seven years of age. Epidemiology 1, 370–375.

Spear, L.P., 2000. The adolescent brain and age-related behavioral manifestations. Neurosci. Biobehav. Rev. 24, 417–463.

Stehr-Green, P., Tull, P., Stellfeld, M., Mortenson, P.B., Simpson, D., 2003. Autism and thimerosal-containing vaccines: lack of consistent evidence for an association. Am. J. Prev. Med. 25 (2), 101–106.

Sugiura, Y., Hojo, Y., Tamai, Y., Tanaka, H., 1976. Selenium protection against mercury toxicity. Binding of methylmercury by the selenohydryl-containing ligand. J. Am. Chem. Soc. 98, 2339–2341.

Sugiura, Y., Tamai, Y., Tanaka, H., 1978. Selenium protection against mercury toxicity: high binding affinity of methylmercury by selenium-containing ligands in comparison with sulfur-containing ligands. Bioinorg. Chem. 9, 167–180.

Suzuki, K.T., Sasakura, C., Yoneda, S., 1998. Binding sites for the (Hg-Se) complex on selenoprotein P. Biochim. Biophys. Acta 1429, 102–112.

Szot, P., Knight, L., Franklin, A., Sikkema, C., Foster, S., Wilkinson, C.W., White, S.S., Raskind, M.A., 2012. Lesioning noradrenergic neurons of the locus coeruleus in C57Bl/6 mice with unilateral 6-hydroxydopamine injection, to assess molecular, electrophysiological and biochemical changes in noradrenergic signaling. Neuroscience 216, 143–157.

Taylor, L.L., DiStefano, V., 1976. Effects of methylmercury on brain biogenic amines in the developing rat pup. Toxicol. Appl. Pharmacol. 38, 489–497.

Teychenne, P.F., Feuerstein, G., Lake, C.R., Ziegler, M.G., 1985. Central catecholamine systems: interaction with neurotransmitters in normal subjects and in patients with selected neurologic diseases. In: The Catecholamines in Psychiatric and Neurologic Disorders. Butterworth-Heinemann, pp. 91–119.

Thurston, S.W., Bovet, P., Myers, G.J., Davidson, P.W., Georger, L.A., Shamlaye, C., Clarkson, T.W., 2007. Does prenatal methylmercury exposure from fish consumption affect blood pressure in childhood? Neurotoxicology 28, 924–930.

Torres, A.D., Rai, A.N., Hardiek, M.L., 2000. Mercury intoxication and arterial hypertension: report of two patients and review of the literature. Pediatrics 105. e34.

Trillo, L., Das, D., Hsieh, W., Medina, B., Moghadam, S., Lin, B., Dang, V., Sanchez, M.M., De Miguel, Z., Ashford, J.W., Salehi, A., 2013. Ascending monoaminergic systems alterations in Alzheimer's disease. Translating basic science into clinical care. Neurosci. Biobehav. Rev. 37, 1363–1379.

Tritsch, N.X., Sabatini, B.L., 2012. Dopaminergic modulation of synaptic transmission in cortex and striatum. Neuron 76, 33–50.

Tsuzuki, Y., 1982. Effect of methylmercury exposure on different neurotransmitter systems in rat brain. Toxicol. Lett. 13, 159–162.

Usuki, F., Fujimura, M., 2016. Decreased plasma thiol antioxidant barrier and selenoproteins as potential biomarkers for ongoing methylmercury intoxication and an individual protective capacity. Arch. Toxicol. 90, 917–926.

VanDuyn, N., Nass, R., 2014. The putative multidrug resistance protein MRP-7 inhibits methylmercury-associated animal toxicity and dopaminergic neurodegeneration in Caenorhabditis elegans. J. Neurochem. 128, 962–974.

Van Enkhuizen, J., Janowsky, D.S., Olivier, B., Minassian, A., Perry, W., Young, J.W., Geyer, M.A., 2015. The catecholaminergic-cholinergic balance hypothesis of bipolar disorder revisited. Eur. J. Pharmacol. 753, 114–126.

Van Os, J., Kapur, S., 2009. Schizophrenia. Lancet 374, 635–645.

Veyrac, A., Sacquet, J., Nguyen, V., Marien, M., Jourdan, F., Didier, A., 2009. Novelty determines the effects of olfactory enrichment on memory and neurogenesis through noradrenergic mechanisms. Neuropsychopharmacology 34, 786–795.

Vilensky, J.A., Redman, K., 2003. British anti-Lewisite (dimercaprol): an amazing history. Ann. Emerg. Med. 41, 378–383.

Volkow, N.D., Wang, G.-J., Baler, R.D., 2011. Reward, dopamine and the control of food intake: implications for obesity. Trends Cogn. Sci. 15, 37–46.

Volkow, N.D., Wang, G.-J., Tomasi, D., Baler, R.D., 2013. Obesity and addiction: neurobiological overlaps. Obes. Rev. 14, 2–18.

Wagner, C., Sudati, J.H., Nogueira, C.W., Rocha, J.B.T., 2010. In vivo and in vitro inhibition of mice thioredoxin reductase by methylmercury. Biometals 23, 1171–1177.

Yamamoto, T., Fujimoto, Y., Shimura, T., Sakai, N., 1995. Conditioned taste aversion in rats with excitotoxic brain lesions. Neurosci. Res. 22, 31–49.

Yoneda, S., Suzuki, K.T., 1997. Equimolar Hg-Se complex binds to selenoprotein P. Biochem. Biophys. Res. Commun. 231, 7–11.

Yoshizawa, K., Rimm, E.B., Morris, J.S., Spate, V.L., Hsieh, C.-C., Spiegelman, D., Stampfer, M.J., Willett, W.C., 2002. Mercury and the risk of coronary heart disease in men. N. Engl. J. Med. 347, 1755–1760.

Zhong, F.X., Ji, X.Q., Tsou, K., 1985. Intrathecal DSP4 selectively depletes spinal noradrenaline and attenuates morphine analgesia. Eur. J. Pharmacol. 116, 327–330.

Zhou, T., Rademacher, D.J., Steinpreis, R.E., Weis, J.S., 1999. Neurotransmitter levels in two populations of larval Fundulus heteroclitus after methylmercury exposure. Comp. Biochem. Physiol. C 124, 287–294.

FURTHER READING

Mutter, J., Curth, A., Naumann, J., Deth, R., Walach, H., 2010. Does inorganic mercury play a role in Alzheimer's disease? A systematic review and an integrated molecular mechanism. J. Alzheimers Dis. 22, 357–374.

CHAPTER THREE

Pesticides and Parkinson's Disease: Current Experimental and Epidemiological Evidence

Samuel M. Goldman*,†, Ruth E. Musgrove‡, Sarah A. Jewell‡, Donato A. Di Monte‡,1

*University of California, San Francisco, CA, United States
†San Francisco Veterans Affairs Health Care System, San Francisco, CA, United States
‡German Center for Neurodegenerative Diseases (DZNE), Bonn, Germany
1Corresponding author: e-mail address: donato.dimonte@dzne.de

Contents

1. INTRODUCTION

Parkinson's disease (PD) is the second most common human neurodegenerative disease affecting as many as 1.5 million people in the United States. Clinically, it is characterized by symptoms and signs of motor dysfunction that include resting tremor, bradykinesia, and postural instability; the pathological basis of these clinical manifestations is the degeneration of dopaminergic neurons in the nigrostriatal pathway. Another cardinal feature of the disease from the pathological standpoint is the accumulation of

Advances in Neurotoxicology, Volume 1
ISSN 2468-7480
http://dx.doi.org/10.1016/bs.ant.2017.07.004

intraneuronal inclusions, called Lewy bodies and Lewy neurites, whose primary component is the protein α-synuclein. As further discussed below (see Section 3.2), progressive spreading of α-synuclein pathology throughout the brain of PD patients is likely to underlie the worsening of clinical course and appearance of a constellation of nonmotor symptoms such as sleep disorders, depression, cognitive impairment, and pain.

Despite the discovery of gene mutations over the past 20 years that are associated with familial forms of parkinsonism, the majority of PD cases are idiopathic in nature and likely develop as a consequence of both genetic and environmental determinants. Environmental hypotheses of PD causation have been proposed for more than 100 years and found significant support from the recognition of parkinsonian syndromes induced by viral agents (i.e., postencephalitic parkinsonism) or the neurotoxicant 1-methyl-4-phenyl-1,2,3,6-tetrahydropyridine (MPTP) (see Section 3.1). The concordance rate of PD in pairs of dizygotic twins was reported to be similar to that in pairs of monozygotic twins, further supporting a role of nongenetic risk factors in disease pathogenesis (Tanner et al., 1999).

Epidemiological investigations have significantly contributed to the identification of specific environmental culprits that may modulate PD risk. In particular, the longstanding recognition that disease rates may be increased in persons living in rural areas with high concentrations of farming-related industries has focused major attention on pesticides as potential neurotoxicants involved in PD development (e.g., Barbeau et al., 1987; Kamel et al., 2007; Svenson et al., 1993; Wan and Lin, 2016). Findings from epidemiological studies have been reinforced by results of experimental investigations that also support a relationship between pesticide exposure, damage to the nigrostriatal system, and induction of α-synuclein pathology in a variety of in vitro and in vivo model systems (Di Monte, 2003). Both epidemiological and experimental evidence linking pesticide exposure to PD will be reviewed and discussed next.

2. EPIDEMIOLOGICAL EVIDENCE

Important caveats should be mentioned when reviewing epidemiological findings concerning PD. PD presents many challenges to epidemiologists. First, there is no diagnostic test specific for the disease, and disorders such as essential tremor and secondary and atypical forms of parkinsonism may be commonly misclassified as PD (Alves et al., 2009; Marttila and Rinne, 1976; Mutch et al., 1986; Taylor et al., 2006). Differences in clinician experience, patient access to care, as well as changing diagnostic criteria over time complicate direct comparisons between studies (Anderson et al., 1998; Benito-León et al., 2004; Twelves et al., 2003). Because PD is predominantly a disease of old age with a long preclinical phase, many individuals with early disease die from other causes before motor symptoms manifest, obscuring risk factor associations (Ross et al., 2004, 2006). Assessing the "dose" and timing of exposure(s) to pesticides poses an additional challenge (see Section 2.3). For example, causal environmental exposures may occur during early life or even in utero, at chronic low levels or in combination, making them extremely difficult to characterize retrospectively.

2.1 Studies of Occupations

Farming-related and horticultural occupations have been associated with PD risk in many studies (Tanner and Goldman, 1996). Farming-related occupations were assessed in a US mortality study that included more than 2.6 million decedents between 1992 and 1998, 33,678 of whom had PD listed as an underlying or contributing cause of death on their death record (Park et al., 2005). In analyses adjusted for age, race, gender, region, and socioeconomic status, PD risk was modestly, but significantly elevated for farming-related occupations (mortality odds ratio, MOR, 1.14, 95% confidence interval, CI, 1.08–1.19), and for horticultural specialists (MOR 1.65, 95% CI 0.92–2.7), though results for more specific farming-related occupations were not reported.

In nations with nationalized health plans, PD diagnostic data from inpatient and/or outpatient records have been linked with census-derived occupational data to provide extensive power to assess occupational risk factors. A large Danish study linked PD diagnoses from hospitalization records with annual census data for 2.3 million persons over a 12-year period from 1981 to 1993 (Tüchsen and Jensen, 2000). Among nearly 1000 persons with PD, risk ratios (RR) were significantly increased for agriculture and horticulture occupations (RR 1.34, 95% CI 1.11–1.56) and for male farmers (RR 1.30, 95% CI 1.03–1.63). Similarly, Li et al. (2009) utilized a Swedish nationwide database linking census-derived occupation and first hospitalizations for PD from 1987 to 2004. In analyses adjusted for age, region, and education, risk was modestly increased among men coded as farmers in one or two consecutive censuses (standardized incidence ratio, SIR, 1.08 and 1.17, respectively), but was not elevated for gardeners or forestry workers.

Many studies have used clinic-based case–control designs to investigate whether agricultural occupations are more common among persons with PD than among control subjects. In general, these results should be interpreted with caution, because socioeconomic factors such as occupation may determine who is likely to be ascertained as a case in any particular clinic population. Farming occupation was associated with a markedly increased risk of PD in an Italian study of 136 PD specialty clinic patients and 272 non-neurodegenerative neurology clinic controls (OR 7.7, 95% CI 1.4–44.1) (Zorzon et al., 2002). In one of the largest clinic-based occupational studies, Goldman et al. (2005) reported a cross-sectional survey of 2072 PD patients drawn from three US movement disorders clinics in New York, Georgia, and the San Francisco Bay Area. Occupational frequencies among patients were compared with the expected frequencies based on Department of Labor regional estimates (BLS, 2003). The observed-to-expected ratio for farming occupation was markedly increased at 3.0 (95% CI 2.1–4.2). Of note, the proportional increase was greatest in the San Francisco Bay Area, whose catchment area includes one of the country's most heavily pesticide-treated regions (http://www.cdpr.ca.gov/).

Other studies of occupation and PD risk have not found significantly increased risk associated with agricultural occupations per se (Ascherio et al., 2006; Baldereschi et al., 2003; Baldi et al., 2003b; Dick et al., 2007b; Hertzman et al., 1994; Tanner et al., 2009). However, it is important to recognize that farming-related environmental exposures differ dramatically depending on geographic location, time-period, and particularly on the type of farming (e.g., field crops, orchard, livestock, etc.). As reviewed below, most of the studies that failed to find significant associations with agricultural occupational titles did in fact identify significant associations when they more directly assessed exposures to pesticides.

2.2 Studies of Exposure to Pesticides

In contrast to studies that focus solely on occupational categories, studies that more directly assess pesticide exposure consistently find an increased risk of PD (Table 1). Over the past 15 years, several metaanalyses of studies of pesticide exposure have been published. Priyadarshi et al. (2000) pooled 19 studies from North America, Europe, and Asia and found a combined odds ratio of 1.94 (95% CI 1.49–2.53) for PD associated with a history of pesticide exposure. More recently, van der Mark pooled results from 39 studies and reported a pooled relative risk of 1.62 (95% CI 1.40–1.88)

Table 1 Selected Epidemiologic Associations of Pesticides and PD

Study	Location	Design	Cases/ Controls	Exposure	Risk Ratio, 95% CI
Hertzman et al. (1990)	Canada	Case–control; interview	57/122	Pesticide spraying Paraquat	6.6, $p=0.031$ nc, $p=0.01$
Semchuk et al. (1992)	Canada	Case–control; interview	130/260	Pesticide application Herbicide use	2.3 (1.3–4.0) 3.0 (1.2–7.3)
Hertzman et al. (1994)	Canada	Case–control; interview	127/245	Occupational pesticide exposure	2.0 (1.0–4.1)
Seidler et al. (1996)	Germany	Case–control; interview	380/755	Insecticide use; Herbicide use Organochlorines Control Group 1 Control Group 2	Duration $p=0.001$ 1.6 (0.4–6.2) 5.8 (1.1–30.4)
Liou et al. (1997)	Taiwan	Case–control; interview	120/240	Exposure to pesticides Paraquat	2.9 (2.3–3.7) 3.2 (2.4–4.3)
Petrovitch et al. (2002)	USA	Prospective cohort; interview	116/7986	Plantation work	1.9 (1.0–3.5)
Baldi et al. (2003a)	France	Prospective cohort; interview	24/1507	Occupational pesticide exposure	5.6 (1.5–22)
Baldereschi et al. (2003)	Italy	Nested case–control; interview	113/4383	Pesticide-use license	3.7 (1.6–8.6)

Continued

Table 1 Selected Epidemiologic Associations of Pesticides and PD—cont'd

Study	Location	Design	Cases/ Controls	Exposure	Risk Ratio, 95% CI
Ascherio et al. (2006)	USA	Prospective cohort; baseline survey	413/143,325	Any pesticide exposure	1.8 (1.3–2.5)
Kamel et al. (2007)	USA	Nested case–control; baseline interview	78/55,931	Professional pesticide use Paraquat DDT 2,4-D	2.3 (1.2–4.5) 1.8 (1.0–3.4) 1.0 (0.6–1.8) 0.9 (0.5–1.8)
Dick et al. (2007a)	Europe	Case–control; interview, JEM	649/1587	Pesticide use (vs none) Low High	1.13 (0.82–1.57) 1.41 (1.06–1.88)
Hancock et al. (2008)	USA	Family study; interview	319/296	Pesticide application Organochlorines	1.6 (1.1–2.3) 2.0 (1.1–3.6)
Gatto et al. (2009)	USA	Case–control; GIS– pesticide usage linkage	368/341	Well water inferred exposure Diazinon Chlorpyrifos Paraquat Propargite	1.6(1.0–2.4) 1.5 (0.9–2.2) 1.1 (0.75–1.6) 1.3 (0.9–2.0)
Tanner et al. (2009)	USA	Case–control; interview	519/511	Occupational pesticide use Paraquat Permethrin 2,4-D	1.9 (1.1–3.2) 2.8 (0.8–9.7) 3.1 (0.7–15.8) 2.6 (1.03–6.5)

Study	Country	Method	Cases/Controls	Exposure	OR (95% CI)
Elbaz et al. (2009)	France	Case–control, professional users; interview	224/557	Pesticide application Organochlorines	1.8 (1.1–3.1) 2.4 (1.2–5.0)
Firestone et al. (2010)	USA	Case–control; interview	404/526	Pesticide worker DDT 2,4-D	1.5 (0.5–4.4) 0.8 (0.4–1.6) 0.8 (0.3–2.0)
Tanner et al. (2011)	USA	Case–control, professional applicators; interview	110/358	Rotenone Paraquat Dieldrin	2.5 (1.3–4.7) 2.5 (1.4–4.7) 1.6 (0.7–3.3)
Wang et al. (2011)	USA	Case–control GIS–pesticide usage linkage	362/341	Ziram + Maneb + Paraquat Occupational Exposure Residential Exposure	3.1 (1.7–5.6) 1.9 (1.1–3.2)
Fitzmaurice et al. (2013)	USA	Case–control; interview and GIS–pesticide usage linkage	360/754	Benomyl (highest quartile) Occupational exposure Residential exposure	2.0 (1.3–3.0) p-trend 0.002 1.2 (0.8–1.9) p-trend 0.07
van der Mark et al. (2014)	Netherlands	Case–control JEM; CEM	444/876	Highest quantile of: Pesticide exposure Insecticide exposure Benomyl Lindane Permethrin 2,4-D	1.28 (0.79–2.1) 1.46 (0.84–2.53) 2.23 (1.01–4.82) 1.26 (0.67–2.38) 1.44 (0.57–3.67) 1.51 (0.74–3.08)

Continued

Table 1 Selected Epidemiologic Associations of Pesticides and PD—cont'd

Study	Location	Design	Cases/ Controls	Exposure	Risk Ratio, 95% CI
Liew et al. (2014)	USA	Case–control; JEM	205/350	Lifetime occupational pesticide exposure (highest quartile)	2.2 (1.3–3.8) p-trend 0.007
Moisan et al. (2015)	France	Case–control; professional users; interview	133/298	Cumulative exposure (highest quartile) Pesticides	2.31 (1.09–4.9) p-trend 0.013
				Insecticides	1.88 (0.98–3.61)
				Fungicides	2.90 (1.22–6.89)
				Herbicides	1.12 (0.57–2.20)
Brouwer et al. (2015)	Netherlands	Case–Cohort mortality study; occupational JEM	402/2098	Pesticide exposure (vs none)	
				Low exposure	1.35 (0.81–2.26)
				High exposure	1.27 (0.86–1.88)

Abbreviations: *nc*=not calculable; *CI*=confidence interval; *JEM*=job-exposure matrix; *CEM*=crop-exposure matrix; *GIS*=geographic information systems.

for use of any pesticide (van der Mark et al., 2012). In a subanalysis of those studies that had information on pesticide functional class, the pooled risk was 1.40 (95% CI 1.08–1.81) for insecticides, 1.50 (95% CI 1.07–2.11) for herbicides, and 0.99 (95% CI 0.71–1.40) for fungicides. Not surprisingly, the greatest between-study heterogeneity was related to differences in the exposure assessment. Similar results were reported by Pezzoli and Cereda (2013) in a metaanalysis of 51 case–control studies. Metaanalytic risk was increased for any pesticide (odds ratio, OR, 1.76, 95% CI 1.56–2.04), herbicides (OR 1.33, 95% CI 1.08–1.65), insecticides (OR 1.53, 95% CI 1.12–2.08), but not for fungicides (OR 0.97, 95% CI 0.69–1.38). Despite this consistency, far fewer studies have assessed PD risk associated with specific pesticides. While pooling compounds on the basis of their functional category (e.g., herbicide, insecticide) seems a logical approach, these compounds may have widely disparate toxicological profiles in humans, and doing so could obscure large effects imparted by a small number of agents within any functional class (Kamel, 2013).

2.3 Exposure Assessment and Studies of Specific Pesticides

Table 1 summarizes selected analytic epidemiologic studies of PD and pesticide exposure. Although pesticide use is consistently associated with an increased risk of PD, most studies assessed exposure in a very general way—relying on self-reported "ever" exposure to "pesticides." Some studies provide greater detail, including functional class or application purpose such as eradication of termites, fleas, or weeds. However, because most nonprofessionals using pesticides in the home or garden do not know which specific compounds they were exposed to, far fewer studies assessed exposure to specific compounds or chemical classes (Brown et al., 2006). Recall bias (the propensity for case subjects to overendorse suspected exposures relative to control subjects) is an important consideration for retrospective studies with self-reported exposure data. Alternative approaches to self-report include the use of job- or crop-exposure matrices that impartially assign exposures based on a priori considerations, and the use of industrial hygienist expert raters to infer exposures while blinded to case status. Biological measurements have been useful for some highly persistent pesticides, such as organochlorines in blood and brain, but few if any biomarkers exist for most pesticides. Finally, some highly specialized populations, such as professional pesticide applicators, are able to provide very accurate recall of their use of specific agents, though these studies may have relatively limited analytic power due to their small size. Despite these limitations, epidemiologic studies have convincingly implicated several specific pesticides, as discussed below.

2.3.1 Rotenone

Derived from the leaves, roots, and seeds of the derris or cubé plant and other members of the pea family, rotenone containing plants and their extracts have been used as insecticides and piscicides (to kill fish) for centuries (Cabras et al., 2002). First registered in the United States in 1947, rotenone has been used broadly as a nonselective insecticide in agriculture and home gardening, and ubiquitously in household pet products such as flea powders. A so-called "organic" insecticide, until recently, its usage was permitted on produce labeled "organic" in the United States. In 2006, its registration with the US EPA was voluntarily withdrawn for livestock, residential, and domestic pet usage. However, it is still commonly used as a piscicide to eradicate invasive fish species in lakes and reservoirs (ATSDR, 2010).

Despite its mechanistic similarity to MPP^+ (the 1-methyl-4-phenylpyridinium metabolite of MPTP, see Section 3.1) and its ability to produce a compelling animal model of parkinsonism, few studies have specifically investigated associations between PD and rotenone usage. Exposure assessment is complicated by the fact that rotenone was sold under a broad range of trade names, the composition of which changed frequently. Tanner et al. (2011) published the most compelling epidemiologic study of rotenone and PD, a case–control study nested within the 84,000-member Agricultural Health Study (AHS), a cohort of professional pesticide applicators (mostly farmers), and their spouses in North Carolina and Iowa. Important strengths of this population include their familiarity with the specific products they utilized, and the documented validity of their self-reported exposures (Blair et al., 2002). AHS members were screened for PD by questionnaire. Suspected cases and randomly selected matched controls underwent in-person confirmatory diagnostic evaluations and completed a detailed risk factor interview that focused on mixing or applying 31 specific pesticides of a priori interest. In analyses of 110 cases and 358 controls adjusted for age, gender, state, and smoking, prior use of rotenone was associated with a significantly increased risk of PD (OR 2.5, 95% CI 1.3–4.7). In the only other published epidemiologic study on rotenone, Dhillon et al. (2008) found a markedly increased risk associated with gardening use of "organic pesticides such as rotenone" in a clinic-based case–control study in Texas (OR 10.9, 95% CI 2.5–48).

2.3.2 Paraquat

Paraquat is a restricted-use bipyridylium herbicide that was first produced commercially in 1961. Used primarily as a nonselective contact defoliant and preemergence dessicant, it is one of the most widely used pesticides in the world (US Centers for Disease Control) despite being banned by the European Union in 2007 (Isenring, 2017). At least five case–control studies have reported increased PD risk associated with exposure (Gatto et al., 2009; Hertzman et al., 1990; Kamel et al., 2007; Liou et al., 1997; Tanner et al., 2009, 2011), while two found no association (Firestone et al., 2010; Hertzman et al., 1994). In particular, risk was markedly increased in the high-quality case–control study by Tanner et al. (2011) nested within the AHS (OR 2.5, 95% CI 1.4–4.7). In addition to the documented validity of self-reported exposures, and in contrast to other studies of paraquat, this study had a relatively large number of exposed individuals. Remarkably, of the 31 a priori pesticides for which specific data were obtained, only rotenone and paraquat—the two compounds with compelling animal models, were significantly associated with PD risk. On a hopeful note (as well as further implicating a true causal association with paraquat exposure), the association with paraquat was markedly attenuated in those applicators who regularly wore protective gloves (Furlong et al., 2015). Gatto et al. (2009) conducted a study of 368 incident PD cases and 341 population controls in the California central valley—an area with very high pesticide usage. Their unique, unbiased design used geographic mapping (GIS) to link residences to California state pesticide application data in order to estimate exposure to paraquat in well water. They found a nonsignificant increased risk associated with high paraquat exposure (OR 1.31, 95% CI 0.79–2.17). This same group used similar methods to assess risk associated with combined ambient exposures to paraquat and the dithiocarbamate pesticide Maneb between 1974 and 1999 (Costello et al., 2009). Exposure to both pesticides within 500 m of a residence was associated with an increased risk of PD (OR 1.75, 95% CI 1.13–2.73), and this was particularly evident in those diagnosed at or before age 60 (OR 5.07, 95% CI 1.75–14.7).

The association of paraquat and PD risk may be substantially greater in combination with specific genetic vulnerability. In the AHS study, use of paraquat was associated with a remarkable 11-fold increased risk of PD (95% CI 3.0–44.6) in those lacking active glutathione-S-transferase T1 due to homozygous deletion of *GSTT1*—a variation present in 20% of the population (Goldman et al., 2012). Similarly, paraquat exposure may also be more harmful in combination with other environmental exposures, such as mild traumatic brain injury (Goldman et al., 2012; Hutson et al., 2011; Lee et al., 2012), or in combination with other pesticide exposures (Costello et al., 2009).

2.3.3 Organochlorine Pesticides

Organochlorines comprise the pesticide class most frequently associated with PD. These include DDT, aldrin, dieldrin, endosulfan, hexachlorocyclohexanes (e.g., lindane), heptachlor, methoxychlor, chlordane, and others. Most of these compounds were commercialized in the period following WWII, and although many have been banned since the 1970s, these highly persistent compounds are still identified in animal and human fatty tissues. A large German case–control study and a US-based family study both found approximate doubling of PD risk in association with use of organochlorine insecticides (Hancock et al., 2008; Seidler et al., 1996). Risk was also significantly increased in professional users of organochlorine insecticides in a large French agricultural worker cohort (OR 2.4, 95% CI 1.2–5.0) (Elbaz et al., 2009). Importantly, risk was markedly higher in association with polymorphisms in the *ABCB1* gene that encodes the transmembrane xenobiotic efflux pump p-glycoprotein (Dutheil et al., 2010). Further supporting a possible causal association, organochlorine pesticide residues in brain were correlated with Lewy pathology in a postmortem study (Ross et al., 2012).

2.3.4 Specific Organochlorines

Dieldrin is a highly persistent organochlorine insecticide that was widely used on a variety of crops for several decades until 1970, and for termite control until 1987 (ATSDR, 2002). In a postmortem study, dieldrin was detected in 6 of 20 brains from PD patients, but in 0 of 14 controls, and importantly in only 1 of 7 with Alzheimer's disease (AD)—demonstrating a specificity for PD (Fleming et al., 1994). Higher levels of dieldrin were also found in PD substantia nigra and caudate relative to controls (Corrigan et al., 1998, 2000). Consistent with studies in brain tissue, Weisskopf et al. (2010) found higher levels of dieldrin in blood from participants of a 60,000-member prospective Finnish cohort who developed PD 20–40 years later (OR per interquartile range 1.7, 95% CI 1.2–2.4). This prospective design mitigates any potential reverse causation—i.e., the possibility that an enhanced catabolic state in PD might result in increased mobilization of dieldrin from fat stores. Dieldrin exposure has been associated with a borderline increased risk in men in the AHS (OR 1.8, 95% CI 0.8–4.0) (Tanner et al., 2011), and residential dieldrin exposure estimated by GIS linkage to pesticide application data was associated with a significantly increased risk (OR 3.08, 95% CI 1.13–8.43) (Fitzmaurice et al., 2014).

Lindane (gamma-hexachlorohexane; HCH) and its byproduct beta-hexachlorohexane have been associated with PD in several studies. Lindane levels were fourfold higher in PD substantia nigra than in control brain (Corrigan et al., 2000). Richardson et al. (2009) detected β-HCH in 76% of PD patients, but only 30% of Alzheimer's and 40% of controls, with odds ratios (OR) of 4.4 and 5.2, respectively. In a larger follow-up study, higher levels of β-HCH were associated with threefold increased risk of PD (Richardson et al., 2011). However, HCH was not associated with PD in a large prospective Finnish study (Weisskopf et al., 2010).

2.3.5 Other Pesticide Associations

Benomyl is a systemic benzimidazole fungicide that has been associated with PD with increasing consistency over the past several years (Fitzmaurice et al., 2013, 2014; Tanner et al., 2011; van der Mark et al., 2014).

Dithiocarbamate pesticides such as maneb, mancozeb, and zineb, alone or in combination with paraquat have been inconsistently associated with increased PD risk (Costello et al., 2009; Seidler et al., 1996; Semchuk et al., 1993; Wang et al., 2011).

Permethrin. Several epidemiologic studies have reported increased risk associated with use of the pyrethroid insecticide permethrin. In addition to its direct use as an insecticide, permethrin is often used to impregnate clothing to repel insects. Like rotenone, permethrin is often considered an "organic" pesticide, because it is historically derived from chrysanthemums, though, more recently, pyrethroids are synthetic. Most studies of permethrin have had limited analytic power. Tanner et al. (2009, 2011) reported a trend toward increased risk associated with use of permethrin in two independent studies, including one in the AHS. More recently, permethrin was associated with modestly increased risk in a large Dutch case–control study (OR 1.44, 95% CI 0.57–3.67) (van der Mark et al., 2014).

2,4-Dichlorophenoxyacetic acid (2,4-D) is a chlorphenoxy herbicide that has been used to kill broadleaf weeds since the 1940s and is still one of the most commonly used herbicides in the world (ATSDR, 1999). It is a primary component of Agent Orange, an herbicide mixture that the Institute of Medicine has recognized as having "limited or suggestive evidence" of a causal association with PD (Institute of Medicine, 2016). Use of 2,4-D was significantly associated with increased risk of PD in a large multicenter US case–control study (OR 2.59, 95% CI 1.03, 6.48) (Tanner et al., 2009), and trended toward significance in several others, including a large Dutch study (OR 1.51, 95% CI 0.74–3.08) (van der Mark et al., 2014), and a French study of farming professionals (OR 1.8, 95% CI 0.9–3.3) (Elbaz et al., 2009).

2.4 Studies of Interaction Between Genes and Environment

Genetically determined variation in pesticide metabolism and response to injury is increasingly recognized as mediating associations between pesticides exposure and PD risk. Importantly, many of these genetic variants have strong biologic functional plausibility and do not appear to increase PD risk in the absence of pesticide exposures (and vice versa).

The *ABCB1* gene encodes the membrane transporter p-glycoprotein (p-gp; also known as MDR1 (multidrug resistance protein)), which binds and extrudes intracellular xenobiotics from a variety of cells, including the endothelial cells of the blood–brain barrier (Miller, 2010). Single-nucleotide polymorphic variants (SNPs) in *ABCB1* have been shown to dramatically affect associations of PD and rotenone (Goldman et al., 2016), organochlorines (Dutheil et al., 2010; Narayan et al., 2015), organophosphates (Narayan et al., 2015), and pesticides (nonspecifically) (Zschiedrich et al., 2009).

Variation in genes involved in metabolic and antioxidant processes have also been shown to modify associations with pesticides. These include *CYP2D6* (cytochrome P450 2D) and organochlorines (Deng et al., 2004; Elbaz et al., 2004), *GSTP1* (glutathione-*S*-transferase P1) and herbicides (Wilk et al., 2006), *NOS1* (nitric oxide synthase) and insecticides and herbicides (Hancock et al., 2008), the marked effect observed with *GSTT1* (glutathione-*S*-transferase T1) and paraquat noted above (Goldman et al., 2012), *NQO1* (NAD(P)H:quinone oxidoreductase 1) (Fong et al., 2007), *SOD2* (manganese superoxide dismutase) (Fong et al., 2007), *PON1* (paraoxonase 1) and organophosphates (Manthripragada et al., 2010), and *ALDH2* (aldehyde dehydrogenase) (Fitzmaurice et al., 2014).

3. EXPERIMENTAL EVIDENCE

The two pathological hallmarks of PD are nigrostriatal degeneration and accumulation of α-synuclein-containing inclusions, raising the critical questions of whether specific pesticides can target the nigrostriatal system and damage nigral dopaminergic neurons, and whether pesticide–α-synuclein interactions occur and are capable of promoting α-synuclein pathology. Experimental evidence supporting a role of pesticides in nigrostriatal injury and α-synuclein pathology mostly derives from studies with rotenone and paraquat, which will be discussed first in this section of the paper. Then, the effects of other pesticides and additional features of pesticide exposures relevant to PD will also be reviewed.

3.1 Pesticides and Nigrostriatal Degeneration: Rotenone and Paraquat

In the early 1980s, the discovery of MPTP as the culprit of a parkinsonian syndrome that affected young individuals who accidentally injected them-selves with an MPTP-contaminated synthetic drug represented a turning point in PD research and, in particular, research on the role of neuro-toxicants in disease pathogenesis (Di Monte et al., 2002). The discovery clearly indicated that mere exposure to a toxic agent was capable of inducing selective damage to the nigrostriatal system, with severe depletion of striatal dopamine, and significant loss of dopaminergic neurons in the substantia nigra pars compacta, two cardinal pathological features of PD. The ensuing clinical syndrome recapitulated motor dysfunctions, such as tremor, rigidity, and bradykinesia, virtually indistinguishable from symptoms and signs typ-ical of idiopathic PD. Based on these considerations, it is not surprising that, following MPTP identification, major efforts focused on the search for potential environmental agents that possessed MPTP-like toxic properties and could therefore act as disease risk factors. Pesticides became primary tar-gets of interest in virtue of the convergence of laboratory findings and epi-demiological clues (Di Monte, 2003).

Extensive laboratory work has elucidated important mechanistic features of MPTP-induced parkinsonism (Di Monte et al., 2002). MPTP does not damage nigrostriatal neurons itself but must undergo a process of metabolic activation in order to become neurotoxic. Within the brain, the enzyme monoamine oxidase type B (MAO-B) converts MPTP to its fully oxidized MPP+ metabolite, which represents the actual mediator of neuronal injury. MPP+ is taken up and accumulated within dopaminergic cells, triggering a series of toxic events that ultimately result in neurodegeneration. It is widely accepted that a primary event underlying MPP+ toxicity is an impairment of mitochondrial oxidative metabolism due to specific inhibition of complex I activity. Dopaminergic neurons in the substantia nigra pars compacta func-tion under conditions of high metabolic demand (due, for example, to their calcium-dependent autonomous pacemaking activity) and are characterized by pronounced arborization of their unmyelinated projections, making them particularly susceptible to failures of energy supplies, such as those caused by MPTP exposure (Chan et al., 2007; Matsuda et al., 2009). Another toxic mechanism likely involved in MPTP/MPP+-induced neurodegeneration is the production of reactive oxygen species (ROS). The MAO-B-mediated conversion of MPTP to MPP+ results in the generation of hydrogen peroxide, and further sources of ROS production include blockage of mitochondrial complex I and MPP+-mediated displacement of dopamine from its normal vesicular storage into the cytosol (Lotharius and O'Malley, 2000). Dopaminergic neurons are uniquely susceptible to oxidative stress mostly due to the prooxidant properties of "free" dopamine (i.e., dopamine that is not contained within vesicles); this distinct vulnerability is likely to contribute to their demise after MPTP exposure and would make them preferential targets of other endogenous or exogenous toxin with ROS-generating ability (Przedborski and Vila, 2003).

3.1.1 Rotenone

Elucidation of mechanisms of MPTP neurotoxicity and susceptibility factors that characterize nigrostriatal dopaminergic neurons guided subsequent investigations into specific classes of environmental agents with toxic properties predictively relevant to PD pathogenesis. A landmark study was published by Betarbet et al. (2000). Capitalizing on knowledge of the relationship between MPTP toxicity and mitochondrial complex I inhibition, this study assessed PD-like pathology and behavior in rats infused intravenously with rotenone, a lipophilic insecticide known for its highly specific ability to block complex I. By titrating rotenone exposure levels, these investigators were able to show that, at relatively low doses, the insecticide caused selective degeneration of dopaminergic terminals and cell bodies in the striatum and substantia nigra, respectively. These pathological outcomes were paralleled by the development of behavioral motor deficits that included hypokinesia and rigidity. The rotenone model, as described in this study, had a few drawbacks. In particular, the route and method of toxicant administration (i.e., infusion via a jugular vein cannula) were relatively cumbersome, and unambiguous evidence of nigrostriatal dopaminergic lesions was reported in only 50% of rats exposed to the insecticide. Subsequent refinements of the model have addressed these weaknesses, revealing that consistent nigrostriatal damage could be achieved, for example, by repeated intraperitoneal injections or intragastric administrations of rotenone (Pan-Montojo et al., 2010; Zharikov et al., 2015).

Over the past 15 years, experimental work using the rotenone model has provided strong support in favor of a link between environmental exposure and PD risk. As importantly, it has also yielded clues on mechanisms of nigrostriatal degeneration of likely translational relevance. For example, one important feature of rotenone neurotoxicity, which is shared by MPTP and other PD-relevant toxicants, is pronounced axonal damage. Investigations using the rotenone model have revealed that microtubule depolymerization may underlie this effect (Ren et al., 2005). Dopaminergic neurons could be particularly vulnerable to perturbations of microtubule stability leading to impaired axonal transport, loss of axonal integrity and, ultimately, retrograde (dying back) cell degeneration. Other important mechanistic clues derived from work with the rotenone model include the role that inhibition of autophagy may play in toxicant-induced dopaminergic cell death and the relationship between complex I deficiency and mitochondrial DNA damage (Sanders et al., 2014).

3.1.2 Paraquat

The chemical similarity between MPP + and the herbicide paraquat was one of the initial reasons for studies on the potential role of paraquat as a dopaminergic neurotoxicant. Chemical similarity, however, does not necessarily imply identical modes of action and, in fact, some properties of paraquat (e.g., its ability to redox cycle with molecular oxygen) are quite distinct from those of MPP + (Di Monte et al., 1986). Nevertheless, as indicated above, one of the intrinsic features that are likely to make nigrostriatal neurons preferential targets of neurodegenerative processes is their great vulnerability to oxidative stress. Paraquat's mode of action involving the generation of ROS via redox-cycling reactions provided (and still represents) a strong rationale for placing it on the list of candidate pesticides that could affect dopaminergic cell integrity and thus contribute to pathogenetic processes in PD (Bonneh-Barkay et al., 2005a). The ability of paraquat to damage and kill dopaminergic neurons has been demonstrated by numerous studies in several different laboratories in which the herbicide was injected into mice or rats, alone or in combination with the fungicide maneb (Kachroo and Schwarzschild, 2014; McCormack et al., 2002; Muthukumaran et al., 2014; see also Section 3.3). In a few instances (Smeyne et al., 2016), the neurotoxic properties of paraquat could not be confirmed but it is quite possible that methodological discrepancies (e.g., the accuracy of separation of the substantia nigra pars compacta from the ventral tegmental area during stereo-logical cell counting, see below) underlie these apparent incongruences. Images in Fig. 1 show a decreased density of dopaminergic (tyrosine hydroxylase-positive) neurons in the substantia nigra pars compacta (delin-eated in red) of paraquat- vs saline-injected mice (panel 1B vs panel 1A). Paraquat-induced nigral cell degeneration was confirmed by stereological counting of tyrosine hydroxylase-immunoreactive and Nissl-stained neu-rons (Fig. 1C).

Features of paraquat-induced neurodegeneration in in vivo experimental models mimic pathological characteristics of PD. Similar to observations in PD brains, the herbicide selectively targets dopaminergic neurons in the sub-stantia nigra pars compacta, while sparing nearby cells in the ventral tegmen-tal area (McCormack et al., 2006). Another important finding with the paraquat model concerns the role that neuroinflammation appears to play in its neurotoxicity. Neurodegeneration induced by the herbicide was found to be associated with robust microglial activation and could be alleviated by treatment with the antiinflammatory agent minocycline. Furthermore, no dopaminergic cell loss was observed when paraquat was injected to mutant mice lacking functional microglial NADPH oxidase (Purisai et al., 2007). Based on these findings, the following series of deleterious events have been proposed to underlie paraquat neurotoxicity. Once systemically adminis-tered, paraquat reaches the brain and, in particular, the ventral mesenceph-alon; its half-life in this brain region has been estimated to be approximately 28 days (Prasad et al., 2007). An initial tissue reaction to paraquat exposure may promote its toxic potential. Indeed, microglial activation and induction of microglial NADPH oxidase could facilitate the one-electron reduction of the herbicide and its subsequent redox cycling with molecular oxygen (Purisai et al., 2007). This redox cycling may occur at the level of microglial plasma membranes and would not necessarily require access of paraquat into the intracellular space (Bonneh-Barkay et al., 2005b). ROS generated through this mechanism, however, could themselves cross cell membranes, including membranes of nearby neurons. Dopaminergic cells would be especially susceptible to the damaging effects of ROS, thus explaining their distinct vulnerability to paraquat-induced degeneration.

Additional experimental evidence linking paraquat exposure to oxida-tive stress and dopaminergic cell injury derives from studies in which para-quat was administered to mice in combination with iron. Findings of these studies revealed that enhanced oral intake of iron during the neonatal period (pups were fed iron daily from postnatal days 10 to 17) resulted in an

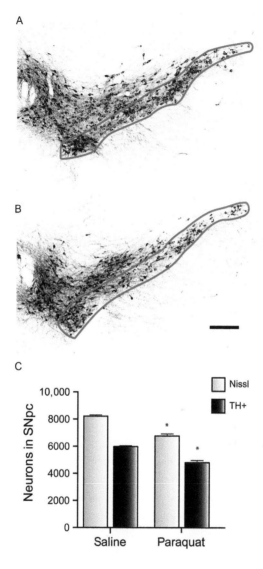

Fig. 1 Mice received two intraperitoneal injections of either saline or paraquat (15 mg/kg), given 1 week apart, and were sacrificed at 10 days after the second injection. (A and B) Coronal midbrain sections were stained with an antibody against tyrosine hydroxylase (TH). In two representative sections, one from a control animal injected with saline (A) and the other from a mouse treated with paraquat (B), TH-positive neurons are shown in the substantia nigra pars compacta (SCpc) delineated in *red*. Scale bar = 200 µm. (C) The number of Nissl-stained and TH-immunoreactive neurons was counted in the SNpc of saline- and paraquat-injected mice ($n = 5$/group) using unbiased stereology. Data are means ± SEM. *$P < 0.001$ (unpaired t-test).

augmented nigral cell loss when animals were exposed to paraquat at 12 and 24 months of age; this synergistic toxic effect was abolished if the synthetic superoxide dismutase/catalase mimetic, EUK-189, was administered prior to paraquat (Peng et al., 2007). The significance of these results cannot be overstated in terms of their translational implications. They underscore the fact that combined exposures rather than a single "hit" are more likely to play a role in the pathogenesis of PD. As importantly, the experimental data are consistent with an exposure model of sequential hits that, albeit separated by a long interval, could still produce additive or synergistic pathology; in this specific paradigm, early-in-life iron exposure predisposed to dopaminergic cell degeneration triggered by paraquat 1 or 2 years later.

3.2 Rotenone, Paraquat, and α-Synuclein

Fifteen years after the discovery of MPTP, another major breakthrough in PD research derived from genetic studies that identified a causal association between a single point mutation in the gene encoding the protein α-synuclein and a familial form of parkinsonism (Polymeropoulos et al., 1997; Ulusoy and Di Monte, 2013). The new spotlight on α-synuclein prompted a series of investigations into the role that this protein could have not only in rare genetic forms of the disease but also in typical sporadic PD. It shortly became clear that α-synuclein is indeed a major player in the pathogenesis of idiopathic PD. A pathognomonic feature of the disease is the accumulation of proteinaceous inclusions, called Lewy bodies and Lewy neurites, within neurons in the substantia nigra and other brain regions, such as the cerebral cortex, the nucleus basalis of Meynert, the locus coeruleus, and the dorsal motor nucleus of the vagus nerve (Forno, 1996). Postmortem histopathological evaluation of PD brains revealed that α-synuclein is a major component of Lewy inclusions (Spillantini et al., 1998); results of laboratory investigations also showed a tendency of α-synuclein to form both soluble and fibrillar aggregates that are likely to constitute the backbone of these intraneuronal deposits (Conway et al., 1998). Braak et al. (2003a) performed an extensive survey of postmortem brains from PD patients as well as individuals with no clinical evidence of the disease mapping, the occurrence of α-synuclein-positive lesions in tissue sections that spanned throughout the entire brain. This study confirmed the extensive diffusion of Lewy pathology in late disease stages, underscoring the fact that PD is not a disorder of the nigrostriatal system but as a "whole brain" pathological entity. Most importantly, however, Braak and colleagues noted a stereotypical pattern of progressive diffusion of α-synuclein inclusions that began in the lower brainstem and olfactory bulbs and proceeded toward pontine, midbrain, and forebrain regions. This groundbreaking observation bore a number of significant implications. First, it led to the classification of PD pathology into six stages, including two stages that precede the clinical onset of motor dysfunction (Braak et al., 2003a). Second, the peculiar sites of initial lesions, i.e., the dorsal motor nucleus of the vagus nerve and the olfactory bulbs (both these tissues are potentially exposed to exogenous noxae), prompted the suggestion that the disease process may actually be initiated outside the central nervous system by a neurotropic pathogen (Braak et al., 2003b). Such a pathogen could, for example, cross the mucosal intestinal barrier and enter postganglionic enteric neurons; from there, it could be transferred transynaptically into preganglionic efferent fibers of the Xth nerve and, traveling within these axons, could ultimately reach neuronal cell bodies in the dorsal motor nucleus of the vagus nerve. A third implication of the pathological staging proposed by Braak and colleagues is that the progressive accumulation of α-synuclein lesions affecting anatomically interconnected brain regions is compatible with the hypothesis of an interneuronal spreading of pathological forms of the protein (Braak et al., 2003b). Experimental evidence has confirmed the ability of α-synuclein to travel long-distance intra- and interneuronally via mechanisms that may include prion-like misfolding and template-dependent protein aggregation (Lee et al., 2010).

The prominent role played by α-synuclein in PD pathology, when considered within the context of environmental disease risk factors, raises the following critical questions: Is there a relationship between pesticide exposure and α-synuclein pathology? Are there mechanisms by which toxicant–α-synuclein interactions could promote or enhance susceptibility to α-synuclein lesions? Experimental work that has explored potential links between pesticide and α-synuclein is described below.

3.2.1 α-Synuclein and Rotenone

When Betarbet et al. (2000) described features of rotenone-induced neurotoxicity in rats, they noted the presence of nigral cytoplasmic inclusions that contained α-synuclein and displayed other immunohistochemical characteristics typical of Lewy inclusions such as their immunoreactivity for ubiquitin. More recently, work from the same laboratory elegantly showed a direct relationship between α-synuclein expression and rotenone-induced neurodegeneration (Zharikov et al., 2015). A short hairpin (sh) RNA targeting the endogenous α-synuclein gene transcript was delivered intranigrally using adeno-associated viral vectors in rotenone-treated rats. Quite remarkably, α-synuclein knockdown significantly reduced dopaminergic cell degeneration and attenuated the motor deficits associated with rotenone exposure in this animal model.

Another intriguing paradigm of rotenone–α-synuclein interactions was developed and characterized by Pan-Montojo et al. (2010, 2012). A chronic regimen of rotenone administered intragastrically to 1-year-old mice resulted in accumulation and aggregation of α-synuclein first within neurons of the enteric nervous system, then within cholinergic neurons of the dorsal motor nucleus of the vagus nerve and finally, after 3 months of treatment, within dopaminergic nigral neurons. α-Synuclein changes were paralleled by evidence of neuroinflammation in the dorsal motor nucleus of the vagus nerve and frank neurodegeneration in the substantia nigra (Pan-Montojo et al., 2010). Consistent with the interpretation that this progressive pathology involved spreading of toxic α-synuclein species from the intestine to the brain, significant neuroprotection could be achieved by hemivagotomy or partial sympathectomy (Pan-Montojo et al., 2012). The potential relevance of this rotenone model to pathogenetic processes in humans is underscored by the results of a recent study in which PD risk was compared between individuals who, for medical reasons, underwent vagotomy vs a control population that did not receive this surgical procedure; data revealed that full truncal vagotomy was associated with a decreased risk for subsequent development of PD (Svensson et al., 2015).

3.2.2 α-Synuclein and Paraquat

Genetic studies have shown that, besides point mutations, multiplication mutations of the α-synuclein gene cause familial parkinsonism, indicating that increased protein expression is itself a trigger of pathological processes (Ulusoy and Di Monte, 2013). It is noteworthy, therefore, that enhanced α-synuclein expression could be a consequence of toxic injury. Initial studies described this effect in MPTP-treated mice and nonhuman primates in which toxicant exposure induced an upregulation of α-synuclein and posttranslational protein modifications (i.e., nitration and phosphorylation) similar to those seen in the brain of PD patients (McCormack et al., 2008; Purisai et al., 2005; Vila et al., 2000). α-Synuclein upregulation was also observed in the brain of mice injected with the herbicide paraquat; protein expression was increased in the substantia nigra but also other brain regions and was accompanied by evidence of intraneuronal α-synuclein aggregation (Manning-Bog et al., 2002). Another interesting feature of enhanced α-synuclein expression in the paraquat model was the accumulation of α-synuclein into neuronal nuclei where it was shown to interact with histones (Goers et al., 2003). Toxicant-induced α-synuclein upregulation, albeit transitory, may have significant pathogenetic implications. Increased levels of intraneuronal α-synuclein could enhance its tendency to aggregate, suggesting a mechanism by which toxic exposures could contribute to the formation of Lewy inclusions (Ulusoy and Di Monte, 2013). Recent experimental findings also indicate that neuron-to-neuron transfer and long-distance propagation of α-synuclein can be prompted by its increased neuronal expression, consistent with the possibility that a toxic insult may result in protein accumulation, generation of deleterious α-synuclein species, and progressive diffusion of harmful α-synuclein throughout the brain (Helwig et al., 2016; Ulusoy et al., 2013).

Investigations into the relationship between paraquat exposure and α-synuclein have provided a number of additional important clues. For example, they emphasize the importance of protein degradation pathways as modulators of α-synuclein's toxic potential. Data indicate that lysosomal pathways, such as chaperone-mediated autophagy, play an important role in counteracting α-synuclein accumulation after paraquat exposure (Mak et al., 2010). Interestingly, evidence in paraquat-injected mice also suggests that the herbicide is capable of impairing both proteasomal and autosomal protein clearing, an effect that may contribute to α-synuclein accumulation and α-synuclein-related pathology (Wills et al., 2012). Epidemiological findings are consistent with a synergistic effect that paraquat exposure and traumatic brain injury may have as PD risk factors (Lee et al., 2012). Changes in α-synuclein toxicity may underlie this observation since both conditions, when reproduced in animal models, were associated with enhanced α-synuclein expression and aggregation (Hutson et al., 2011; Uryu et al., 2003). The precise molecular mechanisms underlying paraquat–α-synuclein interactions remain relatively unknown. Nevertheless, experimental evidence suggests an important role of NADPH oxidase 1, since paraquat-induced increase in α-synuclein expression and aggregation could be significantly reversed by shRNA-mediated knockdown of this enzyme both in vitro and in rats (Cristovao et al., 2012).

3.3 Additional Features of PD-Relevant Pesticides

Beyond the evidence discussed above, other intriguing clues on mechanisms that may contribute to the development of PD pathology have emerged from studies with paraquat and rotenone. Pesticides other than rotenone and paraquat have also been used experimentally to assess their potential role as PD risk factors and to elucidate mechanisms linking toxicant exposures to disease pathogenesis. Important findings from some of these studies are reviewed in the following paragraphs.

3.3.1 Long-Lasting Effects and Toxicant Interactions

Human pesticide exposures rarely involve one single agent. Therefore, experimental models in which PD-like pathology results from or is enhanced by combined toxicant administrations are likely to bear particular translational relevance. As reviewed above (Section 3.1), evidence of toxicant interactions, even after a long latent interval between exposures, derives from experiments in mice treated with iron and paraquat. Further experimental data supporting additive or synergistic effects of combined pesticide exposures are provided by in vivo studies using paraquat in combination with maneb (Kachroo and Schwarzschild, 2014; Thiruchelvam et al., 2000). Of note, Cory-Slechta et al. (2005) developed two models of paraquat/maneb exposures consistent with the hypothesis that developmental insults may enhance the risk for nigrostriatal degeneration in PD. In the first model, administration of the two pesticides from postnatal days 5 to 19 damaged the nigrostriatal system to a greater extent than adult-only exposures. In the second paradigm, treatment with maneb from gestational days 10 to 17 markedly predisposed to paraquat vulnerability during adulthood.

The possibility of long-lasting effects of toxicant exposures would be increased for agents that are retained in body tissues for prolonged periods of time and, in this respect, organochlorine insecticides represent a prototypic class of environmental agents; because of their lipophilicity, insecticides such as dieldrin can be accumulated in the brain, underlying their potential involvement in neurodegenerative processes (see Section 2.3). Laboratory work has revealed that chronic administration of low levels of dieldrin to young adult mice caused striatal changes in redox homeostasis, increased α-synuclein concentration, and decreased expression of the dopamine transporter (Hatcher et al., 2007). Developmental exposures to organochlorine pesticides also had deleterious long-lasting consequences. Treatment of mice with dieldrin, heptachlor, or endosulfam during gestation and lactation not only affected markers of nigrostriatal function in pesticide-exposed offspring but also significantly enhanced their susceptibility to the neurotoxic effects of subsequently administered MPTP (Richardson et al., 2006, 2008; Wilson et al., 2014).

3.3.2 Gender Differences and Genetic Determinants of Pesticide Neurotoxicity

An interesting feature of several models of pesticide-induced nigrostriatal toxicity is a gender specificity. For example, when the offspring of dieldrin- or heptachlor-exposed dams were treated with MPTP, toxicant-induced nigrostriatal injury was greater in male than female mice (McCormack et al., 2008; Richardson et al., 2006). Similarly, female mice were relatively protected from the toxic effects of maneb and paraquat, when administered together or in succession (see Section 3.3.1) (Cory-Slechta et al., 2005). The precise mechanisms underlying this gender differences remain to be fully elucidated. However, potential contributors of this effect may be estrogen secretion and/or expression of specific estrogen receptors, since estrogen-mediated neuroprotection has been shown in models of nigrostriatal degeneration (Bourque et al., 2012).

The toxic action of pesticides appears to be modulated by specific gene mutations/polymorphisms. Particularly relevant are interactions between pesticides and genes linked to familial forms of parkinsonism. In the model of neonatal iron intake followed by paraquat exposure (see Section 3.1), herbicide-induced dopaminergic neurodegeneration was significantly more pronounced in transgenic mice expressing A53T mutant human α-synuclein (Peng et al., 2010). *Parkin* mutations cause autosomal recessive parkinsonism and, interestingly, susceptibility to rotenone was found to be enhanced in neurons from parkin knockout mice (Casarejos et al., 2006). Mutations in the *DJ-1* gene underlie another form of recessive familial parkinsonism. DJ-1 acts as a redox sensor, and its loss of function would be expected to increase vulnerability to oxidative stress. Experiments in *Drosophila* revealed that the fly DJ-1 homolog, DJ-1β, became modified as a result of paraquat-induced oxidative stress; data also showed that lack of DJ-1 function made flies markedly more sensitive to the toxic effects of paraquat and rotenone (Muelener et al., 2005). As a final example of gene–pesticide interactions, PD-linked mutations in leucine-rich repeat kinase 2 (LRRK2) enhanced the loss of dopaminergic markers caused by rotenone treatment in *Caenorhabditis elegans* (Saha et al., 2009).

3.3.3 Pesticide-Induced Transcriptional Alterations

One of the mechanisms by which pesticide exposure could result in long-lasting changes predisposing to PD pathology is via transcriptional dys-regulation. Convincing evidence in support of this possibility derives from the work of Ryan et al. (2013) using human-induced pluripotent stem cells.

Data revealed that treatment with rotenone, paraquat, or paraquat plus maneb caused redox-dependent posttranslational modifications (e.g., S-nitrosylation) of myocyte enhancer factor 2C (MEF2C) and consequent inhibition of the MEF2C-PGC1α (peroxisome proliferator-activated receptor-γ coactivator-1α) transcriptional network. This transcriptional dysfunction contributed to toxin-induced mitochondrial impairment and apoptotic cell death. Other studies linking pesticide exposure to transcriptional changes include work by Slotkin and Seidler (2011) that analyzed the expression of PD-related genes (e.g., *snca*, *pink1*, *uchl1*, and *lrrk2*) in differentiating PC12 cells exposed to organochlorine or organophosphate agents. Results showed significant and specific effects. Indeed, of the two organophosphates used in this study, chlorpyrifos evoked both up- and downregulation whereas diazinon uniformly reduced gene expression. Treatment with the organochlorine dieldrin yielded no significant changes. The concept that pesticide exposure may affect the expression of specific genes relevant to neurodegenerative processes is reinforced by further experimental evidence. This includes the finding that transcriptional regulation of neurogenesis-related genes was modified in the mouse hippocampus after combined exposure to paraquat and maneb; gene dysregulation was magnified in transgenic animals overexpressing human wild-type α-synuclein (Desplats et al., 2012).

4. FINAL REMARKS

Evidence described in this chapter is directly relevant to three fundamental questions: Are environmental agents likely to play a role in PD etiology? Can pesticide usage and/or exposure represent a risk factor for the disease? Do specific pesticides act as neurotoxicants and affect key pathogenetic processes in PD? Current knowledge may not provide final answers to these questions. Nevertheless, a few important considerations emerge from the reviewed data. A large number of epidemiologic studies has found evidence for a relationship between pesticides and PD within and outside the context of occupational exposures. Similarly, pesticide administration has been extensively used to reproduce neurochemical, pathological, and behavioral features of PD in animal models. Although results of these investigations do not always concur and, in some instances, may even be at odds with each other, an overall assessment of current literature indicates that the majority of data is indeed compatible with a role of pesticides in PD pathogenesis. When drawing conclusions on this topic and, in particular, when the involvement of specific pesticides is evaluated, it must also be recognized that definite proof of a link between a single risk factor and the etiology of a complex disease, such as PD, may be difficult to attain. Despite this limitation, current observations on specific pesticides or specific classes of agricultural chemicals are quite compelling if one considers that epidemiological and experimental studies have independently yielded similar conclusions; for example, the deleterious effects of paraquat exposure are suggested by a large body of evidence from both population-based and laboratory research. Taken together, data from epidemiological and experimental investigations also underscore the need for future work further assessing multiple interactions of PD risk factors (e.g., pesticide–gene–aging interactions) and mechanisms of "amplification" of the effects of pesticide exposures on PD pathogenesis (e.g., long-lasting alterations of transcriptional processes). A more integrated approach to the evaluation of pesticides in PD, involving closer collaborations between clinicians, epidemiologists, and basic researchers should be applied to future investigations; this would undoubtedly result in the acquisition of more precise information and in a more direct translation of research findings into public health policies.

ACKNOWLEDGMENT

This work was supported by the Paul Foundation and the Centers of Excellence in Neurodegeneration Research (CoEN).

REFERENCES

Agency for Toxic Substances and Disease Registry, 1999. Toxicological Profile for Chlorophenols. ATSDR, US Department of Health and Human Services, Atlanta, USA.

Agency for Toxic Substances and Disease Registry, 2002. Toxicological Profile for Aldrin/Dieldrin. ATSDR, Preventative Health Services, US Department of Health and Human Services, Atlanta, USA.

Agency for Toxic Substances and Disease Registry, 2010. Toxicological Profile for Taxophene. ATDSR, Preventative Health Services, US Department of Health and Human Services, Atlanta, USA.

Alves, G., Müller, B., Herlofson, K., HogenEsch, I., Telstad, W., Aarsland, D., et al., 2009. Incidence of Parkinson's disease in Norway: the Norwegian ParkWest study. J. Neurol. Neurosurg. Psychiatry 80, 851–857.

Anderson, D.W., Rocca, W.A., de Rijk, M.C., Grigoletto, F., Melcon, M.O., Breteler, M.M., et al., 1998. Case ascertainment uncertainties in prevalence surveys of Parkinson's disease. Mov. Disord. 13, 626–632.

Ascherio, A., Chen, H., Weisskopf, M.G., O'Reilly, E., McCullough, M.L., Calle, E.E., et al., 2006. Pesticide exposure and risk for Parkinson's disease. Ann. Neurol. 60, 197–203.

Baldereschi, M., Di Carlo, A., Vanni, P., Ghetti, A., Carbonin, P., Amaducci, L., et al., 2003. Lifestyle-related risk factors for Parkinson's disease: a population-based study. Acta Neurol. Scand. 108, 239–244.

Baldi, I., Cantagrel, A., Lebailly, P., Tison, F., Dubroca, B., Chrysostome, V., et al., 2003a. Association between Parkinson's disease and exposure to pesticides in southwestern France. Neuroepidemiology 22, 305–310.

Baldi, I., Lebailly, P., Mohammed-Brahim, B., Letenneur, L., Dartigues, J.F., Brochard, P., 2003b. Neurodegenerative diseases and exposure to pesticides in the elderly. Am. J. Epidemiol. 157, 409–414.

Barbeau, A., Roy, M., Bernier, G., Campanella, G., Paris, S., 1987. Ecogenetics of Parkinson's disease: prevalence and environmental aspects in rural areas. Can. J. Neurol. Sci. 14, 36–41.

Benito-León, J., Bermejo-Pareja, F., Morales-González, J.M., Porta-Etessam, J., Trincado, R., Vega, S., et al., 2004. Incidence of Parkinson disease and parkinsonism in three elderly populations of central Spain. Neurology 62, 734–741.

Betarbet, R., Sherer, T.B., MacKenzie, G., Garcia-Osuna, M., Panov, A.V., Greenamyre, J.T., 2000. Chronic systemic pesticide exposure reproduces features of Parkinson's disease. Nat. Neurosci. 3, 1301–1306.

Blair, A., Tarone, R., Sandler, D., Lynch, C.F., Rowland, A., Wintersteen, W., et al., 2002. Reliability of reporting on life-style and agricultural factors by a sample of participants in the Agricultural Health Study from Iowa. Epidemiology 13, 94–99.

Bonneh-Barkay, D., Langston, W.J., Di Monte, D.A., 2005a. Toxicity of redox cycling pesticides in primary mesencephalic cultures. Antioxid. Redox Signal. 7, 649–653.

Bonneh-Barkay, D., Reaney, S.H., Langston, W.J., Di Monte, D.A., 2005b. Redox cycling of the herbicide paraquat in microglial cultures. Brain Res. Mol. Brain Res. 134, 52–56.

Bourque, M., Dluzen, D.E., Di Paolo, T., 2012. Signaling pathways mediating the neuroprotective effects of sex steroids and SERMs in Parkinson's disease. Front. Neuroendocrinol. 33, 169–178.

Braak, H., Del Tredici, K., Rüb, U., de Vos, R.A., Jansen Steur, E.N., Braak, E., 2003a. Staging of brain pathology related to sporadic Parkinson's disease. Neurobiol. Aging 24, 197–211.

Braak, H., Rüb, U., Gai, W.P., Del Tredici, K., 2003b. Idiopathic Parkinson's disease: possible routes by which vulnerable neuronal types may be subject to neuroinvasion by an unknown pathogen. J. Neural Transm. 110, 517–536.

Brouwer, M., Koeman, T., van den Brandt, P.A., Kromhout, H., Schouten, L.J., Peters, S., et al., 2015. Occupational exposures and Parkinson's disease mortality in a prospective Dutch cohort. Occup. Environ. Med. 72, 448–455.

Brown, T.P., Rumsby, P.C., Capleton, A.C., Rushton, L., Levy, L.S., 2006. Pesticides and Parkinson's disease—is there a link? Environ. Health Perspect. 114, 156–164.

Bureau of Labor Statistics, 2003. Occupational Employment and Wages. Department of Labor, Washington, DC.

Cabras, P., Caboni, P., Cabras, M., Angioni, A., Russo, M., 2002. Rotenone residues on olives and in olive oil. J. Agric. Food Chem. 50, 2576–2580.

Casarejos, M.J., Menéndez, J., Solano, R.M., García de Yébenes, J., Mena, M.A., 2006. Susceptibility to rotenone is increased in neurons from parkin null mice and is reduced by minocycline. J. Neurochem. 97, 934–946.

Chan, C.S., Guzman, J.N., Ilijic, E., Mercer, J.N., Rick, C., Tkatch, T., et al., 2007. "Rejuvination" protects neurons in mouse models of Parkinson's disease. Nature 447, 1081–1086.

Conway, K.A., Harper, J.D., Lansbury, P.T., 1998. Accelerated in vitro fibril formation by a mutant α-synuclein linked to early-onset Parkinson's disease. Nat. Med. 4, 1318–1320.

Corrigan, F.M., Murray, L., Wyatt, C.L., Shore, R.F., 1998. Diorthosubstituted polychlorinated biphenyls in caudate nucleus in Parkinson's disease. Exp. Neurol. 150, 339–342.

Corrigan, F.M., Wienburg, C.L., Shore, R.F., Daniel, S.E., Mann, D., 2000. Organochlorine insecticides in substantia nigra in Parkinson's disease. J. Toxicol. Environ. Health A 59, 229–234.

Cory-Slechta, D.A., Thiruchelvam, M., Barlow, B.K., Richfield, E.K., 2005. Developmental pesticide models of the Parkinson disease phenotype. Environ. Health Perspect. 113, 1263–1270.

Costello, S., Cockburn, M., Bronstein, J., Zhang, X., Ritz, B., 2009. Parkinson's disease and residential exposure to maneb and paraquat from agricultural applications in the central valley of California. Am. J. Epidemiol. 169, 919–926.

Cristovao, A.C., Guhathakurta, S., Bok, E., Je, G., Yoo, S.D., Choi, D.H., et al., 2012. NADPH oxidase 1 mediates α-synucleinopathy in Parkinson's disease. J. Neurosci. 32, 14465–14477.

Deng, Y., Newman, B., Dunne, M.P., Silburn, P.A., Mellick, G.D., 2004. Further evidence that interactions between CYP2D6 and pesticide exposure increase risk for Parkinson's disease. Ann. Neurol. 55, 897.

Desplats, P., Patel, P., Kosberg, K., Mante, M., Patrick, C., Rockenstein, E., et al., 2012. Combined exposure to Maneb and Paraquat alters transcriptional regulation of neurogenesis-related genes in mice models of Parkinson's disease. Mol. Neurodegener. 7, 49.

Dhillon, A.S., Tarbutton, G.L., Levin, J.L., Plotkin, G.M., Lowry, L.K., Nalbone, J.T., et al., 2008. Pesticide/environmental exposures and Parkinson's disease in East Texas. J. Agromed. 13, 37–48.

Di Monte, D., Sandy, M.S., Ekström, G., Smith, M.T., 1986. Comparative studies on the mechanisms of paraquat and 1-methyl-4-phenylpyridine (MPP+) cytotoxicity. Biochem. Biophys. Res. Commun. 137, 303–309.

Di Monte, D.A., 2003. The environment and Parkinson's disease: is the nigrostriatal system preferentially targeted by neurotoxins? Lancet Neurol. 2, 531–538.

Di Monte, D.A., Lavasani, M., Manning-Bog, A.B., 2002. Environmental factors in Parkinson's disease. Neurotoxicology 23, 487–502.

Dick, F.D., De Palma, F., Ahmadi, A., Scott, N.W., Prescott, G.J., Bennett, J., et al., 2007a. Environmental risk factors for Parkinson's disease and parkinsonism; the Geoparkinson study. Occup. Environ. Med. 64, 666–672.

Dick, S., Semple, S., Dick, F., Seaton, A., 2007b. Occupational titles as risk factors for Parkinson's disease. Occup. Med. 57, 50–56.

Dutheil, F., Beaune, P., Tzourio, C., Loriot, M.A., Elbaz, A., 2010. Interaction between ABCB1 and professional exposure to organochlorine insecticides in Parkinson disease. Arch. Neurol. 67, 739–745.

Elbaz, A., Clavel, J., Rathouz, P.J., Moisan, F., Galanaud, J.P., Delemotte, B., et al., 2009. Professional exposure to pesticides and Parkinson disease. Ann. Neurol. 66, 494–504.

Elbaz, A., Levecque, C., Clavel, J., Vidal, J.S., Richard, F., Amouyel, P., et al., 2004. CYP2D6 polymorphism, pesticide exposure, and Parkinson's disease. Ann. Neurol. 55, 430–434.

Firestone, J.A., Lundin, J.I., Powers, K.M., Smith-Weller, T., Franklin, G.M., Swanson, P.D., et al., 2010. Occupational factors and risk of Parkinson's disease: a population-based case–control study. Am. J. Ind. Med. 53, 217–223.

Fitzmaurice, A.G., Rhodes, S.L., Lulla, A., Murphy, N.P., Lam, H.A., O'Donnell, K.C., et al., 2013. Aldehyde dehydrogenase inhibition as a pathogenic mechanism in Parkinson disease. Proc. Natl. Acad. Sci. U.S.A. 110, 636–641.

Fitzmaurice, A.G., Rhodes, S.L., Cockburn, M., Ritz, B., Bronstein, J.M., 2014. Aldehyde dehydrogenase variation enhances effect of pesticides associated with Parkinson disease. Neurology 82, 419–426.

Fleming, L., Mann, J.B., Bean, J., Briggle, T., Sanchez-Ramos, J.R., 1994. Parkinson's disease and brain levels of organochlorine pesticides. Ann. Neurol. 36, 100–103.

Fong, C.S., Wu, R.M., Shieh, J.C., Chao, Y.T., Fu, Y.P., Kuao, C.L., et al., 2007. Pesticide exposure on southwestern Taiwanese with MnSOD and NQO1 polymorphisms is associated with increased risk of Parkinson's disease. Clin. Chim. Acta 378, 136–141.

Forno, L.S., 1996. Neuropathology of Parkinson's disease. J. Neuropathol. Exp. Neurol. 55, 259–272.

Furlong, M., Tanner, C.M., Goldman, S.M., Bhudhikanok, G.S., Blair, A., Chade, A., et al., 2015. Protective glove use and hygiene habits modify the associations of specific pesticides with Parkinson's disease. Environ. Int. 75, 144–150.

Gatto, N.M., Cockburn, M., Bronstein, J., Manthripragada, A.D., Ritz, B., 2009. Well-water consumption and Parkinson's disease in rural California. Environ. Health Perspect. 117, 1912–1918.

Goers, J., Manning-Bog, A.B., McCormack, A.L., Millett, I.S., Doniach, S., Di Monte, D.A., et al., 2003. Nuclear localization of α-synuclein and its interaction with histones. Biochemistry 42, 8465–8471.

Goldman, S.M., Kamel, F., Meng, C., Korell, M., Umbach, D.M., Hoppin, J., et al., 2016. Rotenone and Parkinson's disease (PD): effect modification by membrane transporter variants. Mov. Disord. 31, S148–S149.

Goldman, S.M., Kamel, F., Ross, G.W., Bhudhikanok, G.S., Hoppin, J.A., Korell, M., et al., 2012. Genetic modification of the association of Paraquat and Parkinson's disease. Mov. Disord. 27, 1652–1658.

Goldman, S.M., Tanner, C.M., Olanow, C.W., Watts, R.L., Field, R.D., Langston, J.W., 2005. Occupation and parkinsonism in three movement disorders clinics. Neurology 65, 1430–1435.

Hancock, D.B., Martin, E.R., Mayhew, G.M., Stajich, J.M., Jewett, R., Stacy, M.A., et al., 2008. Pesticide exposure and risk of Parkinson's disease: a family-based case–control study. BMC Neurol. 8, 6.

Hatcher, J.M., Richardson, J.R., Guillot, T.S., McCormack, A.L., Di Monte, D.A., Jones, D.P., et al., 2007. Dieldrin exposure induces oxidative damage in the mouse nigrostriatal dopamine system. Exp. Neurol. 204, 619–630.

Helwig, M., Klinkenberg, M., Rusconi, R., Musgrove, R.E., Majbour, N.K., El-Agnaf, O.M., et al., 2016. Brain propagation of transduced α-synuclein involves non-fibrillar protein species and is enhanced in α-synuclein null mice. Brain 139, 856–870.

Hertzman, C., Wiens, M., Bowering, D., Snow, B., Calne, D., 1990. Parkinson's disease: a case–control study of occupational and environmental risk factors. Am. J. Ind. Med. 17, 349–355.

Hertzman, C., Wiens, M., Snow, B., Kelly, S., Calne, D., 1994. A case–control study of Parkinson's disease in a horticultural region of British Columbia. Mov. Disord. 9, 69–75.

Hutson, C.B., Lazo, C.R., Mortazavi, F., Giza, C.C., Hovda, D., Chesselet, M.F., 2011. Traumatic brain injury in adult rats causes progressive nigrostriatal dopaminergic cell loss and enhanced vulnerability to the pesticide paraquat. J. Neurotrauma 28, 1783–1801.

Institute of Medicine, 2016. Veterans and Agent Orange: Update 2014. National Academy of Science, Washington, DC.

Isenring, R., 2017. Adverse Health Effects Caused by Paraquat, A Bibliography of Documented Evidence. Public Eye, Switzerland, Pesticide Action Network Asia Pacific and Pesticide Action Network, UK.

Kachroo, A., Schwarzschild, M.A., 2014. Allopurinol reduces levels of urate and dopamine but not dopaminergic neurons in a dual pesticide model of Parkinson's disease. Brain Res. 1563, 103–109.

Kamel, F., 2013. Epidemiology. Paths from pesticides to Parkinson's. Science 341, 722–723.

Kamel, F., Tanner, C., Umbach, D., Hoppin, J., Alavanja, M., Blair, A., et al., 2007. Pesticide exposure and self-reported Parkinson's disease in the agricultural health study. Am. J. Epidemiol. 165, 364–374.

Lee, P.C., Bordelon, Y., Bronstein, J., Rirz, B., 2012. Traumatic brain injury, paraquat exposure, and their relationship to Parkinson's disease. Neurology 79, 2061–2066.

Lee, S.J., Desplats, P., Sigurdson, C., Tsigelny, I., Masliah, E., 2010. Cell-to-cell transmission of non-prion protein aggregates. Nat. Rev. Neurol. 6, 702–706.

Li, X., Sundquist, J., Sundquist, K., 2009. Socioeconomic and occupational groups and Parkinson's disease: a nationwide study based on hospitalizations in Sweden. Int. Arch. Occup. Environ. Health 82, 235–241.

Liew, Z., Wang, A., Bronstein, J., Ritz, B., 2014. Job exposure (JEM)-derived estimates of lifetime occupational pesticide exposure and risk for Parkinson's disease. Arch. Environ. Occup. Health 69, 241–251.

Liou, H.H., Tsai, M.C., Chen, C.J., Jeng, J.S., Chang, Y.C., Chen, S.Y., et al., 1997. Environmental risk factors and Parkinson's disease: a case–control study in Taiwan. Neurology 48, 1583–1588.

Lotharius, J., O'Malley, J.K., 2000. The parkinsonism-inducing drug 1-methyl-4-phenylpyridinium triggers intracellular dopamine oxidation. A novel mechanism of toxicity. J. Biol. Chem. 275, 38581–38588.

Mak, S.K., McCormack, A.L., Manning-Bog, A.B., Cuervo, A.M., Di Monte, D.A., 2010. Lysosomal degradation of α-synuclein in vivo. J. Biol. Chem. 285, 13621–13629.

Manning-Bog, A.B., McCormack, A.L., Li, J., Uversky, V.N., Fink, A.L., Di Monte, D.A., 2002. The herbicide paraquat causes up-regulation and aggregation of α-synuclein in mice. J. Biol. Chem. 277, 1641–1644.

Manthripragada, A.D., Costello, S., Cockburn, M.G., Bronstein, J.M., Ritz, B., 2010. Paraoxonase 1, agricultural organophosphate exposure, and Parkinson disease. Epidemiology 21, 87–94.

Marttila, R.J., Rinne, U.K., 1976. Epidemiology of Parkinson's disease in Finland. Acta Neurol. Scand. 53, 81–102.

Matsuda, W., Furuta, T., Nakamura, K.C., Hioki, H., Fujiyama, F., Arai, R., et al., 2009. Single nigrostriatal dopaminergic neurons form widely spread and highly dense axonal arborizations in the neostriatum. J. Neurosci. 29, 444–453.

McCormack, A.L., Atienza, J.G., Langston, J.W., Di Monte, D.A., 2006. Decreased susceptibility to oxidative stress inderlies the resistance of specific dopaminergic cell populations to paraquat-induced degeneration. Neuroscience 141, 929–937.

McCormack, A.L., Mak, S.K., Shenasa, M., Langston, W.J., Forno, L.S., Di Monte, D.A., 2008. Pathologic modifications of α-synuclein in 1-methyl-4-phenyl-1,2,3,6-tetrahydropyridine (MPTP)-treated squirrel monkeys. J. Neuropathol. Exp. Neurol. 67, 793–802.

McCormack, A.L., Thiruchelvam, M., Manning-Bog, A.B., Thiffault, C., Langston, W.J., Cory-Slechta, D.A., et al., 2002. Environmental risk factors and Parkinson's disease: selective degeneration of nigral dopaminergic neurons caused by the herbicide paraquat. Neurobiol. Dis. 10, 119–127.

Miller, D.S., 2010. Regulation of P-glycoprotein and other ABC drug transporters at the blood-brain barrier. Trends Pharmacol. Sci. 31, 246–254.

Moisan, F., Spinosi, J., Delabre, L., Gourlet, V., Mazurie, J.L., Benatru, I., et al., 2015. Association of Parkinson's disease and its subtypes with agricultural pesticide exposure in men: a case–control study in France. Environ. Health Perspect. 123, 1123–1129.

Muelener, M., Whitworth, A.J., Armstrong-Gold, C.E., Rizzu, P., Heutink, P., Wes, P.D., et al., 2005. Drosophila DJ-1 mutants are selectively sensitive to environmental toxins associated with Parkinson's disease. Curr. Biol. 15, 1572–1577.

Mutch, W.J., Dingwall-Fordyce, I., Downie, A.W., Paterson, J.G., Roy, S.K., 1986. Parkinson's disease in a Scottish city. Br. Med. J. (Clin. Res. Ed.) 292, 534–536.

Muthukumaran, K., Leahy, S., Harrison, K., Sikorska, M., Sandhu, J.K., Cohen, J., et al., 2014. Orally delivered water soluble coenzyme Q10 (Ubisol-Q10) blocks on-going neurodegeneration in rats exposed to paraquat: potential for therapeutic application in Parkinson's disease. BMC Neurosci. 31, 15–21.

Narayan, S., Sinsheimer, J.S., Paul, K.C., Liew, Z., Cockburn, M., Bronstein, J.M., et al., 2015. Genetic variability in ABCB1, occupational pesticide exposure, and Parkinson's disease. Environ. Res. 143, 98–106.

Pan-Montojo, F., Anichtchik, O., Dening, Y., Knels, K., Pursche, S., Jung, R., et al., 2010. Progression of Parkinson's disease pathology is reproduced by intragastric administration of rotenone in mice. PLoS One 5, e8762.

Pan-Montojo, F., Schwarz, M., Winkel, C., Arnhold, M., O'Sullivan, G.A., Pal, A., et al., 2012. Environmental toxins trigger PD-like progression via increased α-synuclein release from enteric neurons in mice. Sci. Rep. 2, 898.

Park, R.M., Schulte, P.A., Bowman, J.D., Walker, J.T., Bondy, S.C., Yost, M.G., et al., 2005. Potential occupational risks for neurodegenerative diseases. Am. J. Ind. Med. 48, 63–77.

Peng, J., Oo, M.L., Andersen, J.K., 2010. Synergistic effects of environmental risk factors and gene mutations in Parkinson's disease accelerate age-related neurodegeneration. J. Neurochem. 115, 1363–1373.

Peng, J., Peng, L., Stevenson, F.F., Doctrow, S.R., Andersen, J.K., 2007. Iron and paraquat as synergistic environmental risk factors in sporadic Parkinson's disease accelerate age-related neurodegeneration. J. Neurosci. 27, 6914–6922.

Petrovitch, H., Ross, G.W., Abbott, R.D., Sanderson, W.T., Sharp, D.S., Tanner, C.M., et al., 2002. Plantation work and risk of Parkinson's disease in a population-based longitudinal study. Arch. Neurol. 59, 1787–1792.

Pezzoli, G., Cereda, E., 2013. Exposure to pesticides or solvents and risk of Parkinson disease. Neurology 80, 2035–2041.

Polymeropoulos, M.H., Lavedan, C., Leroy, E., Ide, S.E., Dehejia, A., Dutra, A., et al., 1997. Mutation in the α-synuclein gene identified in families with Parkinson's disease. Science 276, 2045–2047.

Prasad, K., Winnik, B., Thiruchelvam, M.J., Buckley, B., Mirochnitchenko, O., Richfield, E.K., 2007. Prolonged toxicokinetics and toxicodynamics of paraquat in mouse brain. Environ. Health Perspect. 115, 1448–1453.

Priyadarshi, A., Khuder, S.A., Schaub, E.A., Shrivastava, S., 2000. A meta-analysis of Parkinson's disease and exposure to pesticides. Neurotoxicology 21, 435–440.

Przedborski, S., Vila, M., 2003. The 1-methyl-4-phenyl-1,2,3,6-tetrahydropyridine mouse model: a tool to explore the pathogenesis of Parkinson's disease. Ann. N. Y. Acad. Sci. 991, 189–198.

Purisai, M.G., McCormack, A.L., Cumine, S., Li, J., Isla, M.Z., Di Monte, D.A., 2007. Microglial activation as a priming event leading to paraquat-induced dopaminergic cell degeneration. Neurobiol. Dis. 25, 392–400.

Purisai, M.G., McCormack, A.L., Langston, W.J., Johnston, L.C., Di Monte, D.A., 2005. α-Synuclein expression in the substantia nigra of MPTP-lesioned non-human primates. Neurobiol. Dis. 20, 898–906.

Ren, Y., Liu, W., Jiang, H., Jiang, Q., Feng, J., 2005. Selective vulnerability of dopaminergic neurons to microtubule depolymerization. J. Biol. Chem. 280, 34105–34112.

Richardson, J.R., Caudle, W.M., Wang, M., Dean, E.D., Pennell, K.D., Miller, G.W., 2006. Developmental exposure to the pesticide dieldrin alters the dopamine system and increases neurotoxicity in an animal model of Parkinson's disease. FASEB J. 20, 1695–1697.

Richardson, J.R., Caudle, W.M., Wang, M., Dean, E.D., Pennell, K.D., Miller, G.W., 2008. Developmental heptachlor exposure increases susceptibility of dopamine neurons to N-methyl-4-phenyl-1,2,3,6-tetrahydropyridine (MPTP) in a gender-specific manner. Neurotoxicology 29, 855–863.

Richardson, J.R., Roy, A., Shalat, S.L., Buckley, B., Winnik, B., Gearing, M., et al., 2011. β-Hexachlorocyclohexane levels in serum and risk of Parkinson's disease. Neurotoxicology 32, 640–645.

Richardson, J.R., Shalat, S.L., Buckley, B., Winnik, B., O'Suilleabhain, P., Diaz-Arrastia, R., et al., 2009. Elevated serum pesticide levels and risk of Parkinson disease. JAMA Neurol. 71, 284–290.

Ross, G.W., Duda, J.E., Abbott, R.D., Pellizzari, E., Petrovitch, H., Miller, D.B., et al., 2012. Brain organochlorines and Lewy pathology: the Honolulu–Asia aging study. Mov. Disord. 27, 1418–1424.

Ross, G.W., Petrovitch, H., Abbott, R.D., Nelson, J., Markesbery, W., Davis, D., et al., 2004. Parkinsonian signs and substantia nigra neuron density in decendents elders without PD. Ann. Neurol. 56, 532–539.

Ross, G.W., Petrovitch, H., Abbott, R.D., Tanner, C.M., White, L.R., 2006. Pre-clinical indicators of Parkinson's disease: recent findings from the Honolulu–Asia aging study. Mov. Disord. 21 (S13), S2.

Ryan, S.D., Dolatabadi, N., Chan, S.F., Zhang, X., Akhtar, M.W., Parker, J., et al., 2013. Isogenic human iPSC Parkinson's model shows nitrosative stress-induced dysfunction in MEF2-PGC1α transcription. Cell 155, 1351–1364.

Saha, S., Guillily, M.D., Ferree, A., Lanceta, J., Chan, D., Ghosh, J., et al., 2009. LRRK2 modulates vulnerability to mitochondrial dysfunction in Caenorhabditis elegans. J. Neurosci. 29, 9210–9218.

Sanders, L.H., McCoy, J., Hu, X., Mastrobernardino, P.G., Dickinson, B.C., Chang, C.J., 2014. Mitochondrial DNA damage: molecular marker of vulnerable nigral neurons in Parkinson's disease. Neurobiol. Dis. 70, 214–223.

Seidler, A., Hellenbrand, W., Robra, B.P., Vieregge, P., Nischan, P., Joerg, J., et al., 1996. Possible environmental, occupational, and other etiologic factors for Parkinson's disease: a case–control study in Germany. Neurology 46, 1275–1284.

Semchuk, K.M., Love, E.J., Lee, R.G., 1992. Parkinson's disease and exposure to agricultural work and pesticide chemicals. Neurology 42, 1328–1335.

Semchuk, K.M., Love, E.J., Lee, R.G., 1993. Parkinson's disease: a test of the multifactorial etiologic hypothesis. Neurology 43, 1173–1180.

Slotkin, T.A., Seidler, F.J., 2011. Developmental exposure to organophosphates triggers transcriptional changes in genes associated with Parkinson's disease in vitro and in vivo. Brain Res. Bull. 86, 340–347.

Smeyne, R.J., Breckenridge, C.B., Back, M., Jiao, Y., Butt, M.T., Wolf, J.C., et al., 2016. Assessment of the effects of MPTP and paraquat on dopeminergic neurons and microglia in the substantia nigra pars compacta of C57BL/6 mice. PLoS One 11, e0164094.

Spillantini, M.G., Crowther, R.A., Jakes, R., Hasegawa, M., Goedert, M., 1998. α-Synuclein in filamentous inclusions of Lewy bodies from Parkinson's disease and dementia with lewy bodies. Proc. Natl. Acad. Sci. U.S.A. 95, 6469–6473.

Svenson, L.W., Platt, G.H., Woodhead, S.E., 1993. Geographic variations in the prevalence of Parkinson's disease in Alberta. Can. J. Neurol. Sci. 20, 307–311.

Svensson, E., Horvath-Puho, E., Thomsen, R.W., Djurhuus, J.C., Pedersen, L., Borghammer, P., et al., 2015. Vagotomy and subsequent risk of Parkinson's disease. Ann. Neurol. 78, 522–529.

Tanner, C.M., Goldman, S.M., 1996. Epidemiology of Parkinson's disease. Neurol. Clin. 14, 317–335.

Tanner, C.M., Kamel, F., Ross, G.W., Hoppin, J.A., Goldman, S.M., Korell, M., et al., 2011. Rotenone, paraquat, and Parkinson's disease. Environ. Health Perspect. 119, 866–872.

Tanner, C.M., Ottman, R., Goldman, S.M., Ellenberg, J., Chan, P., Mayeux, R., et al., 1999. Parkinson disease in twins: an etiologic study. JAMA 281, 341–346.

Tanner, C.M., Ross, G.W., Jewell, S.A., Hauser, R.A., Jankovic, J., Factor, S.A., et al., 2009. Occupation and risk of parkinsonism: a multicenter case–control study. Arch. Neurol. 66, 1106–1113.

Taylor, K.S., Counsell, C.E., Harris, C.E., Gordon, J.C., Smith, W.C., 2006. Pilot study of the incidence and prognosis of degenerative parkinsonian disorders in Aberdeen, United Kingdom: methods and preliminary results. Mov. Disord. 21, 976–982.

Thiruchelvam, M., Richfield, E.K., Baggs, R.B., Tank, A.W., Cory-Slechta, D.A., 2000. The nigrostriatal dopaminergic system as a preferential target of repeated exposures to combined paraquat and maneb: implications for Parkinson's disease. J. Neurosci. 20, 9207–9214.

Tüchsen, F., Jensen, A.A., 2000. Agricultural work and the risk of Parkinson's disease in Denmark, 1981–1993. Scand. J. Work Environ. Health 26, 359–362.

Twelves, D., Perkins, K.S., Counsell, C., 2003. Systematic review of incidence studies of Parkinson's disease. Mov. Disord. 18, 19–31.

Ulusoy, A., Di Monte, D.A., 2013. α-Synuclein elevation in human neurodegenerative diseases: experimental, pathogenetic, and therapeutic implications. Mol. Neurobiol. 47, 784–794.

Ulusoy, A., Rusconi, R., Perez-Revuelta, B.I., Musgrove, R.E., Helwig, M., Winzen-Reichert, B., et al., 2013. Caudo-rostral brain spreading of α-synuclein through vagal connections. EMBO Mol. Med. 5, 1119–1127.

Uryu, K., Giasson, B.I., Longhi, L., Martinez, D., Murray, I., Conte, V., et al., 2003. Age-dependent synuclein pathology following traumatic brain injury in mice. Exp. Neurol. 184, 214–224.

van der Mark, M., Brouwer, M., Kromhout, H., Nijssen, P., Huss, A., Vermeulen, R., 2012. Is pesticide use related to Parkinson disease? Some clues to heterogeneity in study results. Environ. Health Perspect. 120, 340–347.

van der Mark, M., Vermeulen, R., Nijssen, P.C., Mulleners, W.M., Sas, A.M., van Laar, T., et al., 2014. Occupational exposure to pesticides and endotoxin and Parkinson disease in the Netherlands. Occup. Environ. Med. 71, 757–764.

Vila, M., Vukosavic, S., Jackson-Lewis, V., Neystat, M., Jakowec, M., Przedborski, S., 2000. α-Synuclein up-regulation in substantia nigra dopaminergic neurons following administration of the parkinsonian toxin MPTP. J. Neurochem. 74, 721–729.

Wan, N., Lin, G., 2016. Parkinson's disease and pesticides exposure; new findings from a comprehensive study in Nebraska, USA. J. Rural Health 32, 303–313.

Wang, A., Costello, S., Cockburn, M., Zhang, X., Bronstein, J., Ritz, B., 2011. Parkinson's disease risk from ambient exposure to pesticides. Eur. J. Epidemiol. 26, 547–555.

Weisskopf, M.G., Knekt, P., O'Reilly, E.J., Lyytinen, J., Reunanen, A., Laden, F., 2010. Persistent organochlorine pesticides in serum and risk of Parkinson disease. Neurology 74, 1055–1061.

Wilk, J.B., Tobin, J.E., Suchowersky, O., Shill, H.A., Klein, C., Wooten, G.F., 2006. Herbicide exposure modifies GSTP1 haplotype association to Parkinson onset age: the GenePD study. Neurology 67, 2206–2210.

Wills, J., Credle, J., Oaks, A.W., Duka, V., Lee, J.H., Jones, J., et al., 2012. Paraquat, but not maneb, induces synucleinopathy and tauopathy in striata of mice through inhibition of proteasomal and autophagic pathways. PLoS One 7, e30745.

Wilson, W.W., Shapiro, L.P., Bradner, J.M., Caudle, W.M., 2014. Developmental exposure to the organochlorine insecticide endosulfan damages the nigrostriatal dopamine system in male offspring. Neurotoxicology 44, 279–287.

Zharikov, A.D., Cannon, J.R., Tapias, V., Bai, Q., Horowitz, M.P., Shah, V., et al., 2015. shRNA targeting α-synuclein prevents neurodegeneration in a Parkinson's disease model. J. Clin. Invest. 125, 2721–2735.

Zorzon, M., Capus, L., Pellegrino, A., Cazzato, G., Zivadinov, R., 2002. Familial and environmental risk factors in Parkinson's disease: a case–control study in north-east Italy. Acta Neurol. Scand. 105, 77–82.

Zschiedrich, K., König, I.R., Brüggemann, N., Kock, N., Kasten, M., Leenders, K.L., 2009. MDR1 variants and risk of Parkinson disease. Association with pesticide exposure? J. Neurol. 256, 115–120.

CHAPTER FOUR

Metals and Circadian Rhythms

Nancy L. Parmalee, Michael Aschner[1]
Albert Einstein College of Medicine, Bronx, NY, United States
[1]Corresponding author: e-mail address: michael.aschner@einstein.yu.edu

Contents

1. INTRODUCTION

Circadian rhythms describe the roughly 24 h oscillation of many physiological properties, behaviors, and gene transcription that has arisen through evolution to coordinate the activities of biological organisms with the periods of light and dark corresponding to day and night (Mohawk et al., 2012; Takahashi et al., 2008). Mechanisms to control circadian rhythms exist in many if not most organisms, from plants to human, with some mechanisms highly conserved and others perhaps evolving convergently. While much work is ongoing regarding the function of sleep in humans, it is clear that sleep is an essential physiological process, the lack of which causes or exacerbates a plethora of conditions ranging from causing a "bad day" to worsening neurodegenerative conditions such as Parkinson's disease (PD) and Alzheimer's disease (AD). In order to coordinate sleep–wake cycles, many biological functions need to be coordinated, including changes in alertness, appetite, body temperature, blood pressure, production of urine, heart rate, and metabolism.

Advances in Neurotoxicology, Volume 1
ISSN 2468-7480
http://dx.doi.org/10.1016/bs.ant.2017.07.003

In humans, circadian control mechanisms exist at several levels. Centrally, the suprachiasmatic nucleus (SCN) in the brain receives signals from the retina and acts as a central clock (Welsh et al., 1995, 2010). In total darkness, absent light cues to indicate daytime, the human circadian clock oscillates with a period slightly longer than 24 h. In order to stay synchronized with the 24 h day, specialized cells in the retina known as intrinsically photosensitive retinal ganglion cells (ipRGCs) sense light for the purpose of setting the circadian clock (Berson et al., 2002; Hattar et al., 2002). These cells do not participate in vision formation, but do express the photopigment melanopsin, which is particularly sensitive to blue light. These cells synapse with cells of the SCN in order to synchronize, or entrain, the central clock to the 24 h day. Peripheral clocks also exist in organs such as the liver, lung, and kidney (Yoo et al., 2004), and circadian oscillations are seen at the level of individual cells. Transcription of a large portion of the genome is controlled in a circadian manner.

The core genes involved in mammalian circadian control include *Clock*, *Bmal1*, *Per1*, *Per2*, *Cry1*, and *Cry2* (Mohawk et al., 2012). CLOCK and BMAL1 dimerize and bind DNA resulting in the transcription of the *Per* and *Cry* genes as well as many others. PER and CRY, in turn, dimerize with each other and interact with the CLOCK–BMAL1 complex in an inhibitory fashion, repressing transcription (Takahashi et al., 2008). Thus, as the products of transcription accumulate they repress their own transcription, resulting in an oscillatory system (Fig. 1).

Several lines of evidence indicate that various metals interact with components of the circadian clock mechanisms resulting in disruptions to circadian rhythms. The data that exist in the literature largely consist of observational correlations in humans, which does not address mechanisms which may be at work in such interactions. We propose that this is a rich area of interest that merits attention from the scientific community. Taking the example of manganese (Mn), a primary focus of our lab: there is a robust literature concerning high-level acute exposures, such as industrial exposures that occur in mining, welding, battery manufacture, and others (Rodier, 1955). There is less data on the effects of cumulative life-long exposures at lower levels that may be encountered in the day-to-day environment. We have written previously about the potential effects of such exposures on neurodegeneration and aging (Parmalee and Aschner, 2016). Notably, circadian disruption is observed in patients with PD, AD, and Huntington's disease (HD), all neurodegenerative diseases that in addition to a genetic component are thought to be influenced by metal

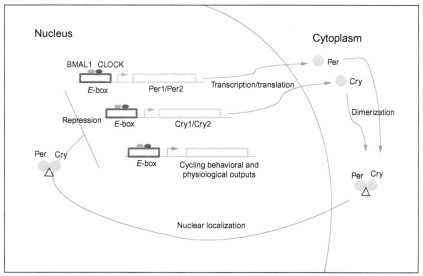

Fig. 1 Circadian oscillations result from the transcription, translation, accumulation, and degradation of genes. This simplified schematic depicts the feedback mechanism of the canonical clock genes that results in circadian oscillation. The proteins BMAL1 and CLOCK dimerize and bind DNA at E-box-binding sites, resulting in transcription of genes under the control of these binding elements. These genes include Per1 and Per2, Cry1 and Cry2, and other genes involved in circadian behavioral and physiological responses. The PER and CRY proteins accumulate in the cytoplasm where they dimerize and are then transported back to the nucleus. The PER–CRY dimer inhibits binding of BMAL1 and CLOCK to DNA, repressing transcription of downstream genes. In this way, the accumulation of PER and CRY represses their own expression, establishing a system that oscillates based on the expression, accumulation, and subsequent repression of these genes.

exposures (Videnovic et al., 2014). As life span increases and metals are present in the environment, these cumulative exposures to a variety of metals are likely to have implications on health. In this review we will present a subset of evidence for interactions between metals and circadian rhythms.

2. MANGANESE

Manganese (Mn) is a metal that occurs commonly in soil and water. It is one of the most abundant elements in the earth's crust. It has been used since antiquity as a pigment in ceramics. It is used as an alloy with other metals, is used in battery manufacture, is present in pesticides, and in the gasoline antiknock additive methylcyclopentadienyl manganese tricarbonyl. Mn is an essential trace element, necessary as a cofactor for many enzymes.

Humans obtain a sufficient amount of Mn through the diet, and Mn deficiency is essentially unknown outside of the laboratory. Low-level exposures occur primarily via well water in regions with a naturally high level of Mn occurring in the soil, where it leaches into the water, and by exposure to contaminated soil in areas nearby to metal refineries. There is evidence for detrimental behavioral and cognitive effects in children exposed to relatively low levels of Mn (Bhang et al., 2013; Carvalho et al., 2014). At higher levels of exposure, especially unregulated industrial exposure, a condition known as manganism can develop that is well characterized and has many features in common with PD (Couper, 1837; Huang et al., 1989; Mena et al., 1967; Rodier, 1955). With excessive exposure, Mn tends to accumulate in the basal ganglia and putamen of the brain, whereas in PD the substantia nigra pars compacta is primarily affected, with degeneration of dopaminergic neurons seen in this area. The similarities and differences between manganism and PD have been reviewed (Guilarte, 2010; Guilarte and Gonzales, 2015). Manganism and PD share the cardinal motor signs: bradykinesia, rigidity, postural instability, and tremor. They also share many nonmotor features, including depression, dementia, loss of sense of smell, and various sleep disorders, including difficulty falling and staying asleep, rapid eye movement (REM) behavior disorder (RBD), and circadian disruptions (Bowler et al., 2006, 2007; Parkinson, 2002; Videnovic and Golombek, 2013). The mechanism of circadian disruption is unknown, but it is a prominent feature of both PD and manganism. Mn toxicity is thought to be mediated by dopamine so it is possible that Mn–dopamine interactions have an adverse effect on the sleep–wake cycle though undetermined mechanisms. Excess Mn is eliminated via bile and excreted in feces. Aside from industrial exposures, patients with liver disease are also known to be at risk for elevated Mn levels. Circadian disruptions have been reported in patients with chronic liver disease, cirrhosis of the liver, fatty liver disease, and hepatitis C (De Cruz et al., 2012), raising the possibility that elevated Mn levels could participate in sleep disruption in these patients.

3. COPPER

Copper (Cu) is also an essential micronutrient which is found bound to many enzymes and unbound in serum. Wilson's disease is a genetic disorder that results in dysregulated copper metabolism. In patients with Wilson's disease copper accumulates in many organs, in the corneas, and in the basal ganglia and putamen of the brain, the same regions affected by Mn accumulation (Nevsimalova et al., 2011). Excess copper can cause parkinsonian movement symptoms, psychiatric disturbances, liver disease, and neurological degeneration. Amyloid precursor protein (*APP*), which is implicated in AD, is a copper-binding protein (Nevsimalova et al., 2011). Patients with Wilson's disease have been reported to have sleep disturbances including RBD, restless leg syndrome, daytime hypersomnolence, excessive nighttime wakefulness, and circadian disruption (Amann et al., 2015; Firneisz et al., 2000; Nevsimalova et al., 2011; Tribl et al., 2016).

Again, the mechanism by which copper overload may lead to circadian disruption is unknown; however, in addition to *APP*, suspect genes that could be involved include a night-specific ATPase *PINA* that is expressed in the pineal gland which participates in circadian biology. Variants in this gene are causative in Wilson's disease, and the gene is expressed at 100-fold greater levels at night than during the day (Borjigin et al., 1999; Firneisz et al., 2000). Additionally, dopamine beta hydroxylase is a copper-binding enzyme that may be functionally affected by copper overload (Kitzberger et al., 2005; Nevsimalova et al., 2011). Copper has also been shown to interact with the vesicular H(+)ATPase, impacting dopamine metabolism (Wimalasena et al., 2007). The regulated cycling of dopamine is critical to maintaining circadian rhythms. In addition to dopamine, the neurotransmitter serotonin (5-hydroxytryptamine, 5-HT) has been shown to interact with unbound copper, resulting in conversion of 5-HT to an intermediate species (Jones et al., 2007). The psychiatric symptoms of Wilson's disease include depression, anxiety, and cognitive changes which could be attributable to changes in the serotonin system, which is also highly circadian. Studies have shown interaction between the prion protein, another copper-binding protein, and 5-HT, both of which are present in the synaptic cleft (Jones et al., 2007; Mouillet-Richard et al., 2005). Notably, symptoms of the human prion diseases Fatal Familial Insomnia and Creutzfeldt–Jakob disease include severe sleep–wake disorders and behavioral disturbance (Jones et al., 2007; Landolt et al., 2006; Wanschitz et al., 2000).

4. ZINC

Zinc (Zn) is another essential metal, necessary for optimal biological functioning. Zinc and copper compete for absorption in the intestine, liver, and kidney. Both copper and zinc are also inhibitory against the NMDA receptor, which is involved in mood, cognitive functions, and sleep (Peters et al., 1987; Song et al., 2012; Vlachova et al., 1996). Song et al.

found that duration of sleep correlated with the ratio of zinc to copper in adult women (Song et al., 2012), suggesting that a balance between these two metals is important for sleep–wake regulation. A similar finding was reported for men (Zhang et al., 2009). In mice, administration of zinc has been shown to decrease locomotion and induce non-REM sleep (Cherasse et al., 2015). Other metals are known to interact with each other, for instance iron and Mn compete for absorption, and iron deficiency is a risk factor for elevated levels of Mn. The balance between zinc and copper appears to be important in the sleep–wake cycle; however, the mechanisms behind this interaction remain to be elucidated.

5. LEAD

While the metals previously discussed are all essential micronutrients that participate in enzymatic activity and are necessary to biological processes, lead (Pb) is not. Pb is an environmental pollutant well characterized as having severe impacts on the central nervous system and cognitive function. Pb does not serve a biological role, but does interact with biological systems; therefore, it is of interest to question what role Pb might play in circadian rhythms and sleep. It is worth noting that sleep studies invariably use *leads*, a homonym for the metal *lead*, confounding searches for the role of Pb in sleep. We suggest that researchers conducting studies related to Pb and sleep employ the elemental nomenclature Pb to allow these studies to be found in the literature.

As previously described, the human circadian system is photoentrained to the 24 h day by sensing daylight via melanopsin-expressing ipRGCs in the retina. Light signals are transmitted synaptically to the SCN, the central clock which coordinates and regulates peripheral and cellular clocks. While ipRGCs are not dopaminergic, they do form syncytial connections through gap junctions with amacrine cells which are dopaminergic. Gestational Pb exposure in mice has been shown to decrease dopamine content in the retina and decrease levels of dopamine by-products 3,4–dihydroxyphenylacetic acid and homovanillic acid (Fox et al., 2011). Many subtypes of amacrine cells exist. Dopaminergic amacrine cells were shown to be depleted, and their processes shortened, while nondopaminergic amacrine cells were unaffected. The dopamine-processing enzyme tyrosine hydroxylase was also depleted with gestational Pb exposure. This study did not address circadian rhythms but raised the possibility that this finding could have a circadian effect.

The early morning production of cortisol is an integral part of the circadian system, responsible for waking and alertness in the morning. In humans, cortisol production is regulated by the hypothalamus–pituitary–adrenal (HPA) axis. Cortisol levels rise dramatically first thing in the morning and decline throughout the day. High levels of stress can result in daytime cortisol production, disrupting this pattern. Braun et al. measured Pb levels and salivary cortisol levels in second trimester pregnant women from Mexico City and evaluated the relationship between Pb and cortisol levels (Braun et al., 2014). Cortisol is also critical for brain development, and gestational Pb exposure is known to carry risk of neurological deficits, possibly resulting from HPA axis effects (Braun et al., 2014; LeWinn et al., 2009). This study found a weak relationship between Pb levels and cortisol awakening response in pregnant women. Diurnal slopes of cortisol production were flatter in those women with the highest Pb levels. Given the known effects of Pb on cognition and the central nervous system, more work remains to be done to understand the role of Pb and circadian rhythms.

6. MERCURY

Mercury (Hg) is another metal that serves no biological purpose, but can accumulate in the body and interact with biological systems. Exposure to mercury can occur in industrial settings, such as gold mining operations and gilding. Mercury is an environmental contaminant that makes its way into waterways where it is taken up by fish and other aquatic animals which then frequently become seafood, resulting in exposure to humans. Because the body does not regulate mercury, exposure is cumulative and bioaccumulation occurs as smaller animals are eaten by larger animals, and eventually by humans. Coastal and island communities, especially indigenous people that traditionally rely on fish and shellfish as a food source, are especially vulnerable to high levels of mercury exposure. This was exemplified by the environmental disaster at Minamata Bay, a fishing village in Japan. In 1956 patients began appearing with what was later recognized as severe mercury poisoning. The cause was large-scale disposal of industrial waste containing high levels of mercury into the bay. The surrounding community was dependent on fish from the region for food, resulting in dire health consequences. While industrial regulation has reduced dumping of mercury-containing waste into waterways, exposure to mercury as a consequence of gold mining operations which use mercury in the extraction process remains an issue in the developing world.

The second oldest mercury mine in the world, the Idrija Mercury Mine, is located in what is now Slovenia and has been in operation since the 1400s. From the 1500s, writers have described the condition of miners at this location as sick, suffering from deformity, paralysis, asthma, and other conditions (Kobal and Grum, 2010). In 1754 Joannes Antonius Scopoli was assigned as physician to the mine and subsequently published the first description of mercury poisoning in Venice in 1761. In addition to tremor, respiratory difficulties, personality changes, difficulty eating and sleeping, and other conditions, Scopoli identified sleep disorders as a prominent sign of mercury toxicity (Kobal and Grum, 2010). Scopoli reported difficulty sleeping, restless sleep, dream disturbances, and what we would now term restless leg syndrome. Sleep disturbance subsequent to mercury exposure was also described by Freeman in 1860 in his description of mercury poisoning in hat makers, the caricature of which was thought to give rise to the character of the Mad Hatter in Lewis Carroll's Through the Looking Glass.

Mercury accumulation has been shown in the pineal gland, which participates in circadian function through the secretion of melatonin and serotonin (Falnoga et al., 2000; Kosta et al., 1975). Methylmercury exposure was shown to result in circadian sleep–wake disruption in rats (Arito et al., 1983). One possible mechanism of sleep disruption is the effect of mercury on glutamate, resulting in increased extracellular glutamate and a possible excitotoxic effect (Aschner et al., 2007). Mercury exposure also results in changes in cytokine production. In children, mercury exposure was found to be associated with reduced levels of TNF-alpha and shorter sleep duration (Gump et al., 2014).

Insomnia and other sleep disorders have been reported subsequent to mercury exposure in worker in fluorescent light bulb plants (Rossini et al., 2000), among sport fishers in Wisconsin (Knobeloch et al., 2006), among dental assistants exposed to mercury-containing amalgam used in fillings (Moen et al., 2008), among gold miners in Indonesia (Bose-O'Reilly et al., 2016), and among Iranian workers gilding a shrine (Vahabzadeh and Balali-Mood, 2016). Additional research is needed to address the mechanism underlying these sleep disruptions observed with mercury exposure.

7. ALUMINUM

Aluminum (Al) is yet another metal that plays no biological role in living systems but has toxic effects in vivo. Aluminum has been implicated as a neurotoxicant and a possible factor in neurodegenerative diseases such as amyotrophic lateral sclerosis (ALS), PD, and AD, which as previously mentioned have strong circadian components (Rodella et al., 2008). While few studies of the effect of aluminum on human sleep and circadian rhythms have been reported, evidence from the fruit fly *Drosophila melanogaster* showed that in addition to shortening the life span of the flies, aluminum exposure influenced the level and rhythm of daily locomotion (Kijak et al., 2014). Given the connection to human neurodegenerative disease with circadian components, this finding in flies suggests that further observation into human aluminum exposure and further research into circadian mechanisms is warranted.

8. CONCLUSION

Many metals are essential for the optimal functional of many physiological processes, including countless enzymatic reactions that use metals as cofactors. Those metals that play a role in biological processes, including but not limited to manganese, copper, and zinc, are tightly regulated at the level of absorption, transport into and out of the cell, and excretion. Even with regulation, excessive levels of these metals can occur, in the case of high levels of environmental exposure, such as in industrial settings, or in the case of disease in which the regulation of the metal is impaired. Given that thousands of genes are under circadian control and involved in maintaining circadian rhythms, it would be surprising if the circadian system was not affected by metals.

Numerous neurological and neurodegenerative diseases have a circadian component, including mood disorders such as depression and bipolar disorder, neurodegenerative diseases such as PD, AD, and HD, and prion diseases such as Fatal Familial Insomnia and Creutzfeldt–Jakob disease. There is evidence that in mood disorders and neurodegenerative diseases that the disease process may initiate the sleep disorder, but dysregulated sleep–wake cycles may worsen the disease process, resulting in a spiraling pattern of dysfunction. Indeed, disturbed sleep patterns are a primary cause of institutionalization in patients suffering from dementia (Hatfield et al., 2004). Patients may benefit from pharmacological interventions to promote normative sleep–wake cycles, which may in turn slow the disease process. While epidemiological studies are useful in observing the correlation between circadian disruption and disease, more research is needed to understand the mechanisms behind these processes.

As the population ages and advances in medicine expand the human life span, healthy aging will be dependent on understanding factors related to neurodegenerative disease. Circadian rhythms are intricately entwined with neurodegenerative disease, as are metals. Understanding the interplay between these factors in health and disease will contribute to healthier aging.

ACKNOWLEDGMENTS

This work has been supported by the National Institutes of Health grant numbers NIEHS R01ES07331, NIEHS R01ES10563, NIEHS R01ES020852, and NIEHS R03ES024849.

REFERENCES

Amann, V.C., Maru, N.K., Jain, V., 2015. Hypersomnolence in Wilson disease. J. Clin. Sleep Med. 11, 1341–1343.

Arito, H., Hara, N., Torii, S., 1983. Effect of methylmercury chloride on sleep-waking rhythms in rats. Toxicology 28, 335–345.

Aschner, M., Syversen, T., Souza, D.O., Rocha, J.B., Farina, M., 2007. Involvement of glutamate and reactive oxygen species in methylmercury neurotoxicity. Braz. J. Med. Biol. Res. 40, 285–291.

Berson, D.M., Dunn, F.A., Takao, M., 2002. Phototransduction by retinal ganglion cells that set the circadian clock. Science 295, 1070–1073.

Bhang, S.Y., Cho, S.C., Kim, J.W., Hong, Y.C., Shin, M.S., Yoo, H.J., Cho, I.H., Kim, Y., Kim, B.N., 2013. Relationship between blood manganese levels and children's attention, cognition, behavior, and academic performance—a nationwide cross-sectional study. Environ. Res. 126, 9–16.

Borjigin, J., Payne, A.S., Deng, J., Li, X., Wang, M.M., Ovodenko, B., Gitlin, J.D., Snyder, S.H., 1999. A novel pineal night-specific ATPase encoded by the Wilson disease gene. J. Neurosci. 19, 1018–1026.

Bose-O'Reilly, S., Schierl, R., Nowak, D., Siebert, U., William, J.F., Owi, F.T., Ir, Y.I., 2016. A preliminary study on health effects in villagers exposed to mercury in a small-scale artisanal gold mining area in Indonesia. Environ. Res. 149, 274–281.

Bowler, R.M., Gysens, S., Diamond, E., Nakagawa, S., Drezgic, M., Roels, H.A., 2006. Manganese exposure: neuropsychological and neurological symptoms and effects in welders. Neurotoxicology 27, 315–326.

Bowler, R.M., Nakagawa, S., Drezgic, M., Roels, H.A., Park, R.M., Diamond, E., Mergler, D., Bouchard, M., Bowler, R.P., Koller, W., 2007. Sequelae of fume exposure in confined space welding: a neurological and neuropsychological case series. Neurotoxicology 28, 298–311.

Braun, J.M., Wright, R.J., Just, A.C., Power, M.C., Tamayo, Y.O.M., Schnaas, L., Hu, H., Wright, R.O., Tellez-Rojo, M.M., 2014. Relationships between lead biomarkers and diurnal salivary cortisol indices in pregnant women from Mexico City: a cross-sectional study. Environ. Health 13, 50.

Carvalho, C.F., Menezes-Filho, J.A., de Matos, V.P., Bessa, J.R., Coelho-Santos, J., Viana, G.F., Argollo, N., Abreu, N., 2014. Elevated airborne manganese and low executive function in school-aged children in Brazil. Neurotoxicology 45, 301–308.

Cherasse, Y., Saito, H., Nagata, N., Aritake, K., Lazarus, M., Urade, Y., 2015. Zinc-containing yeast extract promotes nonrapid eye movement sleep in mice. Mol. Nutr. Food Res. 59, 2087–2093.

Couper, J., 1837. On the effects of black oxide manganese when inhaled into the lungs. Br. Ann. Med. Pharm. Vital Stat. Gen. Sci. 1, 41–42.

De Cruz, S., Espiritu, J.R., Zeidler, M., Wang, T.S., 2012. Sleep disorders in chronic liver disease. Semin. Respir. Crit. Care Med. 33, 26–35.

Falnoga, I., Tusek-Znidaric, M., Horvat, M., Stegnar, P., 2000. Mercury, selenium, and cadmium in human autopsy samples from Idrija residents and mercury mine workers. Environ. Res. 84, 211–218.

Firneisz, G., Szalay, F., Halasz, P., Komoly, S., 2000. Hypersomnia in Wilson's disease: an unusual symptom in an unusual case. Acta Neurol. Scand. 101, 286–288.

Fox, D.A., Hamilton, W.R., Johnson, J.E., Xiao, W., Chaney, S., Mukherjee, S., Miller, D.B., O'Callaghan, J.P., 2011. Gestational lead exposure selectively decreases retinal dopamine amacrine cells and dopamine content in adult mice. Toxicol. Appl. Pharmacol. 256, 258–267.

Guilarte, T.R., 2010. Manganese and Parkinson's disease: a critical review and new findings. Environ. Health Perspect. 118, 1071–1080.

Guilarte, T.R., Gonzales, K.K., 2015. Manganese-induced parkinsonism is not idiopathic Parkinson's disease: environmental and genetic evidence. Toxicol. Sci. 146, 204–212.

Gump, B.B., Gabrikova, E., Bendinskas, K., Dumas, A.K., Palmer, C.D., Parsons, P.J., MacKenzie, J.A., 2014. Low-level mercury in children: associations with sleep duration and cytokines TNF-alpha and IL-6. Environ. Res. 134, 228–232.

Hatfield, C.F., Herbert, J., van Someren, E.J., Hodges, J.R., Hastings, M.H., 2004. Disrupted daily activity/rest cycles in relation to daily cortisol rhythms of home-dwelling patients with early Alzheimer's dementia. Brain 127, 1061–1074.

Hattar, S., Liao, H.W., Takao, M., Berson, D.M., Yau, K.W., 2002. Melanopsin-containing retinal ganglion cells: architecture, projections, and intrinsic photosensitivity. Science 295, 1065–1070.

Huang, C.C., Chu, N.S., Lu, C.S., Wang, J.D., Tsai, J.L., Tzeng, J.L., Wolters, E.C., Calne, D.B., 1989. Chronic manganese intoxication. Arch. Neurol. 46, 1104–1106.

Jones, C.E., Underwood, C.K., Coulson, E.J., Taylor, P.J., 2007. Copper induced oxidation of serotonin: analysis of products and toxicity. J. Neurochem. 102, 1035–1043.

Kijak, E., Rosato, E., Knapczyk, K., Pyza, E., 2014. Drosophila melanogaster as a model system of aluminum toxicity and aging. Insect Sci. 21, 189–202.

Kitzberger, R., Madl, C., Ferenci, P., 2005. Wilson disease. Metab. Brain Dis. 20, 295–302.

Knobeloch, L., Steenport, D., Schrank, C., Anderson, H., 2006. Methylmercury exposure in Wisconsin: a case study series. Environ. Res. 101, 113–122.

Kobal, A.B., Grum, D.K., 2010. Scopoli's work in the field of mercurialism in light of today's knowledge: past and present perspectives. Am. J. Ind. Med. 53, 535–547.

Kosta, L., Byrne, A.R., Zelenko, V., 1975. Correlation between selenium and mercury in man following exposure to inorganic mercury. Nature 254, 238–239.

Landolt, H.P., Glatzel, M., Blattler, T., Achermann, P., Roth, C., Mathis, J., Weis, J., Tobler, I., Aguzzi, A., Bassetti, C.L., 2006. Sleep-wake disturbances in sporadic Creutzfeldt-Jakob disease. Neurology 66, 1418–1424.

LeWinn, K.Z., Stroud, L.R., Molnar, B.E., Ware, J.H., Koenen, K.C., Buka, S.L., 2009. Elevated maternal cortisol levels during pregnancy are associated with reduced childhood IQ. Int. J. Epidemiol. 38, 1700–1710.

Mena, I., Marin, O., Fuenzalida, S., Cotzias, G.C., 1967. Chronic manganese poisoning. Clinical picture and manganese turnover. Neurology 17, 128–136.

Moen, B., Hollund, B., Riise, T., 2008. Neurological symptoms among dental assistants: a cross-sectional study. J. Occup. Med. Toxicol. 3, 10.

Mohawk, J.A., Green, C.B., Takahashi, J.S., 2012. Central and peripheral circadian clocks in mammals. Annu. Rev. Neurosci. 35, 445–462.

Mouillet-Richard, S., Pietri, M., Schneider, B., Vidal, C., Mutel, V., Launay, J.M., Kellermann, O., 2005. Modulation of serotonergic receptor signaling and cross-talk by prion protein. J. Biol. Chem. 280, 4592–4601.

Nevsimalova, S., Buskova, J., Bruha, R., Kemlink, D., Sonka, K., Vitek, L., Marecek, Z., 2011. Sleep disorders in Wilson's disease. Eur. J. Neurol. 18, 184–190.

Parkinson, J., 2002. An essay on the shaking palsy. 1817. J. Neuropsychiatry Clin. Neurosci. 14, 223–236 (discussion 222).

Parmalee, N.L., Aschner, M., 2016. Manganese and aging. Neurotoxicology 56, 262–268.

Peters, S., Koh, J., Choi, D.W., 1987. Zinc selectively blocks the action of N-methyl-D-aspartate on cortical neurons. Science 236, 589–593.

Rodella, L.F., Ricci, F., Borsani, E., Stacchiotti, A., Foglio, E., Favero, G., Rezzani, R., Mariani, C., Bianchi, R., 2008. Aluminium exposure induces Alzheimer's disease-like histopathological alterations in mouse brain. Histol. Histopathol. 23, 433–439.

Rodier, J., 1955. Manganese poisoning in Moroccan miners. Br. J. Ind. Med. 12, 21–35.

Rossini, S.R., Reimao, R., Lefevre, B.H., Medrado-Faria, M.A., 2000. Chronic insomnia in workers poisoned by inorganic mercury: psychological and adaptive aspects. Arq. Neuropsiquiatr. 58, 32–38.

Song, C.H., Kim, Y.H., Jung, K.I., 2012. Associations of zinc and copper levels in serum and hair with sleep duration in adult women. Biol. Trace Elem. Res. 149, 16–21.

Takahashi, J.S., Hong, H.K., Ko, C.H., McDearmon, E.L., 2008. The genetics of mammalian circadian order and disorder: implications for physiology and disease. Nat. Rev. Genet. 9, 764–775.

Tribl, G.G., Trindade, M.C., Bittencourt, T., Lorenzi-Filho, G., Cardoso Alves, R., Ciampi de Andrade, D., Fonoff, E.T., Bor-Seng-Shu, E., Machado, A.A., Schenck, C.H., Teixeira, M.J., Barbosa, E.R., 2016. Wilson's disease with and without rapid eye movement sleep behavior disorder compared to healthy matched controls. Sleep Med. 17, 179–185.

Vahabzadeh, M., Balali-Mood, M., 2016. Occupational metallic mercury poisoning in gilders. Int. J. Occup. Environ. Med. 7, 116–122.

Videnovic, A., Golombek, D., 2013. Circadian and sleep disorders in Parkinson's disease. Exp. Neurol. 243, 45–56.

Videnovic, A., Lazar, A.S., Barker, R.A., Overeem, S., 2014. 'The clocks that time us'—circadian rhythms in neurodegenerative disorders. Nat. Rev. Neurol. 10, 683–693.

Vlachova, V., Zemkova, H., Vyklicky Jr., L., 1996. Copper modulation of NMDA responses in mouse and rat cultured hippocampal neurons. Eur. J. Neurosci. 8, 2257–2264.

Wanschitz, J., Kloppel, S., Jarius, C., Birner, P., Flicker, H., Hainfellner, J.A., Gambetti, P., Guentchev, M., Budka, H., 2000. Alteration of the serotonergic nervous system in fatal familial insomnia. Ann. Neurol. 48, 788–791.

Welsh, D.K., Logothetis, D.E., Meister, M., Reppert, S.M., 1995. Individual neurons dissociated from rat suprachiasmatic nucleus express independently phased circadian firing rhythms. Neuron 14, 697–706.

Welsh, D.K., Takahashi, J.S., Kay, S.A., 2010. Suprachiasmatic nucleus: cell autonomy and network properties. Annu. Rev. Physiol. 72, 551–577.

Wimalasena, D.S., Wiese, T.J., Wimalasena, K., 2007. Copper ions disrupt dopamine metabolism via inhibition of V-H+-ATPase: a possible contributing factor to neurotoxicity. J. Neurochem. 101, 313–326.

Yoo, S.H., Yamazaki, S., Lowrey, P.L., Shimomura, K., Ko, C.H., Buhr, E.D., Siepka, S.M., Hong, H.K., Oh, W.J., Yoo, O.J., Menaker, M., Takahashi, J.S., 2004. PERIOD2::LUCIFERASE real-time reporting of circadian dynamics reveals persistent circadian oscillations in mouse peripheral tissues. Proc. Natl. Acad. Sci. U.S.A. 101, 5339–5346.

Zhang, H.Q., Li, N., Zhang, Z., Gao, S., Yin, H.Y., Guo, D.M., Gao, X., 2009. Serum zinc, copper, and zinc/copper in healthy residents of Jinan. Biol. Trace Elem. Res. 131, 25–32.

CHAPTER FIVE

Aluminum and Neurodegenerative Diseases

Stephen C. Bondy*[,1], Arezoo Campbell[†]

*Environmental Toxicology Program, Center for Occupational and Environmental Health, University of California, Irvine, CA, United States
[†]Western University of Health Sciences, Pomona, CA, United States
[1]Corresponding author: e-mail address: scbondy@uci.edu

Contents

1. INTRODUCTION

Aluminum (Al) is a common element found in large amounts in the earth's crust (Priest et al., 1988). Aluminum-containing minerals are present in relatively inert rock types, especially in igneous formations, such as granite and quartz. Laterization of various silicate rocks weathering into finer particles results in the formation of sedimentary bauxite, where together with iron, Al is present largely as the oxide. It is as bauxite that Al is generally mined and second only to iron, Al is the most widely used metal (Hetherington, 2007).

Despite its commonality, Al has no known beneficial biological roles and is not an essential element for any organism. Aluminum-containing minerals are rather unreactive, and this is also true for metallic aluminum, as this is quickly oxidized in air and thus coated by a very thin but robust layer of the oxide. This apparent inertness has led to the concept that aluminum may not constitute a health hazard. Consequently a wide range of Al compounds have been added as stabilizers in many processed foods. Alum, which is any trivalent Al-containing salt, is the oldest and most commonly used vaccine adjuvant. Recent findings indicate that the effectiveness of the adjuvant relies on both its immunomodulatory as well as inflammatory properties. Al salts have also found utility in water clarifying processes by effecting precipitation of organic particulate matter. Growing incidence of acidic rain has led to greater solubilization of aluminum salts from their insoluble form in rocks. This has led to an elevated Al content in many water reserves used for residential supply. Thus, human exposure to more soluble forms of Al in water and foodstuffs has grown.

Reports from both biological laboratories and from study of human population health indicate that prolonged aluminum ingestion can result in neurological abnormality. Accumulating indications strongly suggest that Al can further the onset and development of neurodegenerative disorders, principally Alzheimer's disease (AD). There are many reports suggesting that Al can provoke excessive inflammatory events in the brain. Superfluous immune reactivity that is not an obvious response to a trauma such as injury or infection is a distinguishing feature of the elderly brain and appears exacerbated in nervous system abnormalities. Most neurodegenerative diseases have no obvious cause and do not have a clear genetic basis. Thus, it is probable that the origin of such diseases lies in unknown environmental influences that interact with the progression of aging. The nature of most of such factors is unknown, but there is growing evidence, indicating that Al is likely to be one of these environmental factors. In this review, reports that point to the conclusion that aluminum are able to speed up the worsening of brain function with age, and potential mechanisms are discussed. It should be noted that acceleration of this process would inevitably increase the prevalence of those specific neurological disorders where age is a con-comitant risk factor.

2. GROWING BIOAVAILABILITY OF ALUMINUM IN THE ENVIRONMENT

Metallic aluminum was first made by Hans Oersted in 1825 by heating aluminum chloride with elemental potassium (Sigel and Sigel, 1988). Al-containing chemicals have many uses. Mixing aluminum sulfate and lime together in water leads to formation of colloidal aluminum hydroxide, and this can bring about precipitation and removal of waterborne organic material. This method for water clarification is widely used. Al-containing additives are also found in many foodstuffs. They are used as emulsifying agents in prep-aration of processed cheese, as crisping agents in pickles, in baking powder, and in a variety of food colorings. Aluminum-containing compounds are also found in cosmetics. Commercial preparations of infant formula can contain significant amounts of the metal (Burrell and Exley, 2010; Dabeka et al., 2011).

High concentrations of soluble Al can be found in the juice resulting from boiling of acidic fruit in aluminum cookware (Fimreite et al., 1997). The aluminum content of city water supplies is variable, but on occasion, concentrations as high as 0.4–1 mg/L have been reported in drinking water. Although the health effects of these levels of the metal on humans are uncer-tain, the Joint Food and Agriculture Organization/World Health Organiza-tion Expert Committee on Food Additives in 2007 recommended a maximum intake of Al less than 1 mg/kg body weight per week. This cor-responds to 63 mg per week for a 140-pound adult. Some commercial pastry products contain Al sulfate and sodium aluminum phosphate levels up to 28 mg in a single serving of muffin or other baked products where baking powder is used (Pennington and Jones, 1989). Alum is a powerful adjuvant in vaccines and performs this role by enhancing the intensity of the immune responses evoked by the vaccine. The interest in determining the mecha-nism by which Al salts produce adjuvant action has led to the understanding that insoluble aluminum salts activate innate immune responses that lead to a T helper 2 (Th2)-type reaction (Marrack et al., 2009). The metal is also pre-sent in most antiperspirants, in buffered aspirin, and in antacids where the Al content can reach 600 mg/tablet.

The most common form of human absorption of Al compounds is by way of the gastrointestinal tract where the degree of uptake can be about

0.2% (Priest et al., 1988). When Al salts are transferred from the stomach to the hepatic portal vein, the metal is largely bound to transferrin (Harris et al., 2003). Al can subsequently reach the brain by means of receptor-mediated endocytosis of this transferrin complex. By this means, around 0.005% of the total aluminum–transferrin complex can cross the blood–brain barrier (Yokel et al., 2001).

Since soluble Al salts are gradually changed into insoluble high-molecular-weight aggregates and the absorption of such Al complexes is limited, environmental Al has often been considered to be harmless. Yet there is evidence that Al-containing materials can be injurious to both plants (Kochian and Jones, 1997) and animals (Sparling and Campbell, 1997). There has been mounting disquiet over the possible toxic effects of Al on humans (LaZerte et al., 1997). While some trepidation about Al harmfulness to humans has been expressed for over 90 years, conventional medical views have generally disregarded such concerns. An example of the dismissive tone used is to be found in an article in JAMA stating, "Propaganda as to possible dangers resulting from the use of aluminum cooking vessels is so persistent that one suspects ulterior motives in its background" (Monier-Williams, 1935). The recent advances in understanding the mechanisms by which adjuvant alum salts lead to cell death and immune activation reopens the concept that aluminum salts may be biologically harmful (Reed et al., 2013).

3. ACUTE EXPOSURE TO HIGH LEVELS OF ALUMINUM CAN LEAD TO ADVERSE NEUROLOGICAL CONSEQUENCES

There is good evidence that relatively high concentrations of Al can be acutely neurotoxic once accumulated. Formerly, hemodialysis of patients suffering from kidney failure often led to toxic levels of Al in the blood. The sources of the metal were from both the tubing used during dialysis, and also the administration of aluminum-containing phosphate binders in patients who already had an impaired ability to excrete Al. This often led to aluminum-induced dialysis encephalopathy, which was attributed to the ability of major amounts of Al to traverse into the brain (Russo et al., 1992). Blood concentrations up to 7 μM Al were found in dialysis patients prior to the onset of obvious dementia (Altmann et al., 1987). This encephalopathy was associated with pathological changes in the brain indicting an inflammatory state. In one case, treatment of a chronic renal failure patient, with phosphate-binding Al gels produced an encephalopathy, which after

9 months, led to death. Postmortem neuropathology showed pronounced proliferation of microglia and astrocytes, indicative of an inflammatory response, in specific brain areas (Shirabe et al., 2002). The clinical status of patients suffering from such encephalopathy has been reported to be improved by administration of deferoxamine, an Al chelator (Erasmus et al., 1995). Encephalopathy due to acute exposure to high levels of Al has also been found in patients suffering from kidney failure, treated by bladder irrigation with 1% alum (Phelps et al., 1999). Neurological derangement involving intellectual deficits, loss of muscle control, tremor, and spinocerebellar degeneration has been described in workers in the aluminum industry (Polizzi et al., 2002).

An abnormal neurological condition has on occasion been found consequent to an intramuscular injection of a vaccine preparation containing alum adjuvant (Couette et al., 2009). The World Health Organization Vaccine Safety Advisory Committee has determined that there is likely to be a subset of individuals who respond undesirably to Al-containing vaccines (Authier et al., 2001).

In the past, inhalation of Al oxide powder was used as a means of protecting against silicotic lung disease in miners (Crombie et al., 1944). This approach was reported to have utility in an animal model of silicosis (Dubois et al., 1988). This inhalation procedure was continued for some years despite the fact that miners suffering from silicosis did not report any benefit from this treatment (Kennedy, 1956). Injurious effects of this procedure upon brain function were ultimately clearly recognized (Rifat et al., 1990), and such administration was terminated. A major accidental discharge of Al sulfate into the drinking water supply of the town of Camelford, UK, took place in 1988. After the spill, authorities initially indicated that the water was safe to drink and suggested the addition of fruit juice to conceal any unpleasant taste. Many acidic fruit juices can also enhance the absorption of Al from the gastrointestinal tract. Evidence emerged later on, of harmful neurological consequences to at least some of the exposed population (Altmann et al., 1999). Postmortem pathology of a person, who was exposed to Al at Camelford and later died of an undetermined neurological disorder, revealed evidence of early-onset beta amyloid angiopathy in the cerebral cortical and leptomeningeal blood vessels. The Al content of some brain areas, especially the cortex, was also strikingly elevated (Exley and Esiri, 2006).

Correlative findings by themselves cannot conclusively demonstrate causation, and it has been suggested that excessive penetrance of Al into the brain is an ancillary event following disruption of the blood–brain barrier and may not in itself effect neurotoxicity. However, in patients suffering from dialysis encephalopathy, chelation therapy using deferoxamine both reduced the Al burden of the brain and improved neurological status. And this suggests a direct causal relation between Al exposure and neurotoxicity (McLachlan et al., 1991). A recent report of the use of chelation with EDTA to effect Al excretion in "Al-intoxicated patients" acknowledges that reduction of levels of other toxic metals by this relatively nonselective chelator might also account for recovery of patients (Fulgenzi et al., 2015). These findings have not been extensively pursued, perhaps in part because these chelators are rather nonspecific and can chelate essential as well as nonessential metals. This can result in many undesirable side effects including muscle pain, nausea, and visual deficits. Another possible hazard of chelation therapy as means of effecting Al removal from the body is that it can mobilize Al from quiescent deposits in bone, and thus lead to high serum levels of Al which can then translocate to the brain. This then may cause emergence of neurological symptoms resembling those found in dialysis dementia (Sherrard et al., 1988).

Other indications of the acute neurotoxicity of Al include a case report where aluminum-containing cement was used in the surgical resection of an acoustic neuroma. Six weeks later the patient suffered from loss of consciousness, myoclonic jerks, and persistent grand mal seizures, clinical symptoms that resembled those of lethal dialysis encephalopathy (Reusche et al., 2001). Overall, there is considerable evidence that acute exposure to large amounts of Al in humans can have harmful effects on cerebral function.

4. BASAL INFLAMMATION WITHIN THE BRAIN INCREASES WITH AGING. MOST NEURODEGENERATIVE DISEASES ARE CHARACTERIZED BY AN EVEN GREATER DEGREE OF INFLAMMATORY ACTIVITY

In order to assemble evidence that ingestion of Al-containing materials can accelerate brain aging, it is necessary to first take into account some of the transformations associated with normal aging of the brain. This is typically attended by evidence of elevated levels of inflammatory activity (David et al., 1997; Sharman et al., 2004). Even in the absence of detectable provocative exogenous immune stimuli, cerebral immune activity becomes increasingly pronounced during normal aging (Lucin and Wyss-Coray, 2009; Sharman et al., 2008).

Following the systemic injection of mice with an inflammogen such as lipopolysaccharide, levels of inflammatory cytokines rapidly increase in many tissues including serum and liver, but these are restored to basal concentrations within a week. However, the response in the brain to such treatment leads to a much more sustained elevation of inflammatory cytokine content. TNF-α remains at high levels for up to 10 months, before reverting to basal levels. This extended response, which continues over a significant portion of the mouse life span, is attended by evidence of glial activation and extended neuronal death (Qin et al., 2007). Responses to acute inflammatory events such as infections are maintained for a long time in the brain (Bilbo et al., 2005; Galic et al., 2008; Shi et al., 2003). In consequence, the brain progressively accumulates changes reflecting a history of adverse systemic events, leading to a persistent and undesirable degree of inflammatory activity. This may account for the many reports of the excessive extent of inflammation present in the aged brain (Bondy and Sharman, 2010). A continuing state of inflammation is likely to contribute toward the development of age-related neurodegenerative changes (Block et al., 2007; Lucin and Wyss-Coray, 2009).

Several age-related neurological disorders are accompanied by the onset of additional elevations of neuroinflammation greater than that present in normal aging (Bondy, 2010). Neurodegenerative diseases where this has been reported include AD, Parkinson's disease (PD), amyotrophic lateral sclerosis (ALS), and multiple sclerosis (MS). This additional inflammation may underlie some of the characteristic pathological changes associated with each disorder. In AD, evidence of astrocytic and microglial activation is most prominent at the site of amyloid plaques, and such activated glia generate inflammatory cytokines and acute-phase proteins (Cullen, 1997; Mrak et al., 1995; Styren et al., 1998). This can also result in the appearance of inflammatory cytokines in brain and cerebrospinal fluid (Sun et al., 2003; Zhao et al., 2003). Some of these prolonged changes reflect altered expression and activation of inflammatory genes, especially in the hippocampus (Colangelo et al., 2002). Aluminum nanoparticles can also reach the CNS. Following deposition on the nasal epithelium such particles may be taken up by exocytosis and then conveyed into the brain by way of the olfactory nerve. This can result in phosphorylation of various kinases leading to signal transduction and altered gene expression (Kwon et al., 2013).

AD is characterized by amyloid beta (Aβ) deposition as senile plaques. Aluminum salts can promote amyloid peptide aggregation in defined in vitro preparations (Bolognin et al., 2011; Bondy and Truong, 1999; Exley, 1997). Exposure of transgenic mice overexpressing amyloid precursor protein (APP) to Al salts by way of their drinking water has been described as promoting Aβ accretion in plaques and leading to evidence of oxidative stress in the cortex (Pratico et al., 2002). The ability of Al to promote Alzheimer-like changes in animal models of AD has, however, been questioned. Administration of Al salts in drinking water to a double transgenic mouse line overexpressing both APP and Tau protein led to no significant changes in Aβ levels or prooxidant activity (Akiyama et al., 2012). Another study, which tested both wild-type and mutant mice, also found no effect on the extent of amyloid deposition or cerebral Al content (Ribes et al., 2012). Finally an experiment involving normal rats receiving Al in drinking water reported negative results (Poirier et al., 2011). Such conflicting results are hard to explain, but study design variables in terms of varying dosage of Al, duration of exposure, age of animals at the initiation of exposure may all play a substantive role.

5. EPIDEMIOLOGICAL STUDIES SUGGEST A RELATIONSHIP BETWEEN ALUMINUM EXPOSURE AND THE INCIDENCE OF NEURODEGENERATIVE DISEASE

The neurotoxicity of Al was originally reported in patients experiencing comparatively brief exposure to high levels of Al notably in renal dialysis patients. More controversially, harmful effects of prolonged exposure to lower levels of Al have been described. Elevated cerebral Al in brains of AD patients has been found. A greater content of Al has also been reported in brains of less common neurological disorders such as Guamanian Parkinsonian–ALS complex and Hallervorden–Spatz disease (Eidelberg et al., 1987; Garruto et al., 1988). This has led to the question as to whether Al may be a factor in the initiation and development of several neurological disorders (Kawahra and Kato-Negishi, 2011).

Several studies have focused on specific types of worker such as welders, exposed to high levels of Al. In an analysis, no association among welders exposed to Al by inhalation and neurobehavioral functioning was apparent (Kiesswetter et al., 2009). However, behavioral deficits in welders, which showed a dose–response relation in proportion to the degree of Al exposure, have been described (Giorgianni et al., 2014). This latter report highlighted the fact that the tests most sensitive to Al exposure involved intricate trials reflecting attention and memorial capacity.

5.1 Alzheimer's Disease

The possibility that Al exposure may advance the progression of AD is strengthened by descriptions of excessive levels of Al in analyses of AD brain tissue postmortem. The first report of this (Perl and Brody, 1980) was questioned because of the difficulty of precise quantitation of Al in brain samples (Bjertness et al., 1996). However, a range of more advanced analytical methods, such as laser microprobe mass analysis (Bouras et al., 1997), neutron activation (Andrasi et al., 2005), upgraded graphite furnace atomic absorption methods (Xu et al., 1992), or energy-dispersive X-ray spectroscopy together with transmission electron microscopy (Yumoto et al., 2009), have all substantially confirmed the original report. Laser microprobe mass analysis revealed Al to be primarily concentrated in the neurofibrillary tangles associated with AD (Bouras et al., 1997). High Al content is also present in the cerebral arteries of AD patients (Bhattacharjee et al., 2013a). The possibility that high Al content in AD brains may be a secondary epiphenomenon, consequent to disruption of the blood–brain barrier must therefore be borne in mind (Guerriero et al., 2016).

The suggestion of a causal relation between Al ingestion and neurodegenerative disease is apparent from the number of studies linking the Al content of drinking water and the incidence of AD. An early study reported the AD prevalence was highest in areas where Al concentrations in the drinking water supply were over 100 μg/L and incidence was directly related to the concentration of Al in the drinking water (McLachlan et al., 1996). A similar finding was made in a study of the elderly with AD and Al content of drinking water (Rondeau et al., 2009). A review assembling data from a range of epidemiological reports suggested that generally, there is a significant relation between usage of Al-containing antacids and AD prevalence (Flaten, 2001). A meta-analysis of nine independent studies where urinary Al concentrations were measured determined that cognitive performance was impaired relative to control (Meyer-Baron et al., 2007). The effects of protracted exposure to low levels of Al on AD incidence are difficult to unambiguously identify. Since AD is largely idiopathic and not of genetic origin, many possible confounding environmental factors exist that may influence the incidence. For instance, AD has also been associated with other metal imbalances such as abnormal copper levels, but the establishment of a causal relation remains unresolved (Akatsu et al., 2012; Exley et al., 2012; Kitazawa et al., 2009).

There are conflicting claims concerning the neurotoxic hazard of the levels of aluminum present in the human environment. These range from claims that "AD is a human form of chronic aluminum neurotoxicity" (Walton, 2014) and "aluminum may be the single most aggravating factor related to AD" (Tomljenovic, 2011), through the more circumspect "exposure to aluminum dust may possibly increase the risk of cardiovascular disease and dementia of the Alzheimer's type" (Peters et al., 2013), to totally dismissive reports of the lack of evidence for any correlation between AD and occupational exposures to aluminum (Santibáñez et al., 2007), with inferences such as "lifetime occupational exposure to Al is not likely to be an important risk factor for AD" (Flaten, 2001). A recent review sums up this view with "consideration of the published research concerning aluminum's role in AD indicates that none of the four Bradford Hill criteria considered necessary to establish causation with respect to Al and neurocognitive disorders has been fulfilled" (Lidsky, 2014).

In an endeavor to resolve this issue, it has been proposed that inconsistent findings may in part be due to a common failure to take silicate levels in drinking water into account. Aluminosilicates do not readily cross the bloodstream from the alimentary tract, and the presence of silicates in water can be protective against the toxic effects of Al (Foglio et al., 2012; Krewski et al., 2007). When the silicate content in drinking water is low, the risk of impairment of brain function by Al is raised (Rondeau et al., 2009), suggesting that silicates may reduce the harmfulness of waterborne Al (Gillette Guyonnet et al., 2007). The use of chelators to enhance Al excretion has been suggested to be beneficial in the treatment of AD (Jansson, 2001; McLachlan et al., 1991). Chelators with a greater selectivity for Al may improve this approach (Shin et al., 2003).

Overall, despite an extensive literature, the issue of the relation between AD and exposure to aluminum remains unresolved. This is partly due to the difficulty in obtaining unambiguous data from epidemiological studies. Furthermore, a clear identification of the molecular basis of Al toxicity is lacking. However, laboratory studies under well-defined conditions, where the number a confounding factors is lower, are generally consonant with epidemiological reports. In this review, a subsequent section summarizes findings from research involving Al-treated animals.

In summary, while mechanisms by which Al acts as a neurotoxicant are unclear, Al exposure has been repeatedly found correlated with neuropathological changes associated with AD. This association between Al and incidence of AD has been much more frequently described compared to other neurological disorders. This may be at least in part because of the high incidence of the disorder, which expedites epidemiological studies.

5.2 Association Between Al Exposure and Neurological Disorders Other Than AD

Evidence of a link between Al and other neurological disorders is less well established. Aluminum-containing salts that enhance the immune response to vaccines are often used as adjuvant constituent of vaccines (Alvarez-Soria et al., 2011; Chang et al., 2010; Girard, 2005; Schoenfeld and Agmon-Levin, 2011; Sutton et al., 2009). Injection of alum into neonatal mice in quantities parallel to those used in childhood vaccination schedules results in behavioral abnormalities which persist into adulthood (Shaw et al., 2013). Administration of aluminum-containing adjuvants led to induction of granuloma, which persisted for a prolonged period in the injected muscle, and Al was ultimately able to be transferred into the brain from this location (Crépeaux et al., 2015). Injection of alum-containing vaccine intramuscularly induced Al deposition in mouse brain in gradually progressive way (Khan et al., 2013). However, there was no penetrance of Al into the CNS following direct infusion of the vaccine into the vascular system. Entry of Al from vaccines into the brain was facilitated by the lymphatic system and monocyte chemoattractant CCL2 (Khan et al., 2013). Administration of an Al-containing vaccine led to behavioral abnormalities in female mice that was associated with microglial activation in the hippocampus (Inbar et al., 2017). Furthermore, the neurobehavioral effects of Al adjuvant occurred at the lowest but not highest doses tested. Thus, it appears that the neurotoxicity of the metal may follow a nonparametric dose–response relationship. It was also noted that while Al levels in the injected muscle resolved in 6 months, the cerebral levels were selectively increased (Crépeaux et al., 2017).

Peripheral administration of aluminum-containing nanoparticles, 30–60 nm in diameter, for 3 weeks, elevated the Al content of mouse brain and also increased prooxidant events. In this study, hippocampal memory-forming processes were impaired and there was an elevated rate of Aβ formation (Shah et al., 2015). Significant amounts of Al are often present in infant formulae at significant levels, 100–756 µg/L, and it may be that this represents a nontrivial developmental hazard (Chuchu et al., 2013). The use of vaccines has been associated with increased incidence of MS, and urinary aluminum content is elevated in MS patients (Exley et al., 2006). Chelation therapy leading to reduced levels of circulating Al has been stated to have therapeutic value in the treatment of MS (Fulgenzi et al., 2015). On the other hand, vaccines with Al-containing adjuvants have been described as protecting mice from developing experimental autoimmune encephalomyelitis (Wållberg et al., 2003). Despite these apparently conflicting reports, there is convincing evidence that nanomolar levels of aluminum can increase expression of the inflammatory biomarker C-reactive protein in isolated endothelial cells derived from microvessels (Alexandrov et al., 2015). The consistent findings that Al can increase markers of neuroinflammation, either directly by microglial activation, or indirectly by influencing the microvasculature, may be an important mechanism by which exposure to the metal can enhance and promote neurodegeneration and subsequent behavioral abnormalities.

There are suggestions linking Al and PD. PD is another rather widespread common neurological disease characterized by elevated levels of oxidative and inflammatory events (Selley, 2005). Such an association has been made based on a relation between gastric ulcers and the incidence of PD, which may reflect the high use of Al-containing antacids by those suffering from ulcers (Altschuler, 1999). Other hints of a possible connection between Al and PD lie in the property of Al salts to bring about activation of monoamine oxidase B. This enzyme is elevated with age and further raised in PD (Zatta et al., 1999). Monoamine oxidase B is able to promote aggregation and fibril formation of α-synuclein, which could account for the reported association between several neurotoxic metals and PD (Uversky et al., 2001). Activated microglia and high levels of inflammatory cytokines within nervous tissue are present in PD (Nagatsu and Sawada, 2005). The activation of NF-κB, a transcription factors leading to induction of a series of inflammatory events, can occur in a synergistic way after treatment of experimental animals with both the dopaminergic neurotoxin, MPTP in conjunction with the presence of low levels of Al in drinking water (Li et al., 2008). Nonsteroidal antiinflammatory drugs may delay the onset and progression of the disease (Hald et al., 2007).

Neuropathological and behavioral modifications resembling those found in ALS have been described in animals after administration of Al salts. Injection of Al-containing adjuvants at levels equivalent to those typically administered to humans, resulted in motoneuron death, impairments in motor performance, reduced retention of spatial memory and increased activation of astrocytes, and microglia in mice (Petrik et al., 2007; Shaw and Petrik,

2009). Thus, exposure to Al may promote an array of neurological impairments. The unique epigenetic and genetic profile of diverse individuals, the dose and duration of exposure to Al, as well as combination of other environmental factors such as coexposure to other metals may all determine the specific type of neurological abnormality that is manifested.

6. RESEARCH FROM ANIMAL MODELS AND IN VITRO SYSTEMS IMPLIES THAT HIGH LEVELS OF ALUMINUM CAN FURTHER THE EVOLUTION OF AGE-RELATED COGNITIVE DEFICITS

Clinical reports of aluminum neurotoxicity are matched by similar findings found in several experimental animal models. Administration of Al salts in such models can result in neuropathological changes resembling those found in human brain aging (Bowdler et al., 1979; Miu et al., 2004). However, several of these studies have involved concentrations of Al not commonly met among human populations. Other studies that better mirror human exposures have been performed using more extended treatment using low levels of Al paralleling those found in environmental exposures. It is apparent that immunomodulation, neuroinflammation, and oxidative stress are mechanisms that have consistently been shown to contribute to neurodegenerative diseases. In this section, the role of Al in exacerbating these events will be further delineated (Fig. 1).

6.1 Immunomodulation and Neuroinflammation

Aluminum salts, used as adjuvants in vaccines, cause necrosis and uric acid production which functions as a danger signal to activate the NLRP3 inflammasome (Eisenbarth et al., 2008). The activation of the NLRP3 inflammasome plays a crucial role in amplifying the inflammatory response (Baroja-Mazo et al., 2014). Furthermore, it has been reported that DNA released from alum-exposed dying cells functions as another danger signal (damage-associated molecular pattern) contributing to the immunomodulatory role of aluminum adjuvants (Marichal et al., 2011). These findings all point to the impression that Al salts are not biologically benign and that prolonged environmental exposure to the metal may lead to cellular stress and subsequent potentiation of inflammatory events. Since an enhanced inflammatory state underlies many disease states, including age-related neurodegenerative diseases, the role of Al in exacerbating the pathology of these disorders should be further examined.

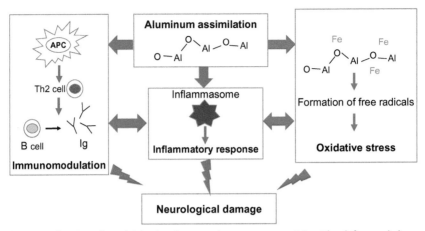

Fig. 1 Mechanisms by which aluminum enhance neurotoxicity. The *left panel* shows how aluminum aggregates cause enhanced activation of antigen-presenting cells (APC) which then provide the signals necessary to promote a T helper 2 (Th2) subset of T cells. Th2 cells are important mediators of B cell activation and antibody (Ig) production. Different isotypes of antibodies cause immunomodulation by their effects on different immune cells. This immunomodulation may result in enhanced neurotoxicity. The *center panel* shows how aluminum aggregates can also activate the NLRP3 inflammasome and by doing so amplify the inflammatory response. Enhanced inflammation has been shown to directly lead to neurotoxicity. The *right panel* shows that aluminum aggregates can function as a platform for redox-active metals such as iron to induce a Fenton reaction. The accelerated formation of free radicals then leads to oxidative stress, another mechanism directly linked to neurotoxicity. Although separate phenomenon, these events may be interrelated and this is indicated in the figure by the two-sided *arrows*.

Low levels of Al in the drinking water of mice led to elevation of indices of inflammation in brain tissue such as increases in levels of inflammatory cytokines and nitric oxide synthase (Campbell et al., 2004). These changes occurred after 3 months of contact with Al salts in the drinking water using concentrations of Al below some of those reported for some residential water reserves. Extended exposure to low levels of Al can also lead to elevated levels of glial fibrillary astrocytic protein an indicator of astrocytic immune activation (Yokel and O'Callaghan, 1998). Other evidence concerning the potential neurotoxicity of Al includes descriptions of cognitive and pathological alterations in aged rats resembling those characteristic of AD, following exposure to Al at levels equivalent to those ingested by some human populations (Walton, 2009a,b, 2012; Walton and Wang, 2009). Exposure to dietary Al also resulted in increased levels of APP in a rodent model (Walton and Wang, 2009). AD-like changes in rats following aluminum exposure have been attributed to induction of α- and β-secretases, leading to increased formation of Aβ from APP (Wang et al., 2014). The evidence for Al salts to induce Aβ formation and its aggregation has been previously reviewed (Zhao et al., 2014).

Changes in expression of specific genes have been described in a transformed neuronal cell line following exposure to an Aβ–aluminum complex, and several of the genes whose expression was increased are among those also elevated in AD (Gatta et al., 2011). After Al treatment of transgenic mouse models of AD, the profile of micro-RNA-mediated gene expression distinctly resembles changes found in AD (Pogue and Lukiw, 2016). These accounts suggest a genetic or epigenetic basis for many of those changes caused by Al treatment that resemble altered expression associated with AD.

If the extended increase in inflammatory activity that characterizes brain aging was worsened in the presence of low Al, this would resemble the doubly enhanced level of inflammation found in many neurodegenerative diseases. Thus, Al may act initially by accelerating the rate of normal brain aging. This could then form a platform that would further the development of a range of more specific neurodegenerative disorders.

Despite the relative inertness of Al salts, there are several potential pathways by which Al could initiate toxic events (Tomljenovic, 2011). The activation of glia and macrophages by Al-containing chemicals and mineral has been described (Evans et al., 1992; Gorell et al., 1999; Platt et al., 2001). Similar to the alum adjuvant, the activation of the NLRP3 inflammasome and consequent amplification of the inflammatory response may play a role in this glial activation. Since Al salts can produce inflammatory reactions in isolated glia as well as in glia of intact animals, it is probable that a direct action on some glial species is implicated (Campbell et al., 2002).

6.2 Oxidative Stress

Al is not a valence-active element and does not have a significant affinity for sulfhydryl groups, but it has the capacity to enhance the production of oxidant free radicals. This property may be due to the enhancement of the redox activity of trace amounts of iron. The ability of aluminum to increase the prooxidant properties of iron is found even in the absence of tissues or organic material (Bondy et al., 1998). The mechanism by which this catalysis is brought about is likely to be by Al complexes in solution forming colloids, upon whose surfaces iron can be loosely sequestered. This partial complexation enables iron to undergo Fenton dynamics and undergo valence redox flux thus causing production of reactive oxygen species (Bondy, 2009; Ruipérez et al., 2012). This can then lead to increased expression of genes associated with inflammation (Alexandrov et al., 2005). An analogous enhancement of the prooxidant properties of iron by an inert mineral is known in the case of silica fibers (Napierska et al., 2012). The binding of transition metals on the surfaces of nanoparticles composed of inert core materials can greatly enhance the toxicity of such particles (Bhattacharjee et al., 2013b). This is likely to have applicability to aluminum salts, which generally exist as colloids in aqueous media.

7. THE NEUROTOXICITY OF ALUMINUM IN AMOUNTS ENCOUNTERED IN THE HUMAN ENVIRONMENT CONTINUES TO BE CONTENTIOUS

The question of Al neurotoxicity has a long history, but no consensus has been reached concerning the hazard posed by environmental exposure. As a result the necessity for increased regulatory action is not regarded as critical. Research on animals is limited by the fact that rodent biology does not completely reflect the human condition, especially as regards to brain function. While there is a range of diverse and often opposing opinions, the preponderance of evidence from both laboratory studies and epidemiology suggests that the issue of Al neurotoxicity should not be cavalierly dispelled. In view of the large number of people ingesting various amounts of this element, this risk should not be dismissed but should remain under careful consideration.

The history of lead toxicity can perhaps give clues as to why there has been failure to reach agreement on the importance of hazards posed by Al, and why there is an inclination to regard these as not of critical concern. Lead has been in use in its metallic form and as salts for more than 3000 years and has been intermittently recognized as poisonous since 700 B.C., and its widespread global presence has risen markedly in the last 200 years. However, only in the last two decades has the harmfulness of lead at low levels been widely accepted. In consequence, increasingly severe legislative measures to curtail lead exposure have been instituted and these appear to be generally effective. A long period of controversy preceded the recognition of the neurotoxicity of lead. Before the universal acceptance of the harmfulness of low levels of lead, the lead industry vigorously fought against the regulation of environmental lead and damaged the reputation of researchers in this area. Prominent scientists evaluating the effects of exposure to low levels of lead on child development were accused of bias and fraud (summarized in Needleman, 2008).

In comparison, Al has only had broad industrial use for a comparatively short time. However, as was the case leading up to recognition of the neurotoxicity of low levels of lead, the harmfulness of low levels of Al is intensely disputed as, once again, major economic forces are involved. Accordingly, no new major efforts to minimize Al levels in food or drinking water are currently being legislatively deliberated. The much shorter history of Al use means that we appear to be at an early stage of concern regarding the dangers posed for human health, than is the case with lead. In common with lead, levels of this metal, once regarded as trivial, are likely to be recognized as potentially hazardous. Also, in common with lead, broad population exposure to ingested Al may cause subtle deficits and vulnerabilities rather than spectacular and specific toxic incidents. It is thus hoped that similar to lead, there will be a growing recognition of the neurotoxicity of environmental aluminum and the introduction of legislation that would protect populations at risk, which are likely to be manifold.

8. SUMMARY

The potential for aluminum ingestion to further the development of neurodegenerative disease is not yet unambiguously accepted. However, several key findings are undisputable. These are as following:
- Al-containing materials have a widespread presence in the environment, and when ingested by humans, some Al salts can reach the brain.
- Brief exposure to high levels of Al can lead to clear evidence of neurological damage.
- The level of basal inflammatory activity in the brain is progressively increased with aging, and this is intensified in several neurodegenerative conditions.
- Administration of amounts of Al to experimental animals in the drinking water that correspond to levels found in some residential water sources can increase inflammatory activity in the brain and are associated with neuropathological changes, resembling those found in AD.

As life expectancy in the United States grows, a greater incidence of slow developing neurodegenerative disorders such as AD, PD, ALS, and MS can be anticipated. These diseases are largely of nongenetic origin and are likely to be initiated by unidentified gene–environmental interactions. As long dormant periods can occur between exposure to an injurious environmental or occupational agent and the manifestation of explicit clinical symptoms, this makes the identification of specific factors that initially begin the disease trajectory difficult. Aging is a critical feature in permitting occurrence of neurodegenerative syndromes. Hastening of normal changes taking place during brain aging could facilitate the incidence of distinctive neurological disorders. A favorable strategy toward alleviation of slowly developing age-related changes might be the recognition of those environmental factors which hasten changes associated with normal brain senescence, and then developing measures to protect against such harmful factors.

The simplest way of explaining much of the research on Al neurotoxicity is the idea that Al can accelerate the development of the inflammatory changes that characterize the normally aging brain. Colloidal aluminum can also exist in a form that promotes the free radical-producing potential of redox-active metals such as copper and iron. Such enhanced free radical generation may also contribute to the inflammatory cascade. This could be a mechanism underlying the impact of Al ingestion upon the promotion of AD. It could also help to account for more tenuous connection suggested for Al and less prevalent age-related neurological diseases. Thus, if Al is able to amplify the inflammatory aspect of normal brain aging, such a chronic state of excessive and ineffective immune function could form a base for the advent and expansion of more specific neurological age-related disorders.

REFERENCES

Akatsu, H., Hori, A., Yamamoto, T., Yoshida, M., Mimuro, M., Hashizume, Y., Tooyama, I., Yezdimer, E.M., 2012. Transition metal abnormalities in progressive dementias. Biometals 25, 337–350.

Akiyama, H., Hosokawa, M., Kametani, F., Kondo, H., Chiba, M., Fukushima, M., Tabira, T., 2012. Long-term oral intake of aluminium or zinc does not accelerate Alzheimer pathology in AβPP and AβPP/tau transgenic mice. Neuropathology 32, 390–397.

Alexandrov, P.N., Zhao, Y., Pogue, A.I., Tarr, M.A., Kruck, T.P., Percy, M.E., Cui, J.G., Lukiw, W.J., 2005. Synergistic effects of iron and aluminum on stress-related gene expression in primary human neural cells. J. Alzheimers Dis. 8, 117–127.

Alexandrov, P.N., Kruck, T.P., Lukiw, W.J., 2015. Nanomolar aluminum induces expression of the inflammatory systemic biomarker C-reactive protein (CRP) in human brain microvessel endothelial cells (hBMECs). J. Inorg. Biochem. 152, 210–213.

Altmann, P., Al-Salihi, F., Butter, K., Cutler, P., Blair, J., Leeming, R., Cunningham, J., Marsh, F., 1987. Serum aluminum levels and erythrocyte dihydropteridine reductase activity in patients on hemodialysis. N. Engl. J. Med. 317, 80–84.

Altmann, P., Cunningham, J., Dhanesha, U., Ballard, M., Thompson, J., Marsh, F., 1999. Disturbance of cerebral function in people exposed to drinking water contaminated with aluminum sulphate: retrospective study of the Camelford water incident. Br. Med. J. 319, 807–811.

Altschuler, E., 1999. Aluminum-containing antacids as a cause of idiopathic Parkinson's disease. Med. Hypotheses 53, 22–23.

Alvarez-Soria, M.J., Hernandez-Gonzalez, A., Carrasco-Garcia de Leon, S., Del Real-Francia, M.A., Gallardo-Alcaniz, M.J., Lopez-Gomez, J.L., 2011. Demyelinating disease and vaccination of the human papillomavirus. Rev. Neurol. 52, 472–476.

Andrasi, E., Pali, N., Molnar, Z., Kosel, S., 2005. Brain aluminum, magnesium and phosphorus contents of control and Alzheimer-diseased patients. J. Alzheimers Dis. 7, 273–284.

Authier, F.J., Cherin, P., Creange, A., Bonnotte, B., Ferrer, X., Abdelmoumni, D., Ranoux, D., Pelletier, J., Figarella-Branger, D., Granel, B., Maisonobe, T., Coquet, M., Degos, J.D., Gherardi, R.K., 2001. Central nervous system disease in patients with macrophagic myofasciitis. Brain 124, 974–983.

Baroja-Mazo, A., Martín-Sánchez, F., Gomez, A.I., Martínez, C.M., Amores-Iniesta, J., Compan, V., Barberà-Cremades, M., Yagüe, J., Ruiz-Ortiz, E., Antón, J., Buján, S., Couillin, I., Brough, D., Arostegui, J.I., Pelegrín, P., 2014. The NLRP3 inflammasome is released as a particulate danger signal that amplifies the inflammatory response. Nat. Immunol. 15, 738–750.

Bhattacharjee, S., Rietjens, I.M., Singh, M.P., Atkins, T.M., Purkait, T.K., Xu, Z., Regli, S., Shukaliak, A., Clark, R.J., Mitchell, B.S., Alink, G.M., Marcelis, A.T., Fink, M.J., Veinot, J.G., Kauzlarich, S.M., Zuilhof, H., 2013a. Cytotoxicity of surface-functionalized silicon and germanium nanoparticles: the dominant role of surface charges. Nanoscale 5, 4870–4883.

Bhattacharjee, S., Zhao, Y., Hill, J.M., Culicchia, F., Kruck, T.P., Percy, M.E., Pogue, A.I., Walton, J.R., Lukiw, W.J., 2013b. Selective accumulation of aluminum in cerebral arteries in Alzheimer's disease (AD). J. Inorg. Biochem. 126, 35–77.

Bilbo, S.D., Biedenkapp, J.C., Der-Avakian, A., Watkins, L.R., Rudy, J.W., Maier, S.F., 2005. Neonatal infection-induced memory impairment after lipopolysaccharide in adulthood is prevented via caspase-1 inhibition. J. Neurosci. 25, 8000–8009.

Bjertness, E., Candy, J.M., Torvik, A., Ince, P., McArthur, F., Taylor, G.A., Johansen, S.W., Alexander, J., Grønnesby, J.K., Bakketeig, L.S., Edwardson, J.A., 1996. Content of brain aluminum is not elevated in Alzheimer disease. Alzheimer Dis. Assoc. Disord. 10, 171–174.

Block, M.L., Zecca, L., Hong, J.S., 2007. Microglia-mediated neurotoxicity: uncovering the molecular mechanisms. Nat. Rev. Neurosci. 8, 57–69.

Bolognin, S., Messori, L., Drago, D., Gabbiani, C., Cendron, L., Zatta, P., 2011. Aluminum, copper, iron and zinc differentially alter amyloid-Aβ(1-42) aggregation and toxicity. Int. J. Biochem. Cell Biol. 43, 877–885.

Bondy, S.C., 2009. Aluminum. In: Squire, L.R. (Ed.), In: Encyclopedia of Neuroscience, vol. 1. Academic Press, Oxford, pp. 253–257.

Bondy, S.C., 2010. The neurotoxicity of environmental aluminum is still an issue. Neurotoxicology 31, 575–581.

Bondy, S.C., Sharman, E.H., 2010. Melatonin, oxidative stress and the aging brain. In: Bondy, S.C., Maiese, K. (Eds.), Oxidative Stress in Basic Research and Clinical Practice: Aging and Age-Related Disorders. Humana Press, Totowa, NJ, pp. 339–357.

Bondy, S.C., Truong, A., 1999. Potentiation of beta-folding of β-amyloid peptide 25-35 by aluminum salts. Neurosci. Lett. 267, 25–35.

Bondy, S.C., Guo-Ross, S.X., Pien, J., 1998. Mechanisms underlying the aluminum-induced potentiation of the pro-oxidant properties of transition metals. Neurotoxicology 19, 65–71.

Bouras, C., Giannakopoulos, P., Good, P.F., Hsu, A., Hof, P.R., Perl, D.P., 1997. A laser microprobe mass analysis of brain aluminum and iron in dementia pugilistica: comparison with Alzheimer's disease. Eur. Neurol. 38, 53–58.

Bowdler, N.C., Beasley, D.S., Fritze, E.C., Goulette, A.M., Hatton, J.D., Hessian, J., Ostman, D.L., Rugg, D.J., Schmittman, C.J., 1979. Behavioural effects of aluminum ingestion on animal and human subjects. Pharmacol. Biochem. Behav. 10, 505–512.

Burrell, S.A., Exley, C., 2010. There is (still) too much aluminium in infant formulas. BMC Pediatr. 10, 63.

Campbell, A., Yang, Y., Tsai-Turton, M., Bondy, S.C., 2002. Pro-inflammatory effects of aluminum in human glioblastoma cells. Brain Res. 933, 62–65.

Campbell, A., Becaria, A., Lahiri, D.K., Sharman, K., Bondy, S.C., 2004. Chronic exposure to aluminum in drinking water increases inflammatory parameters selectively in the brain. J. Neurosci. Res. 75, 565–572.

Chang, J., Campagnolo, D., Vollmer, T.L., Bomprezzi, R., 2010. Demyelinating disease and polyvalent human papilloma virus vaccination. J. Neurol. Neurosurg. Psychiatry 9, 1–3.

Chuchu, N., Patel, B., Sebastian, B., Exley, C., 2013. The aluminium content of infant formulas remains too high. BMC Pediatr. 13, 162.

Colangelo, V., Schurr, J., Ball, M.J., Pelaez, R.P., Bazan, N.G., Lukiw, W.J., 2002. Gene expression profiling of 12633 genes in Alzheimer hippocampal CA1: transcription and neurotrophic factor down-regulation and up-regulation of apoptotic and pro-inflammatory signaling. J. Neurosci. Res. 70, 462–473.

Couette, M., Boisse, M.F., Maison, P., Brugieres, P., Cesaro, P., Chevalier, X., Gherardi, R.K., Bachoud-Levi, A.C., Authier, F.J., 2009. Long-term persistence of vaccine-derived aluminum hydroxide is associated with chronic cognitive dysfunction. J. Inorg. Biochem. 103, 1571–1578.

Crépeaux, G., Eidi, H., David, M., Tzavara, E., Giros, B., Exley, C., Curmi, P.A., Christopher, A., Shawe, C.A., Romain, K., Gherardi, K., Cadusseau, J., 2015. Highly delayed systemic translocation of aluminum-based adjuvant in CD1 mice following intramuscular injections. J. Inorg. Biochem. 152, 199–205.

Crépeaux, G., Eidi, H., David, M., Baba-Amer, Y., Tzavara, E., Giros, B., Authier, F.J., Exley, C., Shaw, C.A., Romain, K., Cadusseau, J., Gherardi, R.K., 2017. Non-linear dose-response of aluminum hydroxide adjuvant particles: selective low dose neurotoxicity. Toxicology 375, 48–57.

Crombie, D.W., Blaisdell, J.L., MacPherson, G., 1944. The treatment of silicosis by aluminum powder. Can. Med. Assoc. J. 50, 318–328.

Cullen, K.M., 1997. Perivascular astrocytes within Alzheimer's disease plaques. Neuroreport 8, 1961–1966.

Dabeka, R., Fouquet, A., Belisle, S., Turcotte, S., 2011. Lead, cadmium and aluminum in Canadian infant formulae, oral electrolytes and glucose solutions. Food Addit. Contam. Part A Chem. Anal. Control Expo. Risk Assess. 28, 744–753.

David, J.P., Ghozali, F., Fallet-Bianco, C., Wattez, A., Delaire, S., Boniface, B., Di Menza, C., Delacourte, A., 1997. Glial reaction in the hippocampal formation is highly concentrated with aging in human brain. Neurosci. Lett. 235, 53–56.

Dubois, F., Bégin, R., Cantin, A., Massé, S., Martel, M., Bilodeau, G., Dufresne, A., Perreault, G., Sébastien, P., 1988. Aluminum inhalation reduces silicosis in a sheep model. Am. Rev. Respir. Dis. 137, 1172–1179.

Eidelberg, D., Sotrel, A., Joachim, C., Selkoe, D.I., Forman, A., Pendlebury, W.W., Perl, D.P., 1987. Adult onset Hallervorden-Spatz disease with neurofibrillary pathology. Brain 110, 993–1013.

Eisenbarth, S.C., Colegio, O.R., O'Connor, W., Sutterwala, F.S., Flavell, R.A., 2008. Crucial role for the Nalp3 inflammasome in the immunomodulatory properties of aluminium adjuvants. Nature 453, 1122–1126.

Erasmus, R.T., Kusnir, J., Stevenson, W.C., Lobo, P., Herman, M.M., Wills, M.R., Savory, J., 1995. Hyperaluminemia associated with liver transplantation and acute renal failure. Clin. Transplant. 9, 307–311.

Evans, P.H., Peterhans, E., Burg, T., Klinowski, J., 1992. Aluminosilicate-induced free radical generation by murine brain glial cells in vitro: potential significance in the aetiopathogenesis of Alzheimer's dementia. Dementia 3, 1–6.

Exley, C., 1997. ATP-promoted amyloidosis of an amyloid β-peptide. Neuroreport 8, 3411–3414.

Exley, C., Esiri, M.M., 2006. Severe cerebral congophilic angiopathy coincident with increased brain aluminum in a resident of Camelford, Cornwall, UK. J. Neurol. Neurosurg. Psychiatry 77, 877–879.

Exley, C., Mamutse, G., Korchazhkina, O., Pye, E., Strekopytov, S., Polwart, A., Hawkins, C., 2006. Elevated urinary excretion of aluminum and iron in multiple sclerosis. Mult. Scler. 12, 533–540.

Exley, C., House, E., Polwart, A., Esiri, M.M., 2012. Brain burdens of aluminum, iron, and copper and their relationships with amyloid-β pathology in 60 human brains. J. Alzheimers Dis. 31, 725–730.

Fimreite, N., Hansen, O.O., Pettersen, H.C., 1997. Aluminum concentrations in selected foods prepared in aluminum cookware, and its implications for human health. Bull. Environ. Contam. Toxicol. 58, 1–7.

Flaten, T.P., 2001. Aluminum as a risk factor in Alzheimer's disease, with emphasis on drinking water. Brain Res. Bull. 55, 187–196.

Foglio, E., Buffoli, B., Exley, C., Rezzani, R., Rodella, L.F., 2012. Regular consumption of a silicic acid-rich water prevents aluminium-induced alterations of nitrergic neurons in mouse brain: histochemical and immunohistochemical studies. Histol. Histopathol. 27, 1055–1066.

Fulgenzi, F.A., De Giuseppe, R., Bamonti, F., Vietti, D., Ferrero, M.E., 2015. Efficacy of chelation therapy to remove aluminium intoxication. J. Inorg. Biochem. 152, 214–218.

Galic, M.A., Riazi, K., Heida, J.G., Mouihate, A., Fournier, N.M., Spencer, S.J., Kalynchuk, L.E., Teskey, G.C., Pittman, Q.J., 2008. Postnatal inflammation increases seizure susceptibility in adult rats. J. Neurosci. 28, 6904–6913.

Garruto, R.M., Shankar, S.K., Yanagihara, R., Salazar, A.M., Amyx, H.L., Gajdusek, D.C., 1988. Low-calcium, high-aluminum diet-induced motor neuron pathology in cynomolgus monkeys. Acta Neuropathol. 78, 210–219.

Gatta, V., Drago, D., Fincati, K., Valenti, M.T., Dalle Carbonare, L., Sensi, S.L., Zatta, P., 2011. Microarray analysis on human neuroblastoma cells exposed to aluminum, β(1–42)-amyloid or the β(1–42)-amyloid aluminum complex. PLoS One 6, e15965.

Gillette Guyonnet, S., Andrieu, S., Vellas, B., 2007. The potential influence of silica present in drinking water on Alzheimer's disease and associated disorders. J. Nutr. Health Aging 11, 119–124.

Giorgianni, C.M., D'Arrigo, G., Brecciaroli, R., Abbate, A., Spatari, G., Tringali, M.A., Gangemi, S., De Luca, A., 2014. Neurocognitive effects in welders exposed to aluminium. Toxicol. Ind. Health 30, 347–356.

Girard, M., 2005. Autoimmune hazards of hepatitis B vaccine. Autoimmun. Rev. 4, 96–100.

Gorell, J.M., Rybicki, B.A., Johnson, C., Peterson, E.L., 1999. Occupational exposure to specific metals (manganese, copper, lead, iron, mercury, zinc, aluminum and others) appears to be a risk factor for Parkinson's disease (PD) in some, but not all, case-control studies. Neuroepidemiology 18, 303–308.

Guerriero, F., Sgarlata, C., Francis, M., Maurizi, N., Faragli, A., Perna, S., Rondanelli, M., Rollone, M., Ricevuti, G., 2016. Neuroinflammation, immune system and Alzheimer disease: searching for the missing link. Aging Clin. Exp. Res. PMID: 27718173 [Epub ahead of print].

Hald, A., Van Beek, J., Lotharius, J., 2007. Inflammation in Parkinson's disease: causative or epiphenomenal? Subcell. Biochem. 42, 249–279.

Harris, W.R., Wang, Z., Hamada, Y.Z., 2003. Competition between transferrin and the serum ligands citrate and phosphate for the binding of aluminum. Inorg. Chem. 42, 3262–3273.

Hetherington, L.E., 2007. World Mineral Production: 2001–2005. British Geological Survey, Keyworth, Nottingham, United Kingdom. ISBN: 978-0-85272-592-4.

Inbar, R., Weiss, R., Tomljenovic, L., Arango, M.T., Deri, Y., Shaw, C.A., Chapman, J., Blank, M., Shoenfeld, Y., 2017. Behavioral abnormalities in female mice following administration of aluminum adjuvants and the human papillomavirus (HPV) vaccine Gardasil. Immunol. Res. 65, 136–149.

Jansson, E.T., 2001. Aluminum exposure and Alzheimer's disease. J. Alzheimers Dis. 3, 541–549.

Kawahra, M., Kato-Negishi, M., 2011. Link between aluminum and the pathogenesis of Alzheimer's disease: the integration of the aluminum and amyloid cascade hypotheses. Int. J. Alzheimers Dis. 2011, 276393.

Kennedy, M.C.S., 1956. Aluminum powder inhalations in the treatment of silicosis of pottery workers and pneumoconiosis of coal-miners. Br. J. Ind. Med. 13, 85–101.

Khan, Z., Combadiere, C., Authier, F.J., Itier, V., Lux, F., Exley, C., Mahrouf-Yorgov, M., Decrouy, X., Moretto, P., Tillement, O., Gherardi, R.K., Cadusseau, J., 2013. Slow CCL2-dependent translocation of biopersistent particles from muscle to brain. BMC Med. 11, 99.

Kiesswetter, E., Schäper, M., Buchta, M., Schaller, K.H., Rossbach, B., Kraus, T., Letzel, S., 2009. Longitudinal study on potential neurotoxic effects of aluminium: II. Assessment of exposure and neurobehavioral performance of Al welders in the automobile industry over 4 years. Int. Arch. Occup. Environ. Health 82, 1191–1210.

Kitazawa, M., Cheng, D., Laferla, F.M., 2009. Chronic copper exposure exacerbates both amyloid and tau pathology and selectively dysregulates cdk5 in a mouse model of AD. J. Neurochem. 108, 1550–1560.

Kochian, L.V., Jones, D.L., 1997. Aluminum toxicity and resistance in plants. In: Yokel, R.A., Golub, M.S. (Eds.), Research Issues in Aluminum Toxicity. Taylor and Francis, Washington, pp. 69–90.

Krewski, D., Yokel, R.A., Nieboer, E., Borchelt, D., Cohen, J., Harry, J., Kacew, S., Lindsay, J., Mahfouz, A.M., Rondeau, V., 2007. Human health risk assessment for aluminium, aluminium oxide, and aluminium hydroxide. J. Toxicol. Environ. Health B Crit. Rev. 10 (Suppl. 1), 1–269.

Kwon, J.T., Seo, G.B., Jo, E., Lee, M., Kim, H.M., Shim, I., Lee, B.W., Yoon, B.I., Kim, P., Choi, K., 2013. Aluminum nanoparticles induce ERK and p38MAPK activation in rat brain. Toxicol. Res. 29, 181–185.

LaZerte, B.D., Van Loon, G., Anderson, B., 1997. Aluminum in water. In: Yokel, R.A., Golub, M.S. (Eds.), Research Issues in Aluminum Toxicity. Taylor and Francis, Washington DC, pp. 17–46.

Li, H., Campbell, A., Ali, S.F., Cong, P., Bondy, S.C., 2008. Chronic exposure to low levels of aluminum alters cerebral cell signaling in response to acute MPTP treatment. Toxicol. Ind. Health 23, 515–524.

Lidsky, T.I., 2014. Is the aluminum hypothesis dead? J. Occup. Environ. Med. 56 (5 Suppl), S73–9.

Lucin, K.M., Wyss-Coray, T., 2009. Immune activation in brain aging and neurodegeneration: too much or too little? Neuron 64, 110–122.

Marichal, T., Ohata, K., Bedoret, D., Mesnil, C., Sabatel, C., Kobiyama, K., Lekeux, P., Coban, C., Akira, S., Ishii, K.J., Bureau, F., Desmet, C.J., 2011. DNA released from dying host cells mediates aluminum adjuvant activity. Nat. Med. 17, 996–1003.

Marrack, P., McKee, A.S., Munks, M.W., 2009. Towards an understanding of the adjuvant action of aluminum. Nat. Rev. Immunol. 9, 287–293.

McLachlan, D.R.C., Dalton, A.J., Kruck, T.P.A., Bell, M.Y., Smith, W.L., Kalow, W., Andrews, D.F., 1991. Intramuscular desferrioxamine in patients with Alzheimer's disease. Lancet 337, 1304–1308.

McLachlan, D.R.C., Bergeron, C., Smith, J.E., Boomer, D., Rifat, S.L., 1996. Risk for neuropathologically confirmed Alzheimer's disease and residual aluminum in municipal drinking water employing weighted residential histories. Neurology 46, 401–405.

Meyer-Baron, M., Schäper, M., Knapp, G., van Thriel, C., 2007. Occupational aluminum exposure: evidence in support of its neurobehavioral impact. Neurotoxicology 28, 1068–1078.

Miu, A.C., Olteanu, A.I., Miclea, M., 2004. A behavioral and ultrastructural dissection of the interference of aluminum with aging. J. Alzheimers Dis. 6, 315–328.

Monier-Williams, G.W., 1935. Aluminum in Food, Report 78 on Public Health and Medical Subjects. Ministry of Health, London.

Mrak, R.E., Sheng, J.G., Griffin, W.S.T., 1995. Glial cytokines in Alzheimer's disease: review and pathogenic implications. Hum. Pathol. 26, 816–823.

Nagatsu, T., Sawada, M., 2005. Inflammatory process in Parkinson's disease: role for cytokines. Curr. Pharm. Des. 11, 999–1016.

Napierska, D., Rabolli, V., Thomassen, L.C., Dinsdale, D., Princen, C., Gonzalez, L., Poels, K.L., Kirsch-Volders, M., Lison, D., Martens, J.A., Hoet, P.H., 2012. Oxidative stress induced by pure and iron-doped amorphous silica nanoparticles in subtoxic conditions. Chem. Res. Toxicol. 25, 828–837.

Needleman, H.L., 2008. The case of Deborah Rice: who is the environmental protection agency protecting? PLoS Biol. 6, e129. http://dx.doi.org/10.1371/journal.pbio.0060129.

Pennington, J.A.T., Jones, J.W., 1989. Dietary intake of aluminum. In: Gitelman, H.J. (Ed.), Aluminum and Health: A Critical Review. Marcel and Dekker, New York, pp. 67–100.

Perl, D.P., Brody, A.R., 1980. Alzheimer's disease: X-ray spectrometric evidence of aluminum accumulation in neurofibrillary tangle-bearing neurons. Science 208, 297–299.

Peters, S., Reid, A., Fritschi, L., de Klerk, N., Musk, A.W., 2013. Long-term effects of aluminium dust inhalation. Occup. Environ. Med. 70, 864–868.

Petrik, M.S., Wong, M.C., Tabata, R.C., Garry, R.F., Shaw, C.A., 2007. Aluminum adjuvant linked to Gulf War illness induces motor neuron death in mice. Neuromolecular Med. 9, 83–100.

Phelps, K.R., Naylor, K., Brien, T.P., Wilbur, H., Haqqie, S.S., 1999. Encephalopathy after bladder irrigation with alum: case report and literature review. Am. J. Med. Sci. 318, 181–185.

Platt, B., Fiddler, G., Riedel, G., Henderson, Z., 2001. Aluminum toxicity in the rat brain: histochemical and immunocytochemical evidence. Brain Res. Bull. 55, 257–267.

Pogue, A.I., Lukiw, W.J., 2016. Aluminum, the genetic apparatus of the human CNS and Alzheimer's disease (AD). Morphologie 100, 56–64.

Poirier, J., Semple, H., Davies, J., Lapointe, R., Dziwenka, M., Hiltz, M., Mujibi, D., 2011. Double-blind, vehicle-controlled randomized twelve-month neurodevelopmental toxicity study of common aluminum salts in the rat. Neuroscience 193, 338–362.

Polizzi, S., Pira, E., Ferrara, M., Bugiani, M., Papaleo, A., Albera, R., Palmi, S., 2002. Neurotoxic effects of aluminum among foundry workers and Alzheimer's disease. Neurotoxicology 23, 761–774.

Pratico, D., Uryu, K., Sung, S., Tang, S., Trojanowski, J.Q., Lee, V.M., 2002. Aluminum modulates brain amyloidosis through oxidative stress in APP transgenic mice. FASEB J. 16, 1138–1140.

Priest, N.D., Talbot, R.J., Newton, D., Day, J.P., King, S.J., Fifield, L.K., 1988. Uptake by man of aluminum in a public water supply. Hum. Exp. Toxicol. 17, 296–301.

Qin, L., Wu, X., Block, M.L., Liu, Y., Breese, G.R., Hong, J.S., Knapp, D.J., Crews, F.T., 2007. Systemic LPS causes chronic neuroinflammation and progressive neurodegeneration. Glia 55, 453–462.

Reed, S.G., Orr, M.T., Fox, C.B., 2013. Key roles of adjuvants in modern vaccines. Nat. Med. 19, 1597–1608.

Reusche, E., Pilz, P., Oberascher, G., Lindner, B., Egensperger, R., Gloeckner, K., Trinka, E., Iglseder, B., 2001. Subacute fatal aluminum encephalopathy after reconstructive otoneurosurgery: a case report. Hum. Pathol. 32, 1136–1140.

Ribes, D., Torrente, M., Vicens, P., Colomina, M.T., Gómez, M., Domingo, J.L., 2012. Recognition memory and β-amyloid plaques in adult Tg2576 mice are not modified after oral exposure to aluminum. Alzheimer Dis. Assoc. Disord. 26, 179–185.

Rifat, S.L., Eastwood, M.R., McLachlan, D.R., Corey, P.N., 1990. Effect of exposure of miners to aluminum powder. Lancet 336, 1162–1165.

Rondeau, V., Jacqmin-Gadda, H., Commenges, D., Helmer, C., Dartigues, J.F., 2009. Aluminum and silica in drinking water and the risk of Alzheimer's disease or cognitive decline: findings from 15-year follow-up of the PAQUID cohort. Am. J. Epidemiol. 169, 489–496.

Ruipérez, F., Mujika, J.I., Ugalde, J.M., Exley, C., Lopez, X., 2012. Pro-oxidant activity of aluminum: promoting the Fenton reaction by reducing Fe(III) to Fe(II). J. Inorg. Biochem. 117, 118–123.

Russo, L.S., Beale, G., Sandroni, S., Ballinger, W.E., 1992. Aluminum intoxication in undialysed adults with chronic renal failure. J. Neurol. Neurosurg. Psychiatry 155, 697–700.

Santibáñez, M., Bolumar, F., García, A.M., 2007. Occupational risk factors in Alzheimer's disease: a review assessing the quality of published epidemiological studies. Occup. Environ. Med. 64, 723–732.

Schoenfeld, Y., Agmon-Levin, N., 2011. 'ASIA' autoimmune/inflammatory syndrome induced by adjuvants. J. Autoimmun. 36, 4–8.

Selley, M.L., 2005. Simvastatin prevents 1-methyl-4-phenyl-1,2,3,6-tetrahydropyridine-induced striatal dopamine depletion and protein tyrosine nitration in mice. Brain Res. 1037, 1–6.

Shah, S.A., Yoon, G.H., Ahmad, A., Ullah, F., Amin, F.U.I., Kim, M.O., 2015. Nanoscale-alumina induces oxidative stress and accelerates amyloid beta (Aβ) production in ICR female mice. Nanoscale 7, 15225.

Sharman, E., Sharman, K.G., Lahiri, D.K., Bondy, S.C., 2004. Age-related changes in murine CNS mRNA gene expression are modulated by dietary melatonin. J. Pineal Res. 36, 165–170.

Sharman, E.H., Sharman, K.Z., Bondy, S.C., 2008. Melatonin causes gene expression in aged animals to respond to inflammatory stimuli in a manner differing from that of young animals. Curr. Aging Sci. 1, 152–158.

Shaw, C.A., Petrik, M.S., 2009. Aluminum hydroxide injections lead to motor deficits and motor neuron degeneration. J. Inorg. Biochem. 103, 1555–1562.

Shaw, C.A., Li, Y., Tomljenovic, L., 2013. Administration of aluminium to neonatal mice in vaccine-relevant amounts is associated with adverse long term neurological outcomes. J. Inorg. Biochem. 128, 237–244.

Sherrard, D.J., Walker, J.V., Boykin, J.L., 1988. Precipitation of dialysis dementia by deferoxamine treatment of aluminum-related bone disease. Am. J. Kidney Dis. 12, 126–130.

Shi, L., Fatemi, S.H., Sidwell, R.W., Patterson, P.H., 2003. Maternal influenza infection causes marked behavioral and pharmacological changes in the offspring. J. Neurosci. 23, 297–302.

Shin, R.W., Kruck, T.P., Murayama, H., Kitamoto, T., 2003. A novel trivalent cation chelator Feralex dissociates binding of aluminum and iron associated with hyperphosphorylated tau of Alzheimer's disease. Brain Res. 961, 139–146.

Shirabe, T., Irie, K., Uchida, M., 2002. Autopsy case of aluminum encephalopathy. Neuropathology 22, 206–210.

Sigel, H., Sigel, A. (Eds.), 1988. Aluminum and its role in biology. In: Metal Ions in Biological Systems, vol. 24. Marcel Dekker, New York.

Sparling, D.W., Campbell, P.G.C., 1997. Ecotoxicology of aluminum to fish and wildlife. In: Yokel, R.A., Golub, M.S. (Eds.), Research Issues in Aluminum Toxicity. Taylor and Francis, Washington, DC, pp. 48–68.

Styren, S.D., Kamboh, M.I., Dekosky, S.T., 1998. Expression of differential immune factors in temporal cortex and cerebellum: the role of α-1-antichymotrypsin, apolipoprotein E, and reactive glia in the progression of Alzheimer's disease. J. Comp. Neurol. 396, 511–520.

Sun, Y.X., Minthon, L., Wallmark, A., Warkentin, S., Blennow, K., Janciauskiene, S., 2003. Inflammatory markers in matched plasma and cerebrospinal fluid from patients with Alzheimer's disease. Dement. Geriatr. Cogn. Disord. 16, 136–144.

Sutton, I., Lahoria, R., Tan, I.L., Clouston, P., Barnett, M.H., 2009. CNS demyelination and quadrivalent HPV vaccination. Mult. Scler. 15, 116–119.

Tomljenovic, L., 2011. Aluminum and Alzheimer's disease: after a century of controversy, is there a plausible link? J. Alzheimers Dis. 23, 567–598.

Uversky, V.N., Li, J., Fink, A.L., 2001. Metal-triggered structural transformations, aggregation, and fibrillation of human alpha-synuclein. A possible molecular link between Parkinson's disease and heavy metal exposure. J. Biol. Chem. 276, 44284–44296.

Wållberg, M., Wefer, J., Harris, R.A., 2003. Vaccination with myelin oligodendrocyte glycoprotein adsorbed to alum effectively protects DBA/1 mice from experimental autoimmune encephalomyelitis. Eur. J. Immunol. 33, 1539–1547.

Walton, J.R., 2009a. Brain lesions comprised of aluminum rich cells that lack microtubules may be associated with the cognitive deficit of Alzheimer's disease. Neurotoxicology 30, 1059–1069.

Walton, J.R., 2009b. Functional impairment in aged rats chronically exposed to human range dietary aluminum equivalents. Neurotoxicology 30, 182–193.

Walton, J.R., 2012. Cognitive deterioration and associated pathology induced by chronic low-level aluminum ingestion in a translational rat model provides an explanation of Alzheimer's disease, tests for susceptibility and avenues for treatment. Int. J. Alzheimers Dis. 2012, 914947.

Walton, J.R., 2014. Chronic aluminum intake causes Alzheimer's disease: applying Sir Austin Bradford Hill's causality criteria. J. Alzheimers Dis. 40, 765–838.

Walton, J.R., Wang, M.X., 2009. APP expression, distribution and accumulation are altered by aluminum in a rodent model for Alzheimer's disease. J. Inorg. Biochem. 103, 1548–1554.

Wang, L., Hu, J., Zhao, Y., Lu, X., Zhang, Q., Niu, Q., 2014. Effects of aluminium on β-amyloid (1-42) and secretases (APP-cleaving enzymes) in rat brain. Neurochem. Res. 39, 1338–1345.

Xu, N., Majidi, V., Markesbery, W.R., Ehmann, W.D., 1992. Brain aluminum in Alzheimer's disease using an improved GFAAS method. Neurotoxicology 13, 735–743.

Yokel, R.A., O'Callaghan, J.P., 1998. An aluminum-induced increase in GFAP is attenuated by some chelators. Neurotoxicol. Teratol. 20, 55–60.

Yokel, R.A., Rhineheimer, S.S., Sharma, P., Elmore, D., McNamara, P.J., 2001. Entry, half-life, and desferrioxamine-accelerated clearance of brain aluminum after a single (26)Al exposure. Toxicol. Sci. 64, 77–82.

Yumoto, S., Kakimi, S., Ohsaki, A., Ishikawa, A., 2009. Demonstration of aluminum in amyloid fibers in the cores of senile plaques in the brains of patients with Alzheimer's disease. J. Inorg. Biochem. 103, 1579–1584.

Zatta, P., Zambenedetti, P., Milanese, M., 1999. Activation of monoamine oxidase type-B by aluminum in rat brain homogenate. Neuroreport 10, 3645–3648.

Zhao, M., Cribbs, D.H., Anderson, A.J., Cummings, B.J., Su, J.H., Wasserman, A.J., Cotman, C.W., 2003. The induction of the TNF alpha death domain signaling pathway in Alzheimer's disease brain. Neurochem. Res. 28, 307–318.

Zhao, Y., Hill, J.M., Bhattacharjee, S., Percy, M.E., Pogue, A.I., Lukiw, W.J., 2014. Aluminum-induced amyloidogenesis and impairment in the clearance of amyloid peptides from the central nervous system in Alzheimer's disease. Front. Neurol. 5, 167.

FURTHER READING

Lukiw, W.J., 2010. Evidence supporting a biological role for aluminum in chromatin compaction and epigenetics. J. Inorg. Biochem. 104, 1010–1012.

Smith, R.W., 1996. Kinetic aspects of aqueous aluminum chemistry: environmental implications. Coord. Chem. Rev. 149, 81–93.

CHAPTER SIX

Manganese Neurodegeneration

David C. Dorman[1]

College of Veterinary Medicine, North Carolina State University, Raleigh, NC, United States
[1]Corresponding author: e-mail address: david_dorman@ncsu.edu

Contents

Advances in Neurotoxicology, Volume 1
ISSN 2468-7480
http://dx.doi.org/10.1016/bs.ant.2017.07.007

1. INTRODUCTION

Manganese is ubiquitous and is naturally found in the air, soil, and water. Manganese is critical to the production of steel and other alloys, with steel production accounting for >90% of the global demand for this metal (U.S. Geological Survey, 2008). Other anthropogenic sources of manganese include welding and metal working, pulp, and paper manufacturing, pesticide use, brick coloration, steel-on-steel abrasion such as subway rails, wheels, cables, and brakes, and use of the octane-enhancing fuel additive methylcyclopentadienyl manganese tricarbonyl (ATSDR, 2000; Chillrud et al., 2005).

Concern has arisen regarding potential exposure to high concentrations of inhaled manganese in the environment due to natural and anthropogenic sources, leading to numerous risk assessments of the compound (ATSDR, 2000; U.S. EPA, 1993; WHO, 2001; Wood and Egyed, 1994). This chapter's initial discussion of manganese neurotoxicity considers the essentiality and pharmacokinetics of manganese since neurotoxicity occurs as a result of much higher exposure than required for normal cell function. The chapter also discusses early literature of manganese neurotoxicity and briefly discusses more contemporary studies that demonstrate neurological effects at lower occupational or environmental exposure. No attempt has been made to provide an exhaustive review of each of the topics addressed, but wherever possible references to review articles are provided.

2. MANGANESE ESSENTIALITY

Manganese is an essential nutrient and as such is present in all tissues including the brain (Aschner et al., 2005; ATSDR, 2000; Sumino et al., 1975; Tipton and Cook, 1963). The body's nutritional requirements for manganese are normally met through dietary intake (food and drinking water), especially whole grains, nuts, leafy vegetables, teas, and other foods enriched in manganese. Manganese is an important enzyme cofactor and is incorporated into various metalloproteins that contribute to normal cell function. For example, manganese is found in manganese superoxide dismutase, glutamine synthetase, arginase, and pyruvate carboxylase (ATSDR, 2000). Manganese deficiency results in decreased growth rates, congenital abnormalities, impaired reproductive performance, ataxia, and abnormal lipid and carbohydrate metabolism (Cotzias, 1958; Hurley,

1981; Takeda, 2003; Wedler and Denman, 1984). The estimated safe and adequate daily dietary intakes (ESADDI) for manganese required to maintain body stores are 0.003–0.06 mg/day in infants, 1.2–2.2 mg/day in children, and 1.6–2.6 mg/day in adults (IOM, 2001). To ensure adequate nutrition in neonates, manganese is often added to infant formula because of their greater need for this element during early growth and development (Aschner and Aschner, 2005).

3. EXPOSURE SOURCES

Neurotoxicity can occur following high dose oral, inhalation, or parenteral exposure to manganese. The development of neurotoxicity following different routes of exposure indicates that the dose to target tissue is the critical determinant of manganese toxicity, regardless of route.

3.1 Inhalation

Air manganese concentrations in rural and urban areas range from 0.01 to 0.07 $\mu g/m^3$ due to the natural weathering of rocks and entrainment of dust in air (ATSDR, 2000; Pellizzari et al., 1999). High dose ($\geq$100 $\mu g/m^3$) inhalation of manganese dust occurs primarily in mining, ore-crushing, and metallurgical operations for iron, steel, ferrous, and nonferrous alloys (ATSDR, 2000). Manufacturing of dry-cell batteries, antiknock gasoline additives, pesticides (e.g., maneb), pigments, dyes, inks, and incendiary devices can also lead to occupational manganese inhalation. Manganese fumes are also produced during certain metallurgical and welding operations.

3.2 Drinking Water, Breastmilk, and Soy-Based Infant Formulas

Water concentrations of manganese typically range from 1 to 100 $\mu g/L$, with most values below 10 $\mu g/L$ (Keen and Zidenberg-Cherr, 1994). The total estimated daily manganese intake in adults from food and drinking water is approximately 115 μg (ATSDR, 2000; Santamaria and Sulsky, 2010). Manganese concentrations in human breast milk can vary from approximately 2 to 10 $\mu g/L$ (Stasny et al., 1984). Much higher manganese concentrations (600 $\mu g/L$) can be found in certain soy-based infant formulas, although health risks associated with these levels remain a topic of debate (Crinella, 2012). On occasion manganese poisoning may arise from high-dose ingestion. For example, Kawamura et al. (1941) and Kondakis et al. (1989) documented outbreaks of manganism in Japan and Greece due to the ingestion of water from wells that were contaminated with extremely high levels of manganese (1.8–14 mg Mn/L).

3.3 Total Parenteral Nutrition (TPN)

Neurological effects and changes in brain magnetic resonance imaging (MRI) consistent with manganese accumulation have also been reported for individuals receiving manganese intravenously as a constituent in TPN (Aschner et al., 2015; Iinuma et al., 2003; Santos et al., 2014). Intravenously administered manganese largely bypasses homeostatic mechanisms regulating manganese absorption. Exposure doses resulting from TPN administration can range from 0.01 to 2.2 mg/day (Iinuma et al., 2003; Wretlind, 1972) with adult exposures of >500 µg/day accounting for most cases of increased brain manganese concentrations (Santos et al., 2014).

3.4 Other Sources

There are several medical applications for manganese. Manganese is paramagnetic and is used as an MRI contrast enhancer (reviewed in Dorman, 2017). Manganese-enhanced MRI (MEMRI) has emerged as an important means to evaluate neuronal function and networks (Inoue et al., 2011; Malheiros et al., 2015). Manganese is also found in the herbicide manganese ethylenebis-dithiocarbamate (maneb). Some impure free-base forms of cocaine (Bazooka) have been contaminated with manganese-carbonate (Ensing, 1985).

4. MANGANESE PHARMACOKINETICS

4.1 Overview

Manganese is present in all tissues with some tissue-to-tissue variability in background manganese concentration. Tissue manganese concentrations reflect developmental stage (neonates are more susceptible), tissue intake, and compensatory mechanisms that restrict manganese intake or retention (e.g., changes in gastrointestinal absorption or hepatobiliary excretion). Thus, at low levels of exposure that modestly exceed typical dietary intake, manganese does not accumulate in the brain or other tissues beyond these normal basal levels. At moderate exposures, homeostatic mechanisms (primarily a compensatory increase in biliary manganese excretion) limit increases in manganese in brain and other target tissues and normal concentrations are often maintained. These homeostatic mechanisms can however become overwhelmed at higher exposures (e.g., repeated inhalation of $\geq 0.1\,mg\,Mn/m^3$) and increased brain and other tissue manganese concentrations result.

A variety of animal pharmacokinetic studies have evaluated the relationship between manganese exposure and tissue manganese concentrations (reviewed in Dorman et al., 2012). These studies show that there are three main mechanisms by which manganese gains access to the brain and other organs: (a) absorption from the gastrointestinal tract with subsequent systemic distribution from blood; (b) absorption across the pulmonary epithelial lining with systemic delivery; and/or (c) direct anterograde axonal delivery via olfactory or trigeminal nerve endings located in the nasal cavity. Schroeter et al. (2012) evaluated short term to chronic manganese exposure of monkeys by inhalation, oral, intravenous, intraperitoneal, and subcutaneous routes. This analysis confirmed that the dose–response relationship for the neurotoxic effects of manganese in monkeys was essentially independent of exposure route.

4.2 Absorption

Gastrointestinal absorption of manganese is under tight homeostatic control (Teeguarden et al., 2007; Yoon et al., 2011). When dietary manganese levels are high, adaptive changes include reduced gastrointestinal absorption of manganese (Abrams et al., 1976; Aschner and Aschner, 2005). Manganese absorption in adults is relatively low (1%–2%) while intestinal neonatal absorption of manganese is much higher (70%–80%) in neonates (ATSDR, 2000).

Multiple transporters, including the divalent metal transporter 1 (DMT1), zinc transporters, dopamine transporter, transferrin receptor (TfR), and calcium channels are involved in the cellular entry of manganese (Chen et al., 2014, 2015; Conrad et al., 2000; Forbes and Gros, 2003). Decreased DMT1 function, as occurs in Belgrade rats for example, results in reduced manganese transport (Chua and Morgan, 1997; Veuthey and Wessling-Resnick, 2014). Other studies however, suggest that DMT1 only plays a role in small intestine uptake of manganese (Canonne-Hergaux et al., 1999; Gunshin et al., 1997; Knöpfel et al., 2005; Roth and Garrick, 2003; Roth et al., 2002; Trinder et al., 2000).

Manganese is absorbed from the lung with more soluble forms (e.g., manganese sulfate) resulting in higher brain manganese concentrations when compared with less-soluble phosphate or oxide forms (Dorman et al., 2001, 2004, 2012; Normandin et al., 2004; Salehi et al., 2003). Like the gastrointestinal epithelium, pulmonary epithelial cells also express DMT1, which plays an important role in iron regulation in the lung. Transferrin binding to manganese (as Mn^{3+}) and subsequent endocytosis by transferrin receptors found on the apical surface of pulmonary epithelial cells may also be involved in manganese clearance from the airway and alveolar fluid (Heilig et al., 2006). Interestingly, experiments with Belgrade rats indicate that impaired DMT1 activity does not alter pulmonary ^{54}Mn absorption.

Manganese (as Mn^{2+}) has a similar ionic radius to that of calcium (Ca^{2+}) and enters neurons through voltage-gated calcium channels (Narita et al., 1990). This observation has prompted studies evaluating whether calcium channels may play a role in the uptake of manganese into heart, liver, and brain (Heilig et al., 2006).

4.3 Distribution

As an essential metal manganese is found in most tissues. Normal manganese concentrations in many tissues often range from 0.1 to 0.5 µg Mn/g tissue wet weight with higher (1–2.5 µg Mn/g) basal levels seen in the kidney, pancreas, and liver (Dorman et al., 2006b). Normal manganese levels in body fluids often range from 4 to 15 µg/L in blood, 1 to 8 µg/L, in urine, and 0.4 to 0.85 µg/L in serum (ATSDR, 2000). As mentioned earlier, high-dose exposure to manganese can overwhelm homeostatic controls and result in a variable degree of tissue manganese concentration increase. For example, a 10-fold increase (vs air-exposed controls) in mean olfactory bulb manganese concentration was observed in monkeys exposed subchronically (65 exposure days) to manganese sulfate at $1.5\,mg\,Mn/m^3$ (Dorman et al., 2006b). In contrast, a threefold to fivefold increase globus pallidus, putamen, and caudate was seen in these monkeys, whereas manganese concentrations in the frontal cortex and cerebellum had less than a threefold increase in manganese concentration. Other tissues including the heart, kidney, liver, bone, pancreas, and skeletal muscle also had increased manganese concentrations following high-dose manganese inhalation (Dorman et al., 2006b). Pseudo steady-state manganese concentrations occur relatively quickly in the rat and monkey striatum, olfactory bulb, and cerebellum following high-dose manganese inhalation (Dorman et al., 2004, 2006b).

Recent in vivo studies suggest that dopamine active transporter (DAT) plays an important role in manganese accumulation in the striatum (Erikson et al., 2005). Inhibition of DAT function in weanling male Sprague–Dawley rats attenuates manganese accumulation in the globus pallidus during chronic exposure (Anderson et al., 2007).

4.4 Elimination

Exposures to increased levels of manganese results in dose-dependent biliary and pancreatic excretion of manganese (Aschner and Dorman, 2006). Upregulation of biliary excretion occurs with increasing manganese intake, independent of dose route (Dorman et al., 2012). Manganese elimination in bile may be due in part to active transport (Crossgrove and Yokel, 2004). The solute carrier 30A10 (SLC30A10) has recently been identified as a putative manganese transporter. An inheritable SLC30A10 deficiency leads to manganese accumulation in the liver and a parkinsonism syndrome in people (Quadri et al., 2012; Tuschl et al., 2013). Other transporters involved in cellular manganese efflux include ATPase 13A2 (ATP13A2), ferroportin, and Ca^{2+}-ATPase 1 (SPCA1) (Chen et al., 2014). Intestinal excretion of manganese is inducible following manganese exposure (Bertinchamps et al., 1966). Manganese demonstrates dose-dependent differences in elimination half times. For example, terminal blood half-time in Chilean manganese-exposed miners was 15 ± 2 days, compared to 37.5 ± 7.5 days for control individuals (Cotzias et al., 1968). Manganese elimination from the brain occurs with an apparent elimination half-life on the order of 45 days or less in animals (Dorman et al., 2004, 2006b).

4.5 Olfactory Transport of Manganese

Inhaled manganese can undergo dose-dependent direct "nose-to-brain" transport (Henriksson et al., 1999; Tjälve and Henriksson, 1999). The propensity of manganese to undergo olfactory transport may contribute to changes in the sense of smell and other neurologic effects (Lucchini et al., 2012a). Interestingly, changes in olfactory function have been seen in people following high-dose manganese exposure and are also reported in individuals with Parkinson's disease and other forms of neurodegeneration (Dorman, 2015).

Initial olfactory transport studies relied on instillation of radiolabeled manganese chloride ($^{54}MnCl_2$) into the nasal cavity of freshwater pike and other animals (Tjälve et al., 1996). Subsequent studies in rats showed that direct delivery of manganese along the olfactory route accounted for nearly all of the ^{54}Mn found in the olfactory bulb and olfactory tract following acute manganese inhalation (Brenneman et al., 2000; Elder et al., 2006;

Fechter et al., 2002). Cross et al. (2004) used unilateral intranasal manganese instillation in rats and brain MRI to demonstrate manganese transport to the olfactory bulb, lateral olfactory tract, and olfactory tubercle ipsilateral to the site of manganese administration. Dorman et al. (2006c) also used brain MRI and measurement of tissue manganese concentrations to demonstrate presumed olfactory transport of inhaled manganese sulfate in rhesus monkeys. Visual evaluation of the monkey's manganese-enhanced neuronal tracts illustrated MRI changes in the olfactory bulb but failed to demonstrate evidence for direct translocation of manganese from the olfactory bulb to the globus pallidus (Dorman et al., 2006c). Sen et al. (2011) used brain MRI to demonstrate that manganese-exposed welders had evidence of manganese accumulation in the olfactory bulb, frontal white matter, globus pallidus, and putamen. These MRI studies do not, however, provide direct evidence that olfactory transport of manganese occurs in humans or other primates.

A working theory to account for olfactory transport postulates that manganese is transported intracellularly within the olfactory sensory neurons (Leavens et al., 2007). Neuronal uptake may occur by passive diffusion, receptor-mediated uptake, or adsorptive endocytosis, followed by axonal transport, which takes several hours to days for manganese to reach the olfactory bulb and other regions of the central nervous system (CNS; Leavens et al., 2007). Thompson et al. (2007) showed that DMT1 may play a role in the olfactory transport of manganese. Pharmacokinetic models indicate a small rate of transfer between the olfactory tract and striatum suggested that delivery of manganese to the striatum from the olfactory transport is minor (<3%) when compared to systemic delivery of manganese in the brain (Leavens et al., 2007).

4.6 Fetal and Neonatal Exposure

Concerns for potential vulnerability to manganese neurotoxicity during fetal and neonatal development have also been raised. Childhood risk factors include higher intestinal absorption of ingested manganese, a lower basal hepatobiliary excretion rate, enhanced delivery of manganese to the neonatal brain, and the use of manganese-supplemented TPN solutions in some infants (Yoon et al., 2011). The rat placenta and fetal liver effectively sequesters manganese limiting delivery to the fetal brain (Dorman et al., 2005b). Pregnant and neonatal rats exposed to comparable levels of inhaled manganese develop qualitatively similar brain manganese concentrations when compared to young male rats, nonpregnant female rats, and senescent rats (Dorman et al., 2004, 2005a, 2012).

4.7 Physiologically Based Pharmacokinetic (PBPK) Models for Manganese

Several PBPK models for manganese have been developed (reviewed in Taylor et al., 2012). These PBPK models effectively simulate manganese tissue kinetics from inhaled, oral, and parenteral manganese intake and represent an important tool for manganese risk assessment (Andersen et al., 2010). This multidose route capability is achieved by incorporating homeostatic control processes, saturable tissue binding capacities, and preferential fluxes in various tissue regions (see Taylor et al., 2012). Available PBPK models have also been used to predict the relationship between brain manganese concentration and neurotoxicity in animals (Schroeter et al., 2012) and offer an important refinement on relying on clinical sign data alone (Gwiazda et al., 2007). These models predict that increased neurologic sign severity occurs in monkeys when peak globus pallidus manganese concentrations are increased twofold to fourfold above normal basal levels (Schroeter et al., 2012). The PBPK models also showed that the dose–response relationship for manganese neurotoxicity in monkeys was independent of exposure route and supports the use of brain manganese concentration as an appropriate dose metric.

5. BRAIN MRI

5.1 Manganese and Brain MRI

Since manganese is paramagnetic it increases the signal intensity seen with T1-weighted MRI (reviewed in Dorman, 2017). Changes in the T1-weighted image correlate with manganese tissue concentration (Dorman et al., 2006c) and allow for identification of tissue sites where manganese accumulates. Brain MRI studies of highly exposed people reveal signal changes in the globus pallidus, striatum, and midbrain consistent with manganese accumulation at these sites (Dorman, 2017; Fitsanakis et al., 2006; Tuschl et al., 2013). Some investigators use MRI to calculate a pallidal index, which is the ratio of the signal intensity in the globus pallidus divided by the signal intensity in the frontal white matter (Baker et al., 2015; Kim et al., 2005; Li et al., 2014; Shin et al., 2007). Studies in monkeys demonstrate that the pallidal index is positively correlated with postmortem globus pallidus manganese concentration (Dorman et al., 2006c). However, increases in white matter manganese concentrations were also seen in manganese-exposed monkeys suggesting that the pallidal index may underestimate globus pallidus manganese accumulation (Dorman et al., 2006c).

Changes in T1 relaxation rate (R1) in manganese-exposed monkeys has been shown to be a better predictor of pallidal manganese concentration when compared with the pallidal index (Dorman et al., 2006c). Similar approaches to evaluate R1 in exposed welders have been used (Lee et al., 2015, 2016; Lewis et al., 2016). Changes in MRI signal intensity can be evaluated longitudinally in manganese workers (Han et al., 2014). Brain MRI studies have also been performed in children. Aschner et al. (2015) reported that infants with greater parenteral manganese exposure had shorter T1 relaxation times, consistent with manganese accumulation, in the basal ganglia when compared with infants with lower parenteral exposures. Children with a SLC39A14 gene defect also develop MRI changes consistent with manganese accumulation in the brain (Dion et al., 2016; Tuschl et al., 2016). Dion et al. (2016) however were unable to detect increased MRI signal intensity in the globus pallidus of children exposed to high (145 µg/L) levels of manganese in drinking water when compared with children consuming a lower level (1 µg/L). Longitudinal MRI studies have shown decreased pallidal hyperintensity in people and monkeys following cessation of manganese exposure (Kim, 2004).

6. MANGANESE NEUROTOXICITY

6.1 Manganism

An association between manganese inhalation and neurotoxicity was first recognized in the mid-19th century in workers at an ore-grinding plant where "black oxide of manganese" was processed (Couper, 1837). Most epidemiologic research on manganese conducted during the late-20th century focused on occupational inhalation exposure. These studies have established a causal association between chronic high dose ($>1 \, mg/m^3$) manganese inhalation and an overt neurotoxic syndrome known as manganism (Cotzias, 1958; Lee, 2000; Lucchini et al., 2009; Pal et al., 1999). The onset of neurological signs in manganese workers may be delayed for several years after exposure (Calne et al., 1994; Huang et al., 1993). Early clinical signs may include emotional instability, compulsive or violent behavior, hallucinations, and other psychiatric symptoms and is referred to as manganese madness or locura manganica (Bouabid et al., 2015). Later hallmarks of manganism include progressive behavioral changes, motor dysfunction including bradykinesia, dystonia, and gait abnormalities, and neurochemical and neuropathological changes in the basal ganglia (Bouabid et al., 2015, Roels et al., 2012; Roth, 2006). These changes are associated with manganese accumulation within the human striatum, globus pallidus, and substantia nigra (Calne et al., 1994; Nagatomo et al., 1999).

6.2 Contemporary Occupational and Environmental Exposures

Subtle and less-consistent neurological effects have been reported in welders and other workers with prolonged inhalation exposures to lower concentrations ($<0.2\,mg/m^3$) of manganese in the workplace (Bast-Pettersen et al., 2004; Beuter et al., 1994; Iregren, 1990; Lucchini et al., 1995, 1999; Mergler et al., 1994; Myers et al., 2003; Park et al., 2009; Roels et al., 1987, 1992, 1999; Young et al., 2005). Manganese-induced cognitive deficits have also been reported in workers with exposure to manganese-based welding fumes (Bowler et al., 2006; Chang et al., 2010; Ellingsen et al., 2008).

Neurological effects have also been reported in environmentally exposed people (Finkelstein and Jerrett, 2007; Lucchini et al., 2007, 2012b; Mergler et al., 1999; Rodríguez-Agudelo et al., 2006). As mentioned earlier, neurotoxicity has been observed in people that consumed drinking water containing high concentrations of manganese (Kondakis et al., 1989; Ljung and Vahter, 2007). Iwami et al. (1994) reported an increased incidence of motor neuron disease in people consuming food with a high manganese content and drinking water with a low magnesium concentration. Decreased olfactory performance has been seen in people living within a manganese mining district when compared with nonexposed subjects living 50 km from the closest source of exposure (Guarneros et al., 2013).

There has also been increasing public health concern regarding the role of environmental manganese exposure and children's health (Sanders et al., 2015; Zoni and Lucchini, 2013). Several community studies have suggested a link between manganese content in drinking water and decreased IQ in children (Bouchard et al., 2011; Khan et al., 2012; Wasserman et al., 2006). Other studies have failed to demonstrate an association between manganese exposure, as assessed using metal concentrations in deciduous teeth, and behavioral deficits in children (Chan et al., 2015). The strength of the associations seen between manganese and changes in IQ and other measures of cognitive performance in children therefore remains uncertain (Lucchini et al., 2009).

6.3 Neurotoxic Mechanisms

Several modes of action have been proposed to explain manganese neurotoxicity. For example, manganese exists in several oxidative states; it facilitates redox reactions; and it may contribute to increased formation of free radicals in the CNS (Erikson et al., 2007; Farina et al., 2013). Manganese plays a role in glutamate metabolism, in particular, manganese is a cofactor for the oxidoreductase, glutamate synthetase, which acts to catalyze glutamine from glutamate (Michalke, 2016). Excess manganese may alter glutamate homeostasis in the basal ganglia (Sidoryk-Wegrzynowicz and Aschner, 2013). Changes in glutamate homeostasis have been associated with excitotoxicity in the CNS (Sidoryk-Wegrzynowicz and Aschner, 2013). Changes in gene expression also occur following manganese exposure, for example, changes in expression of glutamate transporter 1 (GLT-1) and glutamate aspartate transporter (GLAST) may contribute to manganese neurotoxicity (Karki et al., 2015). Monkeys exposed subchronically to inhaled manganese had decreased olfactory cortical GLT-1 and GLAST protein levels (Erikson et al., 2007). Welders exposed to manganese have changes in methylation patterns in the DNA that codes for inducible nitric oxide synthase (Searles Nielsen et al., 2015). To date, the molecular initiating events associated with manganese neurotoxicity remain incompletely understood.

6.4 Neurochemical Changes

Symptoms seen in people with manganese neurotoxicity can be attributed in part to changes in basal ganglia neurochemistry (reviewed in Bouabid et al., 2015; Guilarte, 2013). To date, changes in dopamine function have garnered the most extensive research focus (Erikson et al., 2004; Yamada et al., 1986), however, growing evidence indicates other neurotransmitter systems are also affected (Bouabid et al., 2015). For example, changes in striatal γ-aminobutyric acid (GABA), norepinephrine, and serotonin function are seen following manganese exposure in rodents (Anderson et al., 2008, 2009; Bouabid et al., 2014; Fordahl et al., 2010; Sidoryk-Wegrzynowicz, 2014). Changes in brain neurochemistry have also been reported following intranasal exposure. Intranasal instillation of manganese causes a dose-dependent reduction in odorant-evoked glutamatergic neurotransmitter release in mice as a result of centrally mediated effects (Moberly et al., 2012).

6.5 Neuropathology

The primary neuropathological target of manganism is the globus pallidus (particularly the internal segment) with sparing of the substantia nigra pars compacta and an absence of Lewy bodies (reviewed in Gonzalez-Cuyar et al., 2014; Pal et al., 1999; Pentschew et al., 1963; Perl and Olanow, 2007; Yamada et al., 1986). Chronic manganism in people is associated with decreased GABA neurons, reduced myelinated fibers, and moderate astrocytic proliferation in the medial segment of the globus pallidus (Yamada et al., 1986). Other lesions seen in manganese miners include increased numbers of microglia in the external and internal segments of the globus pallidus (Gonzalez-Cuyar et al., 2014). Banta and Markesbery (1977) reported neuritic plaques and neurofibrillary tangles in the cortical frontal lobe of a patient who displayed dementia and extrapyramidal syndrome secondary to manganese exposure.

Neuropathological changes have also been seen in manganese-exposed animals. Manganese exposure results in neurodegeneration in the mouse globus pallidus and striatum (Liu et al., 2006). Similarly, Eriksson et al. (1987) found reduced numbers of pallidal neurons in manganese-exposed monkeys. Villalobos et al. (2009) showed that the mouse olfactory bulb develops neuron degeneration and myelin sheath disorganization following high-dose intraperitoneal manganese injection. Neuronal cell loss was observed in the globus pallidus and the caudate putamen in rats exposed to manganese by inhalation (Salehi et al., 2006). Colin-Barenque et al. (2011) reported that manganese inhalation was associated with ultrastructural changes including disruption of organelle membranes, cytoplasm vacuolation, increased lipofuscin deposits, and cell death in the olfactory bulb granule cell. Neuronal loss was not seen in rats following intranasal administration of manganese despite the presence of decreased glial fibrillary acidic protein (GFAP) and S-100b in the olfactory cortex, hypothalamus, thalamus, and hippocampus (Henriksson and Tjälve, 2000).

6.6 Species Differences

Significant differences have been widely documented regarding the species-specific sensitivity to manganese neurotoxicity (Aschner et al., 2005; Dorman, 2015; Gwiazda et al., 2007; Newland, 1999). Rodents generally fail to develop a behavioral syndrome or neuropathological lesion comparable to that seen in manganese-poisoned humans. In contrast, nonhuman primates best replicate the neurotoxic effects observed in humans.

Manganese-exposed monkeys develop MRI changes consistent with elevated brain manganese concentrations localized to the striatum, globus pallidus, and substantia nigra (Dorman et al., 2006c; Guilarte et al., 2006; Park et al., 2007; Sung et al., 2007) and gait and other motor abnormalities that mimic those observed in affected humans (Guilarte, 2013; Kim et al., 2013; Olanow et al., 1996). Monkeys also develop reduced levels of striatal and pallidal dopamine and 3,4-dihydroxyphenylacetic acid in conjunction with loss of dopaminergic neurons (Bird et al., 1984; Neff et al., 1969; Struve et al., 2007) without PET imaging apparent loss of dopamine terminals in the caudate and putamen (Guilarte et al., 2006, 2008). Monkeys also develop deficits in spatial and nonspatial working memory as well as effects on visuospatial-paired associate learning (Schneider et al., 2006, 2009, 2013). Histological assessment of the frontal cortex from manganese-exposed monkeys has also shown the presence of cells with apoptotic stigmata and astrocytosis in both the gray and white matter and α-synuclein aggregation in the frontal cortex gray and white matter (Verina et al., 2013). Neuropathologic description of lesions in the globus pallidus of manganese-exposed monkeys is largely lacking (Ulrich et al., 1979) although biomarkers consistent with oxidative stress are present at this location in monkeys following manganese inhalation (Erikson et al., 2007).

7. MANGANESE AND OTHER NEURODEGENERATION SYNDROMES

7.1 Parkinson's Disease and Manganese

The hallmark clinical features of Parkinson's disease include resting tremor, bradykinesia, postural instability, and other signs of motor dysfunction. Motor symptoms generally appear when greater than 60% of dopaminergic neurons are lost and dopamine content in the striatum is reduced to below 20% of normal levels (Jankovic, 2008; Wirdefeldt et al., 2011). Loss of dopaminergic neurons occurs with aging and the prevalence of Parkinson's disease increases dramatically in elderly populations. This observation has led to the hypothesis that Parkinson's disease may reflect an inability of the aging nervous system to remove reactive oxygen species and repair oxidative damage to proteins, lipids, and nucleic acids (Liddell et al., 2010). The development of the neuronal aggregates of α-synuclein and other proteins (i.e., Lewy bodies) seen in Parkinson's disease patients (Meredith, 2005) is consistent with this oxidative stress hypothesis (Sekigawa et al., 2015).

Research investigating toxicological risk factors for Parkinson's disease has largely focused on pesticides, solvents, and other chemicals that induce mitochondrial dysfunction and promote aggregation of α-synuclein and other proteins (Chin-Chan et al., 2015). Since manganese shares many of the pathophysiologic and clinical features of Parkinson's disease it has also garnered attention as a risk factor for Parkinson's disease (Chin-Chan et al., 2015; Kwakye et al., 2015). There is growing epidemiologic literature suggesting that chronic exposure to manganese may predispose an individual to acquire an earlier onset of Parkinson's disease-like signs (Andruska and Racette, 2015; Gorell et al., 1999; Kim et al., 2002; Racette et al., 2001). The epidemiologic evidence in support of a causal association between manganese exposure and Parkinson's disease, however remains largely inconclusive (Flynn and Susi, 2009; Mortimer et al., 2012; Wirdefeldt et al., 2011).

7.2 Hepatic Encephalopathy and Manganese

Experimental studies in animals show that animals with abnormal hepatobiliary function secondary to portacaval anastomosis develop higher brain manganese concentrations when compared with normal animals following manganese inhalation (Salehi et al., 2001). Patients with chronic hepatic failure also develop MRI changes (e.g., hyperintense globus pallidus, putamen, and red nucleus) consistent with manganese accumulation in the CNS (Das et al., 2008; Hauser et al., 1994, 1996; Rovira et al., 2008; Spahr et al., 1996). Increased brain manganese concentrations are also seen in human patients with liver cirrhosis (Rose et al., 1999). Similar MRI changes are seen in the lentiform nuclei of dogs with portosystemic shunts—and in one case a fourfold elevation in brain manganese concentration (vs normal dogs) was associated with MRI hyperintensity (Torisu et al., 2008). Although manganese is considered to be the cause of MRI hyperintense globus pallidus in patients with chronic hepatic failure other metals including copper and iron may also play a role in these MRI changes (Maeda et al., 1997).

7.3 Other Neurological Syndromes

A link between manganese exposure and amyotrophic lateral sclerosis (ALS) has also been reported in manganese workers (Penalver, 1957). However, recent studies have failed to establish an association between blood manganese concentration and ALS (Peters et al., 2016) whereas others have shown positive correlations with hair manganese concentration (Kihira et al., 2015). Other studies have shown an association between manganese exposure as assessed using biomonitoring methods (e.g., hair manganese concentration) and the incidence of attention-deficit/hyperactivity disorder (ADHD) in children (Shin et al., 2015). However, other studies using umbilical blood manganese concentrations found no association and ADHD (Ode et al., 2015).

REFERENCES

Abrams, E., Lassiter, J.W., Miller, W.J., Neathery, M.W., Gentry, R.P., Scarth, R.D., 1976. Absorption as a factor in manganese homeostasis. J. Anim. Sci. 42, 630–636.

Andersen, M.E., Dorman, D.C., Clewell 3rd, H.J., Taylor, M.D., Nong, A., 2010. Multi-dose-route, multi-species pharmacokinetic models for manganese and their use in risk assessment. J. Toxicol. Environ. Health A 73, 217–234.

Anderson, J.G., Cooney, P.T., Erikson, K.M., 2007. Inhibition of DAT function attenuates manganese accumulation in the globus pallidus. Environ. Toxicol. Pharmacol. 23, 179–184.

Anderson, J.G., Fordahl, S.C., Cooney, P.T., Weaver, T.L., Colyer, C.L., Erikson, K.M., 2008. Manganese exposure alters extracellular GABA, GABA receptor and transporter protein and mRNA levels in the developing rat brain. Neurotoxicology 29, 1044–1053.

Anderson, J.G., Fordahl, S.C., Cooney, P.T., Weaver, T.L., Colyer, C.L., Erikson, K.M., 2009. Extracellular norepinephrine, norepinephrine receptor and transporter protein and mRNA levels are differentially altered in the developing rat brain due to dietary iron deficiency and manganese exposure. Brain Res. 1281, 1–14.

Andruska, K.M., Racette, A.B., 2015. Neuromythology of manganism. Curr. Epidemiol. Rep. 2, 143–148.

Aschner, J.L., Aschner, M., 2005. Nutritional aspects of manganese homeostasis. Mol. Aspects Med. 26, 353–362.

Aschner, M., Dorman, D.C., 2006. Manganese: pharmacokinetics and molecular mechanisms of brain uptake. Toxicol. Rev. 25, 147–154.

Aschner, M., Erikson, K.M., Dorman, D.C., 2005. Manganese dosimetry: species differences and implications for neurotoxicity. Crit. Rev. Toxicol. 35, 1–32.

Aschner, J.L., Anderson, A., Slaughter, J.C., Aschner, M., Steele, S., Beller, A., Mouvery, A., Furlong, H.M., Maitre, N.L., 2015. Neuroimaging identifies increased manganese deposition in infants receiving parenteral nutrition. Am. J. Clin. Nutr. 102, 1482–1489.

ATSDR, 2000. Toxicological Profile for Manganese. Agency for Toxic Substances and Disease Registry, Atlanta, GA. Available at:http://www.atsdr.cdc.gov/toxprofiles/tp151.html.

Baker, M.G., Criswell, S.R., Racette, B.A., Simpson, C.D., Sheppard, L., Checkoway, H., Seixas, N.S., 2015. Neurological outcomes associated with low-level manganese exposure in an inception cohort of asymptomatic welding trainees. Scand. J. Work Environ. Health 41, 94–101.

Banta, R.G., Markesbery, W.R., 1977. Elevated manganese levels associated with dementia and extrapyramidal signs. Neurology 27, 213–216.

Bast-Pettersen, R., Ellingsen, D.G., Hetland, S.M., Thomassen, Y., 2004. Neuropsychological function in manganese alloy plant workers. Int. Arch. Occup. Environ. Health 77, 277–287.

Bertinchamps, A.J., Miller, S.T., Cotzias, G.C., 1966. Interdependence of routes excreting manganese. Am. J. Physiol. 211, 217–224.

Beuter, A., Mergler, D., de Geoffroy, A., Carriere, L., Belanger, S., Sreekumar, J., Gauthier, S., 1994. Diadochokinesimetry: a study of patients with Parkinson's disease and manganese exposed workers. Neurotoxicology 15, 655–664.

Bird, E.D., Anton, A.H., Bullock, B., 1984. The effect of manganese inhalation on basal ganglia dopamine concentrations in rhesus monkey. Neurotoxicology 5, 59–65.

Bouabid, S., Delaville, C., De Deurwaerdère, P., Lakhdar-Ghazal, N., Benazzouz, A., 2014. Manganese-induced atypical parkinsonism is associated with altered basal ganglia activity and changes in tissue levels of monoamines in the rat. PLoS One 9 (6), e98952.

Bouabid, S., Tinakoua, A., Lakhdar-Ghazal, N., Benazzouz, A., 2015. Manganese neurotoxicity: behavioral disorders associated with dysfunctions in the basal ganglia and neurochemical transmission. J. Neurochem. http://dx.doi.org/10.1111/jnc.13442.

Bouchard, M.F., Sauvé, S., Barbeau, B., Legrand, M., Brodeur, M.È., Bouffard, T., Limoges, E., Bellinger, D.C., Mergler, D., 2011. Intellectual impairment in school-age children exposed to manganese from drinking water. Environ. Health Perspect. 119, 138–143.

Bowler, R.M., Gysens, S., Diamond, E., Nakagawa, S., Drezgic, M., Roels, H.A., 2006. Manganese exposure: neuropsychological and neurological symptoms and effects in welders. Neurotoxicology 2, 315–326.

Brenneman, K.A., Wong, B.A., Buccellato, M.A., Costa, E.R., Gross, E.A., Dorman, D.C., 2000. Direct olfactory transport of inhaled manganese (^{54}MnCl$_2$) to the rat brain: toxicokinetic investigations in a unilateral nasal occlusion model. Toxicol. Appl. Pharmacol. 169, 238–248.

Calne, D.B., Chu, N.S., Huang, C.C., Lu, C.S., Olanow, W., 1994. Manganism and idiopathic parkinsonism: similarities and differences. Neurology 44, 1583–1586.

Canonne-Hergaux, F., Gruenheid, S., Ponka, P., Gros, P., 1999. Cellular and subcellular localization of the Nramp2 iron transporter in the intestinal brush border and regulation by dietary iron. Blood 1999 (93), 4406–4417.

Chan, T.J., Gutierrez, C., Ogunseitan, O.A., 2015. Metallic burden of deciduous teeth and childhood behavioral deficits. Int. J. Environ. Res. Public Health 12, 6771–6787.

Chang, Y., Lee, J.J., Seo, J.H., Song, H.J., Kim, J.H., Bae, S.J., Ahn, J.H., Park, S.J., Jeong, K.S., Kwon, Y.J., Kim, S.H., Kim, Y., 2010. Altered working memory process in the manganese-exposed brain. Neuroimage 53, 1279–1285.

Chen, P., Parmalee, N., Aschner, M., 2014. Genetic factors and manganese-induced neurotoxicity. Front. Genet. 5, 265.

Chen, P., Chakraborty, S., Mukhopadhyay, S., Lee, E., Paoliello, M.M., Bowman, A.B., Aschner, M., 2015. Manganese homeostasis in the nervous system. J. Neurochem. 134, 601–610.

Chillrud, S.N., Grass, D., Ross, J.M., Coulibaly, D., Slavkovich, V., Epstein, D., Sax, S.N., Pedersen, D., Johnson, D., Spengler, J.D., Kinney, P.L., Simpson, H.J., Brandt-Rauf, P., 2005. Steel dust in the New York City subway system as a source of manganese, chromium, and iron exposures for transit workers. J. Urban Health Bull. N.Y. Acad. Med. 82, 33–42.

Chin-Chan, M., Navarro-Yepes, J., Quintanilla-Vega, B., 2015. Environmental pollutants as risk factors for neurodegenerative disorders: Alzheimer and Parkinson diseases. Front. Cell. Neurosci. 9, 124.

Chua, A.C., Morgan, E.H., 1997. Manganese metabolism is impaired in the Belgrade Laboratory rat. J. Comp. Physiol. B 167, 361–369.

Colin-Barenque, L., Souza-Gallardo, L.M., Fortoul, T.I., 2011. Toxic effects of inhaled manganese on the olfactory bulb: an ultrastructural approach in mice. J. Electron Microsc. (Tokyo) 60, 73.

Conrad, M.E., Umbreit, J.N., Moore, E.G., Hainsworth, L.N., Porubcin, M., Simovich, M.J., Nakada, M.T., Dolan, K., Garrick, M.D., 2000. Separate pathways for cellular uptake of ferric and ferrous iron. Am. J. Physiol. Gastrointest. Liver Physiol. 279, G767–774.

Cotzias, G.C., 1958. Manganese in health and disease. Physiol. Rev. 38, 503.

Cotzias, G.C., Horiuchi, K., Fuenzalida, S., Mena, I., 1968. Chronic manganese poisoning. Clearance of tissue manganese concentrations with persistance of the neurological picture. Neurology 18, 376–382.

Couper, J., 1837. On the effects of black oxide of manganese when inhaled into the lungs. Br. Ann. Med. Pharm. Vital Stat. Gen. Sci. 1, 41–42.

Crinella, F.M., 2012. Does soy-based infant formula cause ADHD? Update and public policy considerations. Expert Rev. Neurother. 12, 395–407.

Cross, D.J., Minoshima, S., Anzai, Y., Flexman, J.A., Keogh, B.P., Kim, Y., Maravilla, K.R., 2004. Statistical mapping of functional olfactory connections of the rat brain in vivo. Neuroimage 23, 1326–1335.

Crossgrove, J.S., Yokel, R.A., 2004. Manganese distribution across the blood-brain barrier III. The divalent metal transporter-1 is not the major mechanism mediating brain manganese uptake. Neurotoxicology 25, 451–460.

Das, K., Singh, P., Chawla, Y., Duseja, A., Dhiman, R.K., Suri, S., 2008. Magnetic resonance imaging of brain in patients with cirrhotic and non-cirrhotic portal hypertension. Dig. Dis. Sci. 53, 2793–2798.

Dion, L.A., Bouchard, M.F., Sauvé, S., Barbeau, B., Tucholka, A., Major, P., Gilbert, G., Mergler, D., Saint-Amour, D., 2016. MRI pallidal signal in children exposed to manganese in drinking water. Neurotoxicology 53, 124–131.

Dorman, D.C., 2015. Extrapyramidal system neurotoxicity: animal models. Handb. Clin. Neurol. 131, 207–223.

Dorman, D.C., 2017. Metal imaging in the brain. In: White, R., Aschner, M., Costa, L.G., Bush, A.I. (Eds.), Biometals in Neurodegenerative Diseases: Mechanisms and Therapeutics. Elsevier Inc., Academic Press, San Diego, CA.

Dorman, D.C., Struve, M.F., James, A.R., Marshall, M.W., Parkinson, C.U., Wong, B.A., 2001. Influence of particle solubility on the delivery of inhaled manganese to the rat brain: manganese sulfate and manganese tetroxide pharmacokinetics following repeated (14-day) exposure. Toxicol. Appl. Pharmacol. 170, 79–87.

Dorman, D.C., McManus, B.E., Parkinson, C.U., Manuel, C.A., McElveen, A.M., Everitt, J.I., 2004. Nasal toxicity of manganese sulfate and manganese phosphate in young male rats following subchronic (13-week) inhalation exposure. Inhal. Toxicol. 16, 481–488.

Dorman, D.C., McElveen, A.M., Marshall, M.W., Parkinson, C.U., James, R.A., Struve, M.F., Wong, B.A., 2005a. Tissue manganese concentrations in lactating rats and their offspring following combined *in utero* and lactation exposure to inhaled manganese sulfate. Toxicol. Sci. 84, 12–21.

Dorman, D.C., McElveen, A.M., Marshall, M.W., Parkinson, C.U., James, R.A., Struve, M.F., Wong, B.A., 2005b. Maternal-fetal distribution of manganese in the rat following inhalation exposure to manganese sulfate. Neurotoxicology 26, 625–632.

Dorman, D.C., Struve, M.F., Marshall, M.W., Parkinson, C.U., James, R.A., Wong, B.A., 2006b. Tissue manganese concentrations in young male rhesus monkeys following subchronic manganese sulfate inhalation. Toxicol. Sci. 92, 201–210.

Dorman, D.C., Struve, M.F., Wong, B.A., Dye, J.A., Robertson, I.D., 2006c. Correlation of brain magnetic resonance imaging changes with pallidal manganese concentrations in rhesus monkeys following subchronic manganese inhalation. Toxicol. Sci. 92, 219–227.

Dorman, D.C., Andersen, M.E., Roper, J.M., Taylor, M.D., 2012. Update on a pharmacokinetic-centric alternative tier II program for MMT-part I: program implementation and lessons learned. J. Toxicol. 2012, 946742.

Elder, A., Gelein, R., Silva, V., Feikert, T., Opanashuk, L., Carter, J., Potter, R., Maynard, A., Ito, Y., Finkelstein, J., Oberdörster, G., 2006. Translocation of inhaled ultrafine manganese oxide particles to the central nervous system. Environ. Health Perspect. 114, 1172–1178.

Ellingsen, D.G., Konstantinov, R., Bast-Pettersen, R., Merkurjeva, L., Chashchin, M., Thomassen, Y., Chashchin, V., 2008. A neurobehavioral study of current and former welders exposed to manganese. Neurotoxicology 29, 48–59.

Ensing, J.G., 1985. Bazooka: cocaine-base and manganese carbonate. J. Anal. Toxicol. 9, 45–46.

Erikson, K.M., Dobson, A.W., Dorman, D.C., Aschner, M., 2004. Manganese exposure and induced oxidative stress in the rat brain. Sci. Total Environ. 334-335, 409–416.

Erikson, K.M., Dorman, D.C., Lash, L.H., Aschner, M., 2005. Persistent alterations in biomarkers of oxidative stress resulting from combined in utero and neonatal manganese inhalation. Biol. Trace Elem. Res. 104, 151–163.

Erikson, K.M., Dorman, D.C., Lash, L.H., Aschner, M., 2007. Manganese inhalation by rhesus monkeys is associated with brain regional changes in biomarkers of neurotoxicity. Toxicol. Sci. 97, 459–466.

Eriksson, H., Mägiste, K., Plantin, L.O., Fonnum, F., Hedström, K.G., Theodorsson-Norheim, E., Kristensson, K., Stålberg, E., Heilbronn, E., 1987. Effects of manganese oxide on monkeys as revealed by a combined neurochemical, histological and neurophysiological evaluation. Arch. Toxicol. 61, 46–52.

Farina, M., Avila, D.S., da Rocha, J.B., Aschner, M., 2013. Metals, oxidative stress and neurodegeneration: a focus on iron, manganese and mercury. Neurochem. Int. 62, 575–594.

Fechter, L.D., Johnson, D.L., Lynch, R.A., 2002. The relationship of particle size to olfactory nerve uptake of a non-soluble form of manganese into brain. Neurotoxicology 23, 177–183.

Finkelstein, M.M., Jerrett, M., 2007. A study of the relationships between Parkinson's disease and markers of traffic-derived and environmental manganese air pollution in two Canadian cities. Environ. Res. 104, 420–432.

Fitsanakis, V.A., Zhang, N., Avison, M.J., Gore, J.C., Aschner, J.L., Aschner, M., 2006. The use of magnetic resonance imaging (MRI) in the study of manganese neurotoxicity. Neurotoxicology 27, 798–806.

Flynn, M.R., Susi, P., 2009. Neurological risks associated with manganese exposure from welding operations—a literature review. Int. J. Hyg. Environ. Health 212, 459–469.

Forbes, J.R., Gros, P., 2003. Iron, manganese, and cobalt transport by Nramp1 (Slc11a1) and Nramp2 (Slc11a2) expressed at the plasma membrane. Blood 102, 1884–1892.

Fordahl, S.C., Anderson, J.G., Cooney, P.T., Weaver, T.L., Colyer, C.L., Erikson, K.M., 2010. Manganese exposure inhibits the clearance of extracellular GABA and influences taurine homeostasis in the striatum of developing rats. Neurotoxicology 31, 639–646.

Gonzalez-Cuyar, L.F., Nelson, G., Criswell, S.R., Ho, P., Lonzanida, J.A., Checkoway, H., Seixas, N., Gelman, B.B., Evanoff, B.A., Murray, J., Zhang, J., Racette, B.A., 2014. Quantitative neuropathology associated with chronic manganese exposure in south African mine workers. Neurotoxicology 45, 260–266.

Gorell, J.M., Rybicki, B.A., Cole Johnson, C., Peterson, E.L., 1999. Occupational metal exposures and the risk of Parkinson's disease. Neuroepidemiology 18, 303–308.

Guarneros, M., Ortiz-Romo, N., Alcaraz-Zubeldia, M., Drucker-Colín, R., Hudson, R., 2013. Nonoccupational environmental exposure to manganese is linked to deficits in peripheral and central olfactory function. Chem. Senses 38, 783–791.

Guilarte, T.R., 2013. Manganese neurotoxicity: new perspectives from behavioral, neuroimaging, and neuropathological studies in humans and non-human primates. Front. Aging Neurosci. 5, 23.

Guilarte, T.R., Chen, M.K., McGlothan, J.L., Verina, T., Zhou, Y., Alexander, M., Pham, L., Griswold, M., Wong, D.F., Syversen, T., Schneider, J.S., 2006. Nigrostriatal dopamine system dysfunction and subtle motor deficits in manganese-exposed non-human primates. Exp. Neurol. 202, 381–390.

Guilarte, T.R., Burton, N.C., McGlothan, J.L., et al., 2008. Impairment of nigrostriatal dopamine neurotransmission by manganese is mediated by pre-synaptic mechanism(s): implications to manganese-induced parkinsonism. J. Neurochem. 107, 1236–1247.

Gunshin, H., Mackenzie, B., Berger, U.V., Gunshin, Y., Romero, M.F., Boron, W.F., Nussberger, S., Gollan, J.L., Hediger, M.A., 1997. Cloning and characterization of a mammalian proton-coupled metal-ion transporter. Nature 388, 482–488.

Gwiazda, R., Lucchini, R., Smith, D., 2007. Adequacy and consistency of animal studies to evaluate the neurotoxicity of chronic low-level manganese exposure in humans. J. Toxicol. Environ. Health A 70, 594–605.

Han, S.H., Ahn, S.W., Youn, Y.C., Shin, H.W., 2014. Reversal of pallidal magnetic resonance imaging T1 hyperintensity in a welder presenting as reversible parkinsonism. Neurol. India 62, 117–118.

Hauser, R.A., Zesiewicz, T.A., Rosemurgy, A.S., Martinez, C., Olanow, C.W., 1994. Manganese intoxication and chronic liver failure. Ann. Neurol. 36, 871–875.

Hauser, R.A., Zesiewicz, T.A., Martinez, C., Rosemurgy, A.S., Olanow, C.W., 1996. Blood manganese correlates with brain magnetic resonance imaging changes in patients with liver disease. Can. J. Neurol. Sci. 23, 95–98.

Heilig, E.A., Thompson, K.J., Molina, R.M., Ivanov, A.R., Brain, J.D., Wessling-Resnick, M., 2006. Manganese and iron transport across pulmonary epithelium. Am. J. Physiol. Lung Cell. Mol. Physiol. 290, L1247–1259.

Henriksson, J., Tjälve, H., 2000. Manganese taken up into the CNS via the olfactory pathway in rats affects astrocytes. Toxicol. Sci. 55, 392–398.

Henriksson, J., Tallkvist, J., Tjälve, H., 1999. Transport of manganese via the olfactory pathway in rats: dosage dependency of the uptake and subcellular distribution of the metal in the olfactory epithelium and the brain. Toxicol. Appl. Pharmacol. 156, 119–128.

Huang, C.C., Lu, C.S., Chu, N.S., Hochberg, F., Lilienfeld, D., Olanow, W., Calne, D.B., 1993. Progression after chronic manganese exposure. Neurology 43, 1479–1483.

Hurley, L.S., 1981. Teratogenic aspects of manganese, zinc and copper nutrition. Physiol. Rev. 61, 249–295.

Iinuma, Y., Kubota, M., Uchiyama, M., Yagi, M., Kanada, S., Yamazaki, S., Murata, H., Okamoto, K., Suzuki, M., Nitta, K., 2003. Whole-blood manganese levels and brain manganese accumulation in children receiving long-term home parenteral nutrition. Pediatr. Surg. Int. 19, 268–272.

Inoue, T., Majid, T., Pautler, R.G., 2011. Manganese enhanced MRI (MEMRI): neurophysiological applications. Rev. Neurosci. 22, 675–694.

IOM, 2001. Manganese. In: Food and Nutrition Board, Institute of Medicine (Eds.), Dietary Reference Intakes for Vitamin A, Vitamin K, Arsenic, Boron, Chromium, Copper, Iodine, Iron, Manganese, Molybdenum, Nickel, Silicon, Vanadium, and Zinc. National Academy Press, pp. 394–419.

Iregren, A., 1990. Psychological test performance in foundry workers exposed to low levels of manganese. Neurotoxicol. Teratol. 12, 673–675.

Iwami, O., Watanabe, T., Moon, C.-S., Nakatsuka, H., Ikeda, M., 1994. Motor neuron disease on the Kii Peninsula of Japan: excess manganese intake from food coupled with low magnesium in drinking water as a risk factor. Sci. Total Environ. 149, 121–135.

Jankovic, J., 2008. Parkinson's disease: clinical features and diagnosis. J. Neurol. Neurosurg. Psychiatry 79, 368–376.

Karki, P., Smith, K., Johnson, J., Aschner, M., Lee, E., 2015. Role of transcription factor yin yang 1 in manganese-induced reduction of astrocytic glutamate transporters: putative mechanism for manganese-induced neurotoxicity. Neurochem. Int. 88, 53–59.

Kawamura, R., Ikuta, H., Fukuzumi, S., Yamada, R., Tsubaki, S., Kodama, T., Kurata, S., 1941. Intoxication by manganese in well water. Kitasato Arch. Exp. Med. 18, 145–169.

Keen, C.L., Zidenberg-Cherr, S., 1994. Should vitamin–mineral supplements be recommended for all women with childbearing potential? Am. J. Clin. Nutr. 59 (2 Suppl), 532S–538S.

Khan, K., Wasserman, G.A., Liu, X., Ahmed, E., Parvez, F., Slavkovich, V., Levy, D., Mey, J., van Geen, A., Graziano, J.H., Factor-Litvak, P., 2012. Manganese exposure from drinking water and children's academic achievement. Neurotoxicology 33, 91–97.

Kihira, T., Sakurai, I., Yoshida, S., Wakayama, I., Takamiya, K., Okumura, R., Iinuma, Y., Iwai, K., Kajimoto, Y., Hiwatani, Y., Kohmoto, J., Okamoto, K., Kokubo, Y., Kuzuhara, S., 2015. Neutron activation analysis of scalp hair from ALS patients and residents in the Kii Peninsula, Japan. Biol. Trace Elem. Res. 164, 36–42.

Kim, Y., 2004. High signal intensities on T1-weighted MRI as a biomarker of exposure to manganese. Ind. Health 42, 111–115.

Kim, Y., Kim, J.M., Kim, J.W., Yoo, C.I., Lee, C.R., Lee, J.H., Kim, H.K., Yang, S.O., Chung, H.K., Lee, D.S., Jeon, B., 2002. Dopamine transporter density is decreased in parkinsonian patients with a history of manganese exposure: what does it mean? Mov. Disord. 17, 568–575.

Kim, E., Kim, Y., Cheong, H.K., Cho, S., Shin, Y.C., Sakong, J., Kim, K.S., Yang, J.S., Jin, Y.W., Kang, S.K., 2005. Pallidal index on MRI as a target organ dose of manganese: structural equation model analysis. Neurotoxicology 26, 351–359.

Kim, C.Y., Sung, J.H., Chung, Y.H., Chung, Y.H., Park, J.D., Han, J.H., Lee, J.S., Heo, J.D., Yu, I.J., 2013. Home cage locomotor changes in non-human primates after prolonged welding-fume exposure. Inhal. Toxicol. 25, 794–801.

Knöpfel, M., Zhao, L., Garrick, M.D., 2005. Transport of divalent transition-metal ions is lost in small-intestinal tissue of b/b Belgrade rats. Biochemistry 44, 3454–3465.

Kondakis, X.G., Makris, N., Leotsinidis, M., Prinou, M., Papapetropoulos, T., 1989. Possible health effects of high manganese concentration in drinking water. Arch. Environ. Health 44, 175–178.

Kwakye, G.F., Paoliello, M.M., Mukhopadhyay, S., Bowman, A.B., Aschner, M., 2015. Manganese-induced parkinsonism and Parkinson's disease: shared and distinguishable features. Int. J. Environ. Res. Public Health 12, 7519–7540.

Leavens, T.L., Rao, D., Andersen, M.E., Dorman, D.C., 2007. Evaluating transport of manganese from olfactory mucosa to striatum by pharmacokinetic modeling. Toxicol. Sci. 97, 265–278.

Lee, J., 2000. Manganese intoxication. Arch. Neurol. 57, 597–599.

Lee, E.Y., Flynn, M.R., Du, G., Lewis, M.M., Fry, R., Herring, A.H., Van Buren, E., Van Buren, S., Smeester, L., Kong, L., Yang, Q., Mailman, R.B., Huang, X., 2015. T1 relaxation rate (R1) indicates nonlinear Mn accumulation in brain tissue of welders with low-level exposure. Toxicol. Sci. 146, 281–289.

Lee, E.Y., Flynn, M.R., Du, G., Li, Y., Lewis, M.M., Herring, A.H., Van Buren, E., Van Buren, S., Kong, L., Fry, R.C., Snyder, A.M., Connor, J.R., Yang, Q.X., Mailman, R.B., Huang, X., 2016. Increased R2* in the caudate nucleus of asymptomatic welders. Toxicol. Sci. 150, 369–377.

Lewis, M.M., Flynn, M.R., Lee, E.Y., Van Buren, S., Van Buren, E., Du, G., Fry, R.C., Herring, A.H., Kong, L., Mailman, R.B., Huang, X., 2016. Longitudinal T1 relaxation rate (R1) captures changes in short-term Mn exposure in welders. Neurotoxicology 57, 39–44.

Li, S.J., Jiang, L., Fu, X., Huang, S., Huang, Y.N., Li, X.R., Chen, J.W., Li, Y., Luo, H.L., Wang, F., Ou, S.Y., Jiang, Y.M., 2014. Pallidal index as biomarker of manganese brain accumulation and associated with manganese levels in blood: a meta-analysis. PLoS One 9, e93900.

Liddell, J.R., Robinson, S.R., Dringen, R., Bishop, G.M., 2010. Astrocytes retain their antioxidant capacity into advanced old age. Glia 58, 1500–1509.

Liu, X., Sullivan, K.A., Madl, J.E., Legare, M., Tjalkens, R.B., 2006. Manganese-induced neurotoxicity: the role of astroglial-derived nitric oxide in striatal interneuron degeneration. Toxicol. Sci. 91, 521–531.

Ljung, K., Vahter, M., 2007. Time to re-evaluate the guideline value for manganese in drinking water? Environ. Health Perspect. 115, 1533–1538.

Lucchini, R., Selis, L., Folli, D., Apostoli, P., Mutti, A., Vanoni, O., Iregren, A., Alessio, L., 1995. Neurobehavioral effects of manganese in workers from a ferroalloy plant after temporary cessation of exposure. Scand. J. Work Environ. Health 21, 143–149.

Lucchini, R., Apostoli, P., Perrone, C., Placidi, D., Albini, E., Migliorati, P., Mergler, D., Sassine, M.P., Palmi, S., Alessio, L., 1999. Long-term exposure to "low levels" of manganese oxides and neurofunctional changes in ferroalloy workers. Neurotoxicology 20, 287–297.

Lucchini, R.G., Albini, E., Benedetti, L., Borghesi, S., Coccaglio, R., Malara, E.C., Parrinello, G., Garattini, S., Resola, S., Alessio, L., 2007. High prevalence of parkinsonian disorders associated to manganese exposure in the vicinities of ferroalloy industries. Am. J. Ind. Med. 50, 788–800.

Lucchini, R.G., Martin, C.J., Doney, B.C., 2009. From manganism to manganese-induced parkinsonism: a conceptual model based on the evolution of exposure. Neuromol. Med. 11, 311–321.

Lucchini, R.G., Dorman, D.C., Elder, A., Veronesi, B., 2012a. Neurological impacts from inhalation of pollutants and the nose–brain connection. Neurotoxicology 33, 838–841.

Lucchini, R.G., Guazzetti, S., Zoni, S., Donna, F., Peter, S., Zacco, A., Salmistraro, M., Bontempi, E., Zimmerman, N.J., Smith, D.R., 2012b. Tremor, olfactory and motor changes in Italian adolescents exposed to historical ferro-manganese emission. Neurotoxicology 33, 687–696.

Maeda, H., Sato, M., Yoshikawa, A., Kimura, M., Sonomura, T., Terada, M., Kishi, K., 1997. Brain MR imaging in patients with hepatic cirrhosis: relationship between high intensity signal in basal ganglia on T1-weighted images and elemental concentrations in brain. Neuroradiology 39, 546–550.

Malheiros, J.M., Paiva, F.F., Longo, B.M., Hamani, C., Covolan, L., 2015. Manganese-enhanced MRI: biological applications in neuroscience. Front. Neurol. 6, 161.

Meredith, S.C., 2005. Protein denaturation and aggregation: cellular responses to denatured and aggregated proteins. Ann. N.Y. Acad. Sci. 1066, 181–221.

Mergler, D., Huel, G., Bowler, R., Iregren, A., Belanger, S., Baldwin, M., Tardif, R., Smargiassi, A., Martin, L., 1994. Nervous system dysfunction among workers with long-term exposure to manganese. Environ. Res. 64, 151–180.

Mergler, D., Baldwin, M., Bélanger, S., Larribe, F., Beuter, A., Bowler, R., Panisset, M., Edwards, R., de Geoffroy, A., Sassine, M.P., Hudnell, K., 1999. Manganese neurotoxicity, a continuum of dysfunction: results from a community based study. Neurotoxicology 20, 327–342.

Michalke, B., 2016. Review about the manganese speciation project related to neurodegeneration: an analytical chemistry approach to increase the knowledge about manganese related parkinsonian symptoms. J. Trace Elem. Med. Biol. 37, 50–61.

Moberly, A.H., Czarnecki, L.A., Pottackal, J., Rubinstein, T., Turkel, D.J., Kass, M.D., McGann, J.P., 2012. Intranasal exposure to manganese disrupts neurotransmitter release from glutamatergic synapses in the central nervous system in vivo. Neurotoxicology 33, 996–1004.

Mortimer, J.A., Borenstein, A.R., Nelson, L.M., 2012. Associations of welding and manganese exposure with Parkinson disease: review and meta-analysis. Neurology 79, 1174–1180.

Myers, J.E., Thompson, M.L., Ramushu, S., Young, T., Jeebhay, M.F., London, L., Esswein, E., Renton, K., Spies, A., Boulle, A., Naik, I., Iregren, A., Rees, D.J., 2003. The nervous system effects of occupational exposure on workers in a South African manganese smelter. Neurotoxicology 24, 885–894.

Nagatomo, S., Umehara, F., Hanada, K., Nobuhara, Y., Takenaga, S., Arimura, K., Osame, M., 1999. Manganese intoxication during total parenteral nutrition: report of two cases and review of the literature. J. Neurol. Sci. 162, 102–105.

Narita, K., Kawasaki, F., Kita, H., 1990. Mn and Mg influxes through Ca channels of motor nerve terminals are prevented by verapamil in frogs. Brain Res. 510, 289–295.

Neff, N.H., Barrett, R.E., Costa, E., 1969. Selective depletion of caudate nucleus dopamine and serotonin during chronic manganese dioxide administration to squirrel monkeys. Experientia 25, 1140–1141.

Newland, M.C., 1999. Animal models of manganese's neurotoxicity. Neurotoxicology 20, 415–432.

Normandin, L., Ann Beaupré, L., Salehi, F., St –Pierre, A., Kennedy, G., Mergler, D., Butterworth, R.F., Philippe, S., Zayed, J., 2004. Manganese distribution in the brain and neurobehavioral changes following inhalation exposure of rats to three chemical forms of manganese. Neurotoxicology 25, 433–441.

Ode, A., Rylander, L., Gustafsson, P., Lundh, T., Källén, K., Olofsson, P., Ivarsson, S.A., Rignell-Hydbom, A., 2015. Manganese and selenium concentrations in umbilical cord serum and attention deficit hyperactivity disorder in childhood. Environ. Res. 137, 373–381.

Olanow, C.W., Good, P.F., Shinotoh, H., Hewitt, K.A., Vingerhoets, F., Snow, B.J., Beal, M.F., Calne, D.B., Perl, D.P., 1996. Manganese intoxication in the rhesus monkey: a clinical, imaging, pathologic, and biochemical study. Neurology 46, 492–498.

Pal, P., Samii, A., Calne, D., 1999. Manganese neurotoxicity: a review of clinical features, imaging and pathology. Neurotoxicology 20, 227–238.

Park, J.D., Kim, K.Y., Kim, D.W., Choi, S.J., Choi, B.S., Chung, Y.H., Han, J.H., Sung, J.H., Kwon, I.H., Mun, J.H., Yu, I.J., 2007. Tissue distribution of manganese in iron-sufficient or iron-deficient rats after stainless steel welding-fume exposure. Inhal. Toxicol. 19, 965–971.

Park, R.M., Bowler, R.M., Roels, H.A., 2009. Exposure–response relationship and risk assessment for cognitive deficits in early welding-induced manganism. J. Occup. Environ. Med. 51, 1125–1136.

Pellizzari, E.D., Clayton, C., Rodes, C.E., Mason, R.E., Piper, L.L., Fort, B.F., Pfeifer, G.D., Lynam, D.R., 1999. Particulate matter and manganese exposures in Toronto, Canada. Atmos. Environ. 33, 721–734.

Penalver, R., 1957. Diagnosis and treatment of manganese intoxication; report of a case. A. M.A. Arch. Ind. Health 16, 64–66.

Pentschew, A., Ebner, F.F., Kovatch, R.M., 1963. Experimental manganese encephalopathy in monkeys: a preliminary report. J. Neuropathol. Exp. Neurol. 22, 488–499.

Perl, D.P., Olanow, C.W., 2007. The neuropathology of manganese-induced parkinsonism. J. Neuropathol. Exp. Neurol. 66, 675–682.

Peters, T.L., Beard, J.D., Umbach, D.M., Allen, K., Keller, J., Mariosa, D., Sandler, D.P., Schmidt, S., Fang, F., Ye, W., Kamel, F., 2016. Blood levels of trace metals and amyotrophic lateral sclerosis. Neurotoxicology 54, 119–126.

Quadri, M., Federico, A., Zhao, T., Breedveld, G.J., Battisti, C., Delnooz, C., Severijnen, L.A., Di Toro Mammarella, L., Mignarri, A., Monti, L., Sanna, A., Lu, P., Punzo, F., Cossu, G., Willemsen, R., Rasi, F., Oostra, B.A., van de Warrenburg, B.P., Bonifati, V., 2012. Mutations in SLC30A10 cause parkinsonism and dystonia with hypermanganesemia, polycythemia, and chronic liver disease. Am. J. Hum. Genet. 90, 467–477.

Racette, B.A., McGee-Minnich, L., Moerlein, S.M., Mink, J.W., Videen, T.O., Perlmutter, J.S., 2001. Welding-related parkinsonism: clinical features, treatment, and pathophysiology. Neurology 56, 8–13.

Rodríguez-Agudelo, Y., Riojas-Rodríguez, H., Ríos, C., Rosas, I., Sabido Pedraza, E., Miranda, J., Siebe, C., Texcalac, J.L., Santos-Burgoa, C., 2006. Motor alterations associated with exposure to manganese in the environment in Mexico. Sci. Total Environ. 368, 542–556.

Roels, H., Lauwerys, R., Buchet, J.P., Genet, P., Sarhan, M.J., Hanotiau, I., de Fays, M., Bernard, A., Stanescu, D., 1987. Epidemiological survey among workers exposed to manganese: effects on lung, central nervous system, and some biological indices. Am. J. Ind. Med. 11, 307–327.

Roels, H.A., Ghyselen, P., Buchet, J.P., Ceulemans, E., Lauwerys, R.R., 1992. Assessment of the permissible exposure level to manganese in workers exposed to manganese dioxide dust. Br. J. Ind. Med. 49, 25–34.

Roels, H.A., Ortega Eslava, M.I., Ceulemans, E., Robert, A., Lison, D., 1999. Prospective study on the reversibility of neurobehavioral effects in workers exposed to manganese dioxide. Neurotoxicology 20, 255–271.

Roels, H.A., Bowler, R.M., Kim, Y., Claus Henn, B., Mergler, D., Hoet, P., Gocheva, V.V., Bellinger, D.C., Wright, R.O., Harris, M.G., Chang, Y., Bouchard, M.F., Riojas-Rodriguez, H., Menezes-Filho, J.A., Téllez-Rojo, M.M., 2012. Manganese exposure and cognitive deficits: a growing concern for manganese neurotoxicity. Neurotoxicology 33, 872–880.

Rose, C., Butterworth, R.F., Zayed, J., Normandin, L., Todd, K., Michalak, A., Spahr, L., Huet, P.M., Pomier-Layrargues, G., 1999. Manganese deposition in basal ganglia structures results from both portal-systemic shunting and liver dysfunction. Gastroenterology 117, 640–644.

Roth, J.A., 2006. Homeostatic and toxic mechanisms regulating manganese uptake, retention, and elimination. Biol. Res. 39, 45–57.

Roth, J.A., Garrick, M.D., 2003. Iron interactions and other biological reactions mediating the physiological and toxic actions of manganese. Biochem. Pharmacol. 66, 1–13.

Roth, J.A., Horbinski, C., Higgins, D., Lein, P., Garrick, M.D., 2002. Mechanisms of manganese-induced rat pheochromocytoma (PC12) cell death and cell differentiation. Neurotoxicology 23, 147–157.

Rovira, A., Alonso, J., Córdoba, J., 2008. MR imaging findings in hepatic encephalopathy. Am. J. Neuroradiol. 29, 1612–1621.

Salehi, F., Carrier, G., Normandin, L., Kennedy, G., Butterworth, R.F., Hazell, A., Therrien, G., Mergler, D., Philippe, S., Zayed, J., 2001. Assessment of bioaccumulation and neurotoxicity in rats with portacaval anastomosis and exposed to manganese phosphate: a pilot study. Inhal. Toxicol. 13, 1151–1163.

Salehi, F., Krewski, D., Mergler, D., Normandin, L., Kennedy, G., Philippe, S., Zayed, J., 2003. Bioaccumulation and locomotor effects of manganese phosphate/sulfate mixture in Sprague–Dawley rats following subchronic (90 days) inhalation exposure. Toxicol. Appl. Pharmacol. 191, 264–271.

Salehi, F., Normandin, L., Krewski, D., Kennedy, G., Philippe, S., Zayed, J., 2006. Neuropathology, tremor and electromyogram in rats exposed to manganese phosphate/sulfate mixture. J. Appl. Toxicol. 26, 419–426.

Sanders, A.P., Claus Henn, B., Wright, R.O., 2015. Perinatal and childhood exposure to cadmium, manganese, and metal mixtures and effects on cognition and behavior: a review of recent literature. Curr. Environ. Health Rep. 2, 284–294.

Santamaria, A.B., Sulsky, S.I., 2010. Risk assessment of an essential element: manganese. J. Toxicol. Environ. Health A 73, 128–155.

Santos, D., Batoreu, C., Mateus, L., Marreilha Dos Santos, A.P., Aschner, M., 2014. Manganese in human parenteral nutrition: considerations for toxicity and biomonitoring. Neurotoxicology 43, 36–45.

Schneider, J.S., Decamp, E., Koser, A.J., Fritz, S., Gonczi, H., Syversen, T., Guilarte, T.R., 2006. Effects of chronic manganese exposure on cognitive and motor functioning in non-human primates. Brain Res. 1118, 222–231.

Schneider, J.S., Decamp, E., Clark, K., Bouquio, C., Syversen, T., Guilarte, T.R., 2009. Effects of chronic manganese exposure on working memory in non-human primates. Brain Res. 1258, 86–95.

Schneider, J.S., Williams, C., Ault, M., Guilarte, T.R., 2013. Chronic manganese exposure impairs visuospatial associative learning in non-human primates. Toxicol. Lett. 221, 146–151.

Schroeter, J.D., Dorman, D.C., Yoon, M., Nong, A., Taylor, M.D., Andersen, M.E., Clewell 3rd, H.J., 2012. Application of a multi-route physiologically based pharmacokinetic model for manganese to evaluate dose-dependent neurological effects in monkeys. Toxicol. Sci. 129, 432–446.

Searles Nielsen, S., Checkoway, H., Criswell, S.R., Farin, F.M., Stapleton, P.L., Sheppard, L., Racette, B.A., 2015. Inducible nitric oxide synthase gene methylation and parkinsonism in manganese-exposed welders. Parkinsonism Relat. Disord. 21, 355–360.

Sekigawa, A., Takamatsu, Y., Sekiyama, K., Hashimoto, M., 2015. Role of α- and β-synucleins in the axonal pathology of Parkinson's disease and related synucleinopathies. Biomolecules 5, 1000–1011.

Sen, S., Flynn, M.R., Du, G., Tröster, A.I., An, H., Huang, X., 2011. Manganese accumulation in the olfactory bulbs and other brain regions of "asymptomatic" welders. Toxicol. Sci. 121, 160–167.

Shin, Y.C., Kim, E., Cheong, H.K., Cho, S., Sakong, J., Kim, K.S., Yang, J.S., Jin, Y.W., Kang, S.K., Kim, Y., 2007. High signal intensity on magnetic resonance imaging as a predictor of neurobehavioral performance of workers exposed to manganese. Neurotoxicology 28, 257–262.

Shin, D.W., Kim, E.J., Lim, S.W., Shin, Y.C., Oh, K.S., Kim, E.J., 2015. Association of hair manganese level with symptoms in attention-deficit/hyperactivity disorder. Psychiatry Investig. 12, 66–72.

Sidoryk-Wegrzynowicz, M., 2014. Impairment of glutamine/glutamate-γ-aminobutyric acid cycle in manganese toxicity in the central nervous system. Folia Neuropathol. 52, 377–382.

Sidoryk-Wegrzynowicz, M., Aschner, M., 2013. Manganese toxicity in the CNS: the glutamine/glutamate-γ-aminobutyric acid cycle. J. Intern. Med. 273, 466–477.

Spahr, L., Butterworth, R.F., Fontaine, S., Bui, L., Therrien, G., Milette, P.C., Lebrun, L.H., Zayed, J., Leblanc, A., Pomier-Layrargues, G., 1996. Increased blood manganese in cirrhotic patients: relationship to pallidal magnetic resonance signal hyperintensity and neurological symptoms. Hepatology 24, 1116–1120.

Stasny, D., Vogel, R.S., Picciano, M.F., 1984. Manganese intake and serum manganese concentration of human milk-fed and formula-fed infants. Am. J. Clin. Nutr. 39, 872–878.

Struve, M.F., McManus, B.E., Wong, B.A., Dorman, D.C., 2007. Basal ganglia neurotransmitter concentrations in rhesus monkeys following subchronic manganese sulfate inhalation. Am. J. Ind. Med. 50, 772–778.

Sumino, K., Hayakawa, K., Shibata, T., 1975. Heavy metals in normal Japanese tissues: amounts of 15 heavy metals in 30 subjects. Arch. Environ. Health 30, 1790–1800.

Sung, J.H., Kim, C.Y., Yang, S.O., Khang, H.S., Cheong, H.K., Lee, J.S., Song, C.W., Park, J.D., Han, J.H., Chung, Y.H., Choi, B.S., Kwon, I.H., Cho, M.H., Yu, I.J., 2007. Changes in blood manganese concentration and MRI t1 relaxation time during 180 days of stainless steel welding-fume exposure in cynomolgus monkeys. Inhal. Toxicol. 19, 47–55.

Takeda, A., 2003. Manganese action in brain function. Brain Res. Rev. 41, 79–87.

Taylor, M.D., Clewell 3rd, H.J., Andersen, M.E., Schroeter, J.D., Yoon, M., Keene, A.M., Dorman, D.C., 2012. Update on a pharmacokinetic-centric alternative tier II program for MMT-part II: physiologically based pharmacokinetic modeling and manganese risk assessment. J. Toxicol. 2012, 791431.

Teeguarden, J.G., Dorman, D.C., Covington, T.R., Clewell 3rd, H.J., Andersen, M.E., 2007. Pharmacokinetic modeling of manganese. I. Dose dependencies of uptake and elimination. J. Toxicol. Environ. Health A 70, 1493–1504.

Thompson, K., Molina, R.M., Donaghey, T., Schwob, J.E., Brain, J.D., Wessling-Resnick, M., 2007. Olfactory uptake of manganese requires DMT1 and is enhanced by anemia. FASEB J. 21, 223–230.

Tipton, I.H., Cook, M.J., 1963. Trace elements in human tissues part II. Adult subjects from the United States. Health Phys. 9, 103–145.

Tjälve, H., Henriksson, J., 1999. Uptake of metals in the brain via olfactory pathways. Neurotoxicology 20, 181–195.

Tjälve, H., Henriksson, J., Tallkvist, J., Larsson, B.S., Lindquist, N.G., 1996. Uptake of manganese and cadmium from the nasal mucosa into the central nervous system via olfactory pathways in rats. Pharmacol. Toxicol. 79, 347–356.

Torisu, S., Washizu, M., Hasegawa, D., Orima, H., 2008. Measurement of brain trace elements in a dog with a portosystemic shunt: relation between hyperintensity on T1-weighted magnetic resonance images in lentiform nuclei and brain trace elements. J. Vet. Med. Sci. 70, 1391–1393.

Trinder, D., Oates, P.S., Thomas, C., Sadleir, J., Morgan, E.H., 2000. Localisation of divalent metal transporter 1 (DMT1) to the microvillus membrane of rat duodenal enterocytes in iron deficiency, but to hepatocytes in iron overload. Gut 46, 270–276.

Tuschl, K., Mills, P.B., Clayton, P.T., 2013. Manganese and the brain. Int. Rev. Neurobiol. 110, 277–312.

Tuschl, K., Meyer, E., Valdivia, L.E., Zhao, N., Dadswell, C., Abdul-Sada, A., Hung, C.Y., Simpson, M.A., Chong, W.K., Jacques, T.S., Woltjer, R.L., Eaton, S., Gregory, A., Sanford, L., Kara, E., Houlden, H., Cuno, S.M., Prokisch, H., Valletta, L., Tiranti, V., Younis, R., Maher, E.R., Spencer, J., Straatman-Iwanowska, A., Gissen, P., Selim, L.A., Pintos-Morell, G., Coroleu-Lletget, W., Mohammad, S.S., Yoganathan, S., Dale, R.C., Thomas, M., Rihel, J., Bodamer, O.A., Enns, C.A., Hayflick, S.J., Clayton, P.T., Mills, P.B., Kurian, M.A., Wilson, S.W., 2016. Mutations in SLC39A14 disrupt manganese homeostasis and cause childhood-onset parkinsonism-dystonia. Nat. Commun. 7, 11601.

U.S. EPA, 1993. Manganese Inhalation Reference Concentration for Chronic Inhalation Exposure (RfC). Integrated Risk Information System (IRIS). www.epa.gov/iris.

U.S. Geological Survey, 2008. Mineral Commodity Summaries. http://minerals.usgs.gov/minerals/pubs/commodity/manganese/. verified July 30, 2008.

Ulrich, C.E., Rinehart, W., Busey, W., Dorato, M.A., 1979. Evaluation of the chronic inhalation toxicity of a manganese oxide aerosol. II. Clinical observations, hematology, clinical chemistry and histopathology. Am. Ind. Hyg. Assoc. J. 40, 322–329.

Verina, T., Schneider, J.S., Guilarte, T.R., 2013. Manganese exposure induces α-synuclein aggregation in the frontal cortex of non-human primates. Toxicol. Lett. 217, 177–183.

Veuthey, T., Wessling-Resnick, M., 2014. Pathophysiology of the Belgrade rat. Front. Pharmacol. 5, 82.

Villalobos, V., Bonilla, E., Castellano, A., Novo, E., Caspersen, R., Giraldoth, D., Medina-Leendertz, S., 2009. Ultrastructural changes of the olfactory bulb in manganese-treated mice. Biocell 33, 187–197.

Wasserman, G.A., Liu, X., Parvez, F., Ahsan, H., Levy, D., Factor-Litvak, P., Kline, J., van Geen, A., Slavkovich, V., LoIacono, N.J., Cheng, Z., Zheng, Y., Graziano, J.H., 2006.

Water manganese exposure and children's intellectual function in Araihazar, Bangladesh. Environ. Health Perspect. 114, 124–129.

Wedler, F.C., Denman, R.B., 1984. Glutamine synthetase: the major Mn(II) enzyme in mammalian brain. Curr. Top. Cell. Regul. 24, 153–169.

WHO, 2001. Air Quality Guidelines, second ed. Regional Office for Europe, Copenhagen, Denmark.

Wirdefeldt, K., Adami, H.O., Cole, P., Trichopoulos, D., Mandel, J., 2011. Epidemiology and etiology of Parkinson's disease: a review of the evidence. Eur. J. Epidemiol. 26 (Suppl. 1), S1–58.

Wood, G., Egyed, M., 1994. Risk Assessment for the Combustion Products of Methylcyclopentadienyl Manganese Tricarbonyl (MMT) in Gasoline. Environmental Health Directorate, Health Canada, Ottawa, Ontario.

Wretlind, A., 1972. Complete intravenous nutrition. Theoretical and experimental background. Nutr. Metab. 14 (Suppl), 1–57.

Yamada, M., Ohno, S., Okayasu, I., Okeda, R., Hatakeyama, S., Watanabe, H., Ushio, K., Tsukagoshi, H., 1986. Chronic manganese poisoning: a neuropathological study with determination of manganese distribution in the brain. Acta. Neuropathol. (Berl) 70, 273–278.

Yoon, M., Schroeter, J.D., Nong, A., Taylor, M.D., Dorman, D.C., Andersen, M.E., Clewell 3rd, H.J., 2011. Physiologically based pharmacokinetic modeling of fetal and neonatal manganese exposure in humans: describing manganese homeostasis during development. Toxicol. Sci. 122, 297–316.

Young, T., Myers, J.E., Thompson, M.L., 2005. The nervous system effects of occupational exposure to manganese—measured as respirable dust—in a South African manganese smelter. Neurotoxicology 26, 993–1000.

Zoni, S., Lucchini, R.G., 2013. Manganese exposure: cognitive, motor and behavioral effects on children: a review of recent findings. Curr. Opin. Pediatr. 25, 255–260.

FURTHER READING

Dorman, D.C., Brenneman, K.A., McElveen, A.M., Lynch, S.E., Roberts, K.C., Wong, B.A., 2002. Olfactory transport: a direct route of delivery of inhaled manganese phosphate to the rat brain. J. Toxicol. Environ. Health A 65, 1493–1511.

Dorman, D.C., Struve, M.F., Clewell III, H.J., Andersen, M.E., 2006a. Application of pharmacokinetic data to the risk assessment of inhaled manganese. Neurotoxicology 27, 752–764.

Foster, M.L., Bartnikas, T.B., Johnson, L.C., Herrera, C., Pettiglio, M.A., Keene, A.M., Taylor, M.D., Dorman, D.C., 2015. Pharmacokinetic evaluation of the equivalency of gavage, dietary, and drinking water exposure to manganese in F344 rats. Toxicol. Sci. 145, 244–251.

Han, J.H., Chung, Y.H., Park, J.D., Kim, C.Y., Yang, S.O., Khang, H.S., Cheong, H.K., Lee, J.S., Ha, C.S., Song, C.W., Kwon, I.H., Sung, J.H., Heo, J.D., Kim, N.Y., Huang, M., Cho, M.H., Yu, I.J., 2008. Recovery from welding-fume-exposure-induced MRI T1 signal intensities after cessation of welding-fume exposure in brains of cynomolgus monkeys. Inhal. Toxicol. 20, 1075–1083.

Penney, D.G., 1990. Acute carbon monoxide poisoning: animal models: a review. Toxicology 62, 123–160.

Silva, A.C., Lee, J.H., Aoki, I., Koretsky, A.P., 2004. Manganese-enhanced magnetic resonance imaging (MEMRI): methodological and practical considerations. NMR Biomed. 17, 532–543.

CHAPTER SEVEN

Roles of Microglia in Inflammation-Mediated Neurodegeneration: Models, Mechanisms, and Therapeutic Interventions for Parkinson's Disease

Hui-Ming Gao*,†,1, Dezhen Tu*,†, Yun Gao*,†, Qiyao Liu*, Ru Yang*, Yue Liu*, Tian Guan*, Jau-Shyong Hong†

*Model Animal Research Center and MOE Key Laboratory of Model Animal for Disease Study, Nanjing University, Nanjing, Jiangsu, China
†Neurobiology Laboratory, National Institute of Environmental Health Sciences/National Institutes of Health, Research Triangle Park, NC, United States
1Corresponding author: e-mail addresses: gaohm@nju.edu.cn; gao2@niehs.nih.gov

Contents

Advances in Neurotoxicology, Volume 1
ISSN 2468-7480
http://dx.doi.org/10.1016/bs.ant.2017.07.005

1. INTRODUCTION

Parkinson's disease (PD), an age-related neurodegenerative movement disorder, affects more than 53 million people worldwide. A progressive degeneration of the nerve terminals in the striatum and the cell bodies of dopamine neurons in the substantia nigra (SN) eventually leads to the development of progressive movement disorders (Jellinger, 2001). PD progresses insidiously for 5–7 years before manifestation of apparent clinical symptoms when 50%–60% of nigral dopamine neurons are lost. PD neurodegeneration continues to worsen even under the symptomatic treatment. Postmortem analysis of the brains of PD patients frequently shows the cytoplasmic inclusions, known as Lewy bodies (Holdorff, 2002; Schiller, 2000) in the remaining dopamine neurons in the SN. Currently, the etiology of PD remains unclear. Epidemiological studies have revealed that most PD cases are sporadic and have a late onset (Tanner, 2003). About 10% of PD cases are characterized by early onset, which primarily occurs in familial clusters (Mizuno et al., 2001) and has been attributed to mutations in several genes, such as parkin, leucine-rich repeat kinase 2 (LRRK2), and α-synuclein (Jiang et al., 2007). At present, available therapeutics only temporarily relieve PD symptoms but fail to stop or slow down neurodegeneration. Therefore, to determine what triggers the disease onset and what drives the chronic, self-propelling neurodegenerative process is critical for the discovery of effective treatments to retard PD progression. For decades, PD has been shown to be strongly linked to environmental exposures such as infectious agents, pesticides, and heavy metals (Betarbet et al., 2000; Di Monte et al., 2002; Gao and Hong, 2011; Guilarte, 2010; Jang et al., 2009, 2012; Rock and Peterson, 2006; Zheng et al., 2011). Currently, development of PD may represent the final outcome of a complex set of interactions among the innate vulnerability of the nigrostriatal dopaminergic system, potential genetic predisposition, and exposure to environmental toxins/toxicants. The recent acceptance of the critical role of neuroinflammation in PD neurodegeneration has spurred hope that the disease course may be altered and effective therapeutic interventions can be developed. However, there remains a dearth of information on the complex mechanisms through which neuroinflammation influences dopamine neuronal survival. Recent advance in understanding how neuroinflammation, as triggered by environmental risk factors, participates in the pathogenesis of PD provides promise to develop novel microglia-based therapies to halt the progression of PD.

2. THE ENVIRONMENTAL CAUSES OF PD

Recent studies uncover genetic defects in 10%–20% of PD cases, most PD cases are idiopathic. The accidental use of 1-methyl-4-phenyl-1,2,3-tetrahydropyridine (MPTP), a by-product of heroin synthesis, causes a Parkinson syndrome that is clinically indistinguishable from PD (Langston et al., 1983). This finding prompted the search for environmental factors as potential causes of PD. Indeed, epidemiological and case–control studies have shown association of rural residence, well-water consumption, pesticide use, and certain occupations (e.g., farming, mining, and welding) with an increased risk of PD (Dhillon et al., 2008; Elbaz et al., 2009; Gorell et al., 1998; Semchuk et al., 1992). Particularly, several specific pesticides including dieldrin (an organochlorine pesticide), maneb (a widely used fungicide), paraquat (a common herbicide with similar structure to 1-methyl-4-phenylpyridinium ion (MPP^+), the active metabolite of MPTP), and pesticide rotenone have recently been suggested to increase the risk for PD by recent epidemiological studies (Brown et al., 2006; Dhillon et al., 2008; Gao and Hong, 2011; Kamel et al., 2007; Liu et al., 2003; Ritz et al., 2009). Miners exposed to high levels of manganese displayed accumulation of this heavy metal in the basal ganglia and PD-like tremors, rigidity, and psychosis (Mergler and Baldwin, 1997). Experimental exposure of animal models to several classes of pesticides, such as paraquat, maneb, rotenone, and dieldrin, leads to dopaminergic neurotoxicity. These pesticides have hence been proposed as potential PD risk factors in humans (Liu et al., 2003).

Gene–environment interactions affect PD susceptibility. Dozens of single-nucleotide polymorphisms (SNPs) in several genes have shown gene–environment interactions in relation to PD. For instance, high levels of exposure to paraquat and maneb and occupational pesticide exposure in carriers of genetic variants in dopamine transporter increased PD risk (Kelada et al., 2006; Ritz et al., 2009). Interactions between heavy solvent exposure (Dick et al., 2007) and GSTM1 (glutathione S-transferase Mu 1)-null genotype and between SNPs in apolipoprotein E and coffee consumption (McCulloch et al., 2008) also affect PD susceptibility. More large-scale human association studies and in vivo experimental investigations will help to identify specific causative environmental toxins/toxicants or infectious agents (DNA or antigens) for PD.

3. NEUROINFLAMMATION AND MICROGLIA

Neuroinflammation (inflammation in the brain) primarily involves the activities of two types of glial cells: microglia and astroglia. Microglia are the resident macrophages of the brain (del Rio-Hortega, 1932). During late embryonic and early postnatal brain remodeling and maturation, microglia participate in the programmed elimination of neural cells (Barron, 1995; Milligan et al., 1991). In mature brains, resting microglia exhibit a characteristic ramified morphology and actively partake in immune surveillance. As the first line of defense in the brain, microglia can rapidly polarize into an activated state in response to brain injuries and immunological stimuli (Kreutzberg, 1996; Liu and Hong, 2003; Streit et al., 1988, 1999). Activated microglia undergo dramatic changes, metamorphosing into an amoeboid morphology (Kreutzberg, 1996) and increasing the expression of many surface molecules (e.g., complement receptors and major histocompatibility complex molecules) (Graeber et al., 1988; Oehmichen and Gencic, 1975). Activated microglia respond to injurious signals by migrating to the site of injury, releasing factors to recruit more cells and phagocytizing the foreign substances. During this process microglia can release a variety of proinflammatory and potentially cytotoxic soluble factors. Microglia can serve neuroprotective and neurotoxic functions, depending on the pathophysiological conditions. In normal or mild inflammatory states, microglia exhibit beneficial immune surveillance and clearance of noxious stimuli. Under strict regulation by a constant, complex interaction between the immune, endocrine, and nervous systems, neuroinflammation is normally self-limiting and is essential for the integrity of the central nervous system. However, escaping from its tight regulation, the immune reaction can become exaggerated and destructive, and turns into chronic persistent inflammation. In chronic inflammatory conditions, microglia can be neurotoxic and significantly contribute to neurodegeneration.

Besides microglia, astroglia are essential to the integrity and function of the brain. Astroglia provide physical contact and nutrition to neurons, maintain ionic homeostasis, buffer excess neurotransmitters, secrete neurotrophic factors, and serve as an important component of the blood–brain barrier. In response to immunologic challenges or brain injuries, astroglia also become activated (Aloisi, 1999; Tacconi, 1998). Activated astroglia produce a host of neurotrophic factors (Lindsay, 1994; Pavlov et al., 2003), which are crucial for the survival of neurons. Balanced interactions among neurons, microglia, and astroglia are critical in maintaining brain immune homeostasis. Further studies on the modulation of neuroinflammation should provide insights promising novel therapies for PD.

4. NEUROINFLAMMATION IN PD

Multiple lines of evidence have indicated involvement of inflammation in PD. Epidemiologic studies and case reports show an association between early-life viral infections and postencephalitic PD or acute Parkinsonism (Duvoisin et al., 1972; Elizan and Casals, 1991; Gao and Hong, 2011; Ghaemi et al., 2000; Liu et al., 2003). A prospective study reveals a correlation between a higher plasma concentration of proinflammatory cytokine interleukin-6 and an increased risk for PD (Chen et al., 2008). Hypothesis-free GWASs lend strong and independent support to the participation of inflammation in PD pathogenesis. For example, DNA polymorphisms of several inflammatory cytokines and genetic variation in the human leukocyte antigen region that contains a large number of human genes related to immune function might become risk factors for PD (Hamza et al., 2010; Wahner et al., 2007). Associations between chronic use of nonsteroidal antiinflammatory drugs (NSAIDs) and PD risk are inconclusive, with a reduced incidence or no association in different studies (Chen et al., 2003, 2005; Powers et al., 2008; Ton et al., 2006). As discussed later, ample experimental evidence strongly supports the participation of brain inflammation in the pathogenesis of PD.

5. INFLAMMATION-ASSOCIATED PD MODELS GENERATED WITH ENVIRONMENTALLY RELEVANT TOXINS

To study the role of chronic neuroinflammation in the pathogenesis of PD, it is essential to develop new models that display delayed, progressive features of PD. Most of the previous animal models of PD failed to recapitulate keys features of PD especially the delayed, progressive degeneration of dopamine neurons in the SN seen in PD patients. To determine how neuroinflammation affects PD neurodegeneration, several inflammation-associated acute, subacute, and chronic PD models have been created. Direct nigral injection of the Gram-negative bacterial endotoxin lipopolysaccharide (LPS) causes robust microglial activation and acute loss of nigral dopamine neurons in rodents (Castano et al., 1998; Liu et al., 2000; Tomas-Camardiel et al., 2004). Chronic nigral infusion of LPS for 2 weeks triggered a rapid activation of microglia, followed by a delayed, selective, and gradual loss of nigral dopamine neurons (Gao et al., 2002). This is the first report that microglial activation induced by chronic exposure to inflammagen could produce a delayed, progressive, and selective degeneration of nigral dopamine neurons. Prenatal exposure to LPS leads to not only lesions of nigrostriatal dopaminergic system in neonates but also increased vulnerability to progressive dopaminergic neurodegeneration elicited by chronic nigral LPS infusion in adult rats (Ling et al., 2002, 2006). Moreover, an intraperitoneal injection of LPS (5 mg/kg) led to significant loss of nigral dopamine neurons in C57BL/6 mice 7 months after the injection; such neurodegeneration progressed further overtime (Qin et al., 2007). This inflammation-elicited chronic PD model recapitulates the delayed, progressive nature of PD and reproduces locomotor impairments in rodents (Gao et al., 2011a; Qin et al., 2007). Clinically, the relevance of our LPS models for PD research is supported by a recent case–control that found a significant correlation between infections and PD patients (Vlajinac et al., 2013).

The delayed and progressive process of degeneration in LPS-elicited chronic PD models provide an opportunity to study the nonmotor symptoms of PD. PD is not merely a movement disorder. There is considerable evidence showing that nonmotor alterations such as changes in olfactory function (hyposmia/anosmia), gastrointestinal disturbances, sleep abnormalities, anxiety, and depression are integral components of PD (Chaudhuri et al., 2006; Lima et al., 2012; Ziemssen and Reichmann, 2007). These nonmotor symptoms often have great impact on disability and quality of life of PD patients. Furthermore, because these nonmotor symptoms usually manifest before the onset of motor symptoms, they are increasingly being recognized as critical events for the early diagnosis and treatment of PD. Although the changes in neuroanatomical and neurochemical features of the nonmotor symptoms in PD are largely unknown, disturbances of nondopaminergic transmitters such as norepinephrine (NE) and serotonin (5-HT) are thought to be involved in nonmotor symptoms (Dauer and Przedborski, 2003; Rommelfanger and Weinshenker, 2007). According to Braak's staging hypothesis of PD, neuronal loss in the brain stem nuclei, such as locus coeruleus (LC; the main NE-containing brain region) and the raphe nuclei (a main 5-HT-containing neuron region), may occur earlier than that of nigral dopamine neurons (Braak et al., 2003). Although information about the role of NE/5-HT on nonmotor symptoms in PD is limited, they are associated with the development of other disorders such as depression, attention deficit, and sleep disorders, many of which are observed in most PD patients. Currently, the research of nonmotor symptoms remains a challenge due to acute neurotoxicity and quick onset of symptoms in widely used PD model created by using MPTP or 6-OHDA (Doty et al., 1992; Jenner and Marsden, 1986). The examination of the temporal loss of neurons in the LC and the SN following a single systemic injection of LPS in C57BL mice showed a time-dependent loss for both dopamine and NE neurons, where NE neurons loss in the LC precedes the loss of nigral dopamine neurons by 3–4 months. Characterization of the correlation between nonmotor symptoms (e.g., olfactory discrimination, delayed gastric emptying, altered sleep latency, anxiety-like behavior, and depressive behaviors) and LC degeneration in LPS-injected mice will provide key evidence for the role of degeneration of NE neurons in the genesis of nonmotor symptoms of PD. Collectively, inflammation-associated chronic PD models not only provide useful tools for studying the critical role of microglia in inflammation-related neurodegeneration but also serve as a good platform for testing the glia-based novel therapies for PD.

6. INFLAMMATION-ASSOCIATED TWO-HIT PD MODELS

Although previous systemic LPS model of PD has been a useful tool, there are limitations to this model. It requires up to 5–6 months for LPS-injected mice to manifest PD-like motor deficits and dopamine neurodegeneration (Qin et al., 2007). Furthermore, this model fails to develop α-synuclein-containing Lewy body-like cytoplasmic inclusions, a pathological hallmark of PD. To study roles of gene–environment interactions in the pathogenesis of PD, a new accelerated PD model has been created by a systemic injection of LPS into transgenic mice overexpressing A53T mutant human α-synuclein gene. This new model shows early onset, pronounced loss of dopamine neurons, and accumulation of pathologically modified insoluble, aggregated α-synuclein as well as formation of Lewy body-like cytoplasmic inclusions (Gao et al., 2011a). α-Synuclein transgenic mice were chosen to generate this two-hit PD model was based on the following three reasons: (1) the multiplications of human α-synuclein gene and missense mutations of α-synuclein gene are responsible for autosomal dominant inherited PD (Kruger et al., 1998; Polymeropoulos et al., 1997; Zarranz et al., 2004); (2) abnormal α-synuclein aggregates are the major components of Lewy bodies in both familial and sporadic PD; and (3) mutant α-synuclein triggers potent microglia-mediated neurotoxicity in neuron–glia cultures (Zhang et al., 2005, 2007).

Another inflammation-associated two-hit PD model has been created by intraperitoneal administration of low dose of LPS twice a week into parkin-null mice for 3–6 months (Frank-Cannon et al., 2008). Loss-of-function mutations in the parkin gene cause early-onset familial PD, but parkin-deficient mice do not display nigrostriatal degeneration. Genetic deficiency in parkin sensitizes nigral dopamine neurons to inflammation-mediated neurotoxicity and subtle fine-motor deficits (Frank-Cannon et al., 2008). Collectively, these inflammation-associated two-hit PD models reproduced the signature lesion of PD, chronic, progressive, and relatively selective degeneration of dopamine neurons in the SN. These models provide a useful tool for exploring mechanisms underlying persistent neuroinflammation in chronic PD progression and provide an excellent opportunity for studying the gene–environment interactions.

7. MECHANISM OF INFLAMMATION-MEDIATED PROGRESSIVE NEURODEGENERATION

Epidemiological and clinical studies indicate a 5–10-year lag between the onset of PD pathogenesis and the appearance of motor symptoms in PD patients. However, the mechanism underlying the progressive nature of this disease remains elusive. With increased awareness of the important contribution of neuroinflammation to PD pathogenesis, it is critical to understand how chronic, low-grade neuroinflammation mediates progressive neurodegeneration. To achieve this, three questions must be addressed: (1) How is neuroinflammation initiated? (2) How does acute neuroinflammation become chronic? (3) How does neuroinflammation trigger progressive neurodegeneration? Based on findings obtained from LPS-elicited PD models, we have formulated a working model whereby interactions between damaged neurons and overactivated microglia create a vicious self-propelling cycle, which leads to uncontrolled, prolonged neuroinflammation that drives the chronic progression of neurodegeneration in PD (Fig. 1).

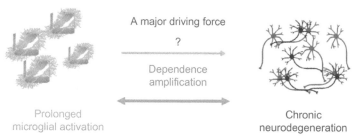

A major driving force

?

Dependence
amplification

Prolonged   Chronic
microglial activation   neurodegeneration

Fig. 1 Interdependency of prolonged microglial activation and chronic, progressive neurodegeneration. Microglial activation and neuronal damages amplify each other to form a vicious cycle. Such positive feedback maintains chronic neuroinflammation thereby mediating progressive neurodegenerative process. Uncontrolled chronic neuroinflammation may be a driving force of progressive neurodegeneration. It is important to determine what factors mediate the constant interaction between damaged neurons and activated microglia thereby triggering and maintaining the formation of the vicious cycle. Inflammation may determine the progression and outcome of PD and other neurodegenerative diseases.

7.1 Microglia Are Essential for Progressive Loss of Dopamine Neurons

A wealth of evidence indicates that overactivation of microglia is an important contributor to the demise of dopamine neurons in PD (Gao and Hong, 2008). To address whether microglial activation is indispensable for the progressive neuronal loss in PD, the survival of dopamine neurons in the presence or absence of microglia after the cultures were stimulated with three toxins widely used to generate PD models: (1) LPS activates microglia and initiates direct immunologic responses that consequently leads to neurodegeneration; (2) 1-methyl-4-phenylpyridinium (MPP^+), the active metabolite of neurotoxin MPTP, directly damages neurons and induces reactive microgliosis (a secondary microglial reactivation) (Gao et al., 2003a); and (3) rotenone, a well-studied pesticide with dual mode of actions: directly damages neuronal function by impairing mitochondrial complex I and activates microglia by increasing the production of superoxide free radicals (Betarbet et al., 2000; Gao et al., 2002). In neuron–glia cultures that contain neurons, astroglia, and microglia, all three toxins, LPS, rotenone, and MPP^+, induced progressive dopamine neuron loss. In contrast, in neuron-enriched cultures, while LPS produced no neurotoxicity, both MPP^+ and rotenone caused acute neurotoxicity, but the neurotoxicity failed to progress further after prolonged treatment. These findings indicate that activated microglia are required for chronic, progressive neurodegeneration,

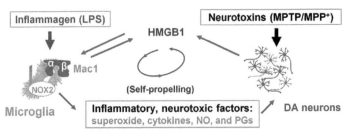

Fig. 2 HMGB1-Mac1-NOX2 signaling axis bridges chronic neuroinflammation and progressive neurodegeneration. Some environmental risk factors such as inflammagen LPS can directly activate microglia. Activated microglia produce and secrete a spectrum of inflammatory mediators, such as cytokines, eicosanoids, chemokines, reactive free radicals, and proteases. Of special interest, activated microglia can also secrete HMGB1 (a DNA-binding nuclear protein). When released in excess quantity, these inflammatory mediators including HMGB1 not only can exaggerate microglial activation but also can damage surrounding neurons. On the other hand, some environmental toxins (e.g., neurotoxin MPTP/MPP$^+$ or pesticide rotenone) can directly trigger neuronal lesions. When neurons are seriously injured and cell membrane breakdown occurs, cytosolic or nuclear compounds (e.g., α-synuclein and HMGB1) will be released or leaked into the extracellular milieu to activate surrounding microglia. Thus, regardless of the type of the initial lesion, neuronal damage and uncontrolled inflammation amplify each other, inducing a vicious self-propagating cycle that causes the chronic progression of neurodegenerative diseases. HMGB1, released from activated microglia and/or degenerating neurons, binds to microglial Mac1 and activates NADPH oxidase (NOX2) to stimulate production of inflammatory and neurotoxic factors such as superoxide, cytokines, nitric oxide, and postglands. Collectively, extracellular HMGB1 may become an important factor mediating constant neuron–microglia interactions and maintaining prolonged neuroinflammation.

regardless of the nature of the initial insult (inflammagens or neurotoxicants) (Gao et al., 2011b; Fig. 2).

To further evaluate the dependence of progressive dopamine neurodegeneration on activated microglia, a reconstituted cell culture system using transwell inserts was utilized to separate neurons from microglia by a permeable membrane that allows soluble factors to diffuse between compartments. In the presence of microglia, LPS treatment for 7 days leads to 50% loss of dopamine neurons, whereas the removal of activated microglia 24 h after LPS treatment significantly attenuated this loss. Furthermore, the removal of activated microglia along with its conditioned media completely abrogated LPS neurotoxicity. These findings support that soluble factors released by microglia are essential in progressive neurodegeneration. Washout of neuron–glia cultures 1 day after LPS treatment or 2 days after MPP$^+$ treatment did not prevent dopamine neurons from additional degeneration,

indicating that continuing presence of the initial stimulants was not necessary for sustained microglial activation and progressive neurodegeneration
(Gao et al., 2011b). These results support the theory that once the activation
of microglia reaches a certain threshold, it can result in sufficient neuronal
damage to propagate reactive microgliosis continuing the cytotoxic inflammatory process even after the initial stimuli have been cleared (Fig. 2). These
findings, combined with in vivo data showing that mice develop delayed and
progressive nigral dopamine neurodegeneration long after LPS has been
cleared from the circulation (Qin et al., 2007), support the notion that neurodegeneration can progress independently in the absence of the initial trigger (e.g., LPS and MPP$^+$). Thus, continued presence of initial stimuli is not
required for progressive neurodegeneration (Gao et al., 2011b).

7.2 Interdependency of Prolonged Microglial Activation and Chronic, Progressive Neurodegeneration

Given the requirement of microglial activation for progressive dopamine
neurodegeneration, it is crucial to address how the activation of microglia
is maintained. LPS treatment caused a long-lasting upregulation of Iba-1
(a marker of microglia) in neuron–glia cultures, which paralleled dopamine
neuronal loss. In contrast, LPS-treated mixed glia cultures (devoid of neurons) produced a short-lived increase of Iba-1. These results suggest an
essential role of ongoing neuronal damage/death for maintaining microglial
activation and chronic neuroinflammation.

A spectrum of noxious endogenous compounds in extracellular milieu
can activate microglia in cultures. These compounds include membrane
breakdown products, abnormally processed, modified or aggregated proteins (e.g., laminin) (Wang et al., 2006), and leaked cytosolic compounds
(e.g., α-synuclein and neuromelanin) (Fig. 3; Zhang et al., 2007, 2011a).
The conditioned media from MPP$^+$-treated dopamine neuron cell line
N27 caused microglial activation and severe dopamine neurotoxicity when
added into primary neuron–glia cultures (Levesque et al., 2010). Thus, activated microglia and damaged neurons formed a vicious cycle mediating progressive neurodegeneration.

7.3 HMGB1-Mac1-NOX2 Signaling Axis Mediates Chronic Neuroinflammation and Subsequent Progressive Neurodegeneration

Determination of neuron-derived activator inducing reactive microgliosis
revealed an important role of nuclear protein high-mobility group box 1

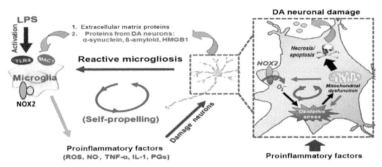

Fig. 3 Dysregulated neuroinflammation drives progression of neurodegeneration. Activated microglia, resulting from either direct activation by inflammagen or a secondary reaction to neuronal injury (termed by reactive microgliosis), can release a variety of proinflammatory factors to damage neurons. A variety of noxious self-compounds in the extracellular milieu, such as membrane breakdown products, abnormally modified, processed, or aggregated proteins (e.g., α-synuclein, β-amyloid, and extracellular matrix proteins), and released or leaked cytosolic/nuclear compounds (e.g., α-synuclein, neuromelanin, and HMGB1) from damaged neurons can induce reactive microgliosis. Sustained microglial activation and continued neuronal death formed a self-propelling, vicious cycle. NOX2-derived oxidants from activated microglia, such as H_2O_2 (nonradical oxidants) and peroxynitrite (a reactive product of superoxide and nitric oxide), can enter neurons and lead to (1) mitochondrial impairment/dysfunction, reduced ATP production, and increased generation of mitochondria-derived ROS; (2) protein oxidation, nitration, aggregation, and accumulation; (3) oxidation of DNA, RNA, and lipids; and (4) impaired redox-sensitive signal transduction. Thus, inflammation-derived oxidative stress leads to another vicious cycle inside the damaged neurons and causes neuronal death. NOX2-derived ROS from activated microglia and damaged neurons mediates chronic oxidative neuronal damage in PD.

(HMGB1) in mediating microglia–neuron interactions. HMGB1, a nonhistone DNA-binding protein, modulates interactions of various transcription factors with DNA in the nucleus. HMGB1 can be passively released from necrotic cells. Interestingly, HMGB1 can also be actively secreted by inflammatory cells to initiate inflammatory responses. Extracellular HMGB1 was detected in neuron–glia culture media after treatment with LPS, MPP$^+$, or rotenone. The concurrent administration of HMGB1 neutralizing antibody with MPP$^+$, rotenone, or LPS attenuated dopamine neuronal loss. These findings indicate that HMGB1 derived from activated microglia or damaged neurons plays a critical role in mediating neuron–microglia interactions and inflammation-associated progressive neurodegeneration.

HMGB1 is known to signal through multiligand receptors, including toll-like receptor 2 (TLR2), TLR4, and RAGE (receptor for advanced glycation end products) in mediating inflammatory responses. Its structural conformation allows HMGB1 to function as a damage-associated molecular pattern molecule (DAMP). DAMPs activate pattern-recognition receptors (PRRs) and signal to the immune system of tissue injury to elicit a non-infectious inflammatory response. Coimmunoprecipitation experiments and a binding analysis of microglial membrane fractions with purified recombinant HMGB1 protein identified macrophage antigen complex 1 (Mac1) as a potential receptor for HMGB1 (Gao et al., 2011b). Mac1 belongs to the β2 integrin family and also functions as a PRR (Fan and Edgington, 1993; Ross and Vetvicka, 1993). The physical interaction between HMGB1 and Mac1 was further supported by functional assay that shows much less production of proinflammatory factors (e.g., tumor necrosis factor-α, interleukin-1β, nitrite oxide, and superoxide) in $Mac1^{-/-}$ microglia than wild-type ($Mac1^{+/+}$) microglia after HMGB1 stimulation. Phagocyte NADPH oxidase (NOX2) is an important downstream effector for Mac1 (Mayadas and Cullere, 2005; Pei et al., 2007). NOX2 is dormant in resting microglia and is separated into individual cytosolic and membrane-bound components. A variety of stimuli including some cellular components (e.g., HMGB1, α-synuclein, and β-amyloid) released from activated microglia and/or damaged neurons can act on microglial Mac1 receptor and activate downstream kinases (e.g., PI3K)—leading to phosphorylation and membrane translocation of the cytosolic subunits of NOX2, $p47^{phox}$ and $p67^{phox}$, leading to the assembly of an active NOX2 and production of the free radical superoxide (Gao et al., 2011b). HMGB1-induced release of extracellular superoxide was only seen in $Mac1^{+/+}$ microglia, but not $Mac1^{-/-}$ or $gp91^{phox-/-}$ microglia. Cotreatment of wild-type microglia with NOX2 inhibitors, diphenyleneiodonium (DPI), or apocynin prevented the superoxide release. Neutralization of HMGB1 and genetic ablation of Mac1 and $gp91^{phox}$ (the catalytic subunit of NOX2) blocked the progressive neurodegeneration. Collectively, HMGB1-Mac1-NOX2 signaling axis bridged chronic neuroinflammation and progressive neurodegeneration (Gao et al., 2011b).

Our recent findings that laminin (Wang et al., 2006), α-synuclein (Zhang et al., 2007), β-amyloid (Zhang et al., 2011b), neuromelanin (Zhang et al., 2011a), matrix metalloproteinase (Kim et al., 2007), and double-stranded RNA (a well-studied viral ligand for TLR3) (Zhou et al., 2013) were able to activate microglia/macrophages through the binding to Mac1 receptor and subsequent activation of the Mac1-NOX2 pathway (Fig. 3). These findings strongly indicate that Mac1 serves as a PRR for a wide spectrum of exogenous/endogenous substances of varying chemical structures and molecular sizes and that the Mac1–NOX2 mediates persistent neuroinflammation, contributing to progressive neurodegeneration.

7.4 NOX2-Derived Oxidative Stress Is a Key Mediator of Inflammation-Induced Progressive Dopamine Neurodegeneration

Oxidative stress is believed to be a major component of age-related diseases such as cancer, strokes, cardiovascular, and neurodegenerative diseases. Given their intimate relationship, oxidative stress and inflammation become integral and inseparable elements of PD pathogenesis. Oxidative stress-related neuronal damage is increased in aging brains and is accelerated in chronic neuroinflammatory conditions. In vivo measurements of dihydroethidium (DHE) oxidation showed an age-dependent (up to 20 months of age) elevation in the levels of reactive oxygen species (ROS) in mouse brain neurons. Interestingly, LPS-injected mice displayed accelerated, intensified increases in the oxidative products of DHE (Qin et al., 2013) and lipid peroxides. Proinflammatory factors generated from activated microglia can promote a progressive increase in neuronal oxidative stress and eventually cause neuronal death in an age-dependent, progressive manner (Fig. 3).

Among different isoforms of NADPH oxidase, NOX2 is accepted as a key enzyme associated with inflammatory oxidative that is highly expressed in macrophages and microglia. Recent experimental evidence has highlighted a pivotal role of overactivated NOX2 in chronic neuroinflammation and progressive neurodegeneration. Deficiency in subunits of NOX2 attenuates neuronal damage induced by diverse insults/stresses relevant to PD (Gao et al., 2012). More importantly, suppression of NOX2 activity correlates with reduced neuronal impairment in PD models. The discovery of NOX2 and nonphagocyte NADPH oxidases in neurons and astroglia further reinforces the crucial role of NADPH oxidases in oxidative stress-mediated chronic neurodegeneration.

Excessive free radicals can damage proteins, lipids, DNA, or RNA, leading to cell dysfunction and eventual cell death. Oxidative stress engages in every aspect of major PD pathological events including protein misfolding and aggregation, mitochondrial dysfunction and impairment, UPS perturbation, aberrant signal transduction, and disruption of intracellular protein homeostasis. Free radicals can oxidize α-synuclein exacerbating its misfolding and accumulation or attack protein surveillance machineries by oxidizing their subunits or key enzymes (e.g., ubiquitin-activating

[E1], –conjugating [E2], and –ligating [E3] enzymes, three enzymes responsible for protein ubiquitination). When the accumulation of misfolded proteins exceeds the capacity for the chaperone, ubiquitin–proteasome, and autophagy systems to detect, to repair, and eventually to destroy faulty proteins, especially when protein surveillance machineries are malfunction under oxidative condition, the buildup of toxic misfolded proteins in the cell will precipitate cellular demise. Oxidative stress appears as a converging point of multiple pathways underlying PD pathogenesis and plays a central role in the chronic progression of PD. Sustained neuronal damage may trigger a self-propelling cycle within neurons to further enhance neuronal NOX2 activity and produce more intracellular ROS. The accumulative oxidative stress in neurons causes damage of mitochondria and leads to neuronal death (Fig. 3).

8. MICROGLIA-BASED NOVEL DISEASE-MODIFYING THERAPIES FOR PD

Increased awareness of the important contribution of neuroinflammation in PD pathogenesis will stimulate research toward new glia-based therapeutics. Since the self-propelling cycle between damaged neurons and activated microglia may become a driving force of progressive neurodegeneration, any interventions that can either halt or dampen this cycle should be effective in preventing further progression of inflammation-mediated neurodegeneration (Gao et al., 2003b). Although encouraging evidence showing that inhibition of inflammation correlates with attenuated neurodegeneration in animal studies (Block et al., 2007), results from clinical studies are less clear. The low success rate of translating antiinflammatory drugs from animal models to human diseases highlights the need for better strategies and use of appropriate animal models for preclinical drug screening. It is critical that animal models used for screening therapeutic drugs should recapitulate the progressive features of PD. In this regard, we believe that the delayed, progressive LPS models are well suited for screening potential disease-modifying drugs. Modulation of microglial activation to controllable level can improve neuronal survival (Fig. 4).

8.1 Antiinflammatory Therapies

Several epidemiological reports indicate that frequent use of the NSAIDs is associated with a lower risk for PD (Chen et al., 2003, 2005). Most NSAIDs are designed to target a limited number of proinflammatory factors released

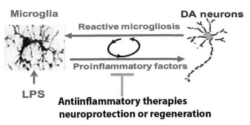

Fig. 4 Microglia-based therapeutic strategies. The modulation of inflammatory processes aimed at preventing or interrupting the self-propelling vicious cycle between damaged neurons and reactive microgliosis has the potential to emerge as a novel therapeutic strategy. Novel antiinflammatory therapies, combined with strategies to improve neuroprotection (e.g., neurotrophic factors) or neuroregeneration, may become promising disease-modifying therapeutics for PD.

from immune cells under inflammatory conditions. For example, COX-2 inhibitors mainly reduce the production of prostaglandins, without affecting other factors. This narrow spectrum of action limits the efficacy of the agent as a general antiinflammatory drug. Nonetheless, given the failure of recent clinical trials, the potential of NSAIDs for the treatment of PD remains unclear. A novel class of antiinflammatory drugs targeting upstream neuro-inflammatory signaling by inhibiting microglial NOX2—which in turn reduces superoxide production and overactivation of microglia and thereby reducing the release of most proinflammatory factors (Qin et al., 2005a). This novel class of antiinflammatory drugs is more efficacious than most of the conventional regimens (Qin et al., 2005b) and thus has received wide attention (Fig. 5).

8.2 Neuroprotective Effects of NOX2 Inhibition in PD Models

Several structurally and functionally different compounds that share common properties of inhibiting NOX2 activity and suppressing inflammation provide significant neuroprotection in PD models. These drugs include dextromethorphan (a widely used antitussive agent), sinomenine (a natural dextrorotatory morphinan analog), squamosamide derivative FLZ, pituitary adenylate cyclase-activating polypeptides, TGF-b1 (transforming growth factor-b1, a known endogenous immune modulator), verapamil, and resveratrol (a nonflavonoid polyphenol with antioxidant and antiinflammatory properties). All these drugs inhibit superoxide release from LPS-stimulated wild-type microglia; more importantly, they fail to exhibit neuroprotection in cultures from NOX2-deficient mice (Liu et al., 2011; Qian et al., 2007a, 2008; Yang et al., 2006; Zhang et al., 2008, 2010; Fig. 6).

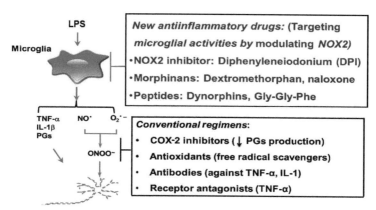

Fig. 5 Novel antiinflammatory therapies. Conventional antiinflammatory therapies are designed to target a limited number of proinflammatory factors released from inflammatory cells. For example, COX-2 inhibitors mainly reduce the production of prostaglandins, without affecting other factors. This narrow spectrum of action limits the efficacy of the agent as a general antiinflammatory drug. A novel class of antiinflammatory drugs targeting upstream neuroinflammatory signaling by inhibiting microglial NOX2—which in turn reduces superoxide production and overactivation of microglia and thereby reducing the release of most proinflammatory factors. This novel class of antiinflammatory drugs is more efficacious than most of the conventional regimens. The main advantages of these novel antiinflammatory drugs are a broad spectrum of action by preventing the overproduction of several proinflammatory factors.

8.3 Subpicomolar DPI, Naloxone, and Dextromethorphan Protect Dopamine Neurons in Neuroinflammation-Mediated Models of PD by Targeting NOX2

Both naloxone and dextromethorphan display antiinflammatory and neuroprotective effects in neuron–glia cultures at the concentration of 10^{-13} or 10^{-14} M (Li et al., 2005; Qin et al., 2005b; Wang et al., 2012). The commonly used NOX2 inhibitor DPI (a widely used and long-acting NOX2 inhibitor that forms a covalent bond with gp91^{phox}) also exhibits specifical inhibition on NOX2 and displays potent neuroprotection both in vitro and in vivo models of PD (Qian et al., 2007b; Wang et al., 2014, 2015). At its recommended micromolar doses, DPI cannot be used clinically due to its nonspecificity and toxicological profile (Aldieri et al., 2008), and at subpicomolar concentrations, DPI protects dopamine neurons from LPS-induced damage in neuron–glial cultures (Qian et al., 2007b; Wang et al., 2014). Posttreatment with subpicomolar DPI also exhibited potent neuroprotection against a variety of toxins (including LPS, MPTP/MPP^{+}, or rotenone) in neuron–glia cultures prepared from wild-type mice

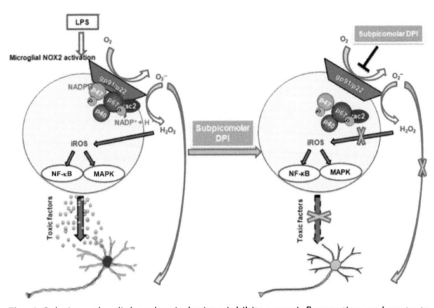

Fig. 6 Subpicomolar diphenyleneiodonium inhibits neuroinflammation and protects dopamine neurons from toxic insults. The phagocyte NADPH oxidase (NOX2) is dormant in resting microglia and is separated into individual cytosolic and membrane-bound components. In activated microglia, translocation of p40^{phox} and phosphorylated p47^{phox} and p67^{phox} and binding to the membrane-bound subunits, p22^{phox}/gp91^{phox} and Rac2, lead to the assembly of an active NOX2 and production of superoxide. Superoxide is biologically very toxic. It can be converted to hydrogen peroxide (H_2O_2) under the enzyme catalyzation by superoxide-scavenging enzyme superoxide dismutase. H_2O_2 can enter neurons inducing oxidative damages or enter microglia to activate NF-κB and MAPK pathways leading to release of more inflammatory and cytotoxic factors to damage neurons. NOX2 inhibitor diphenyleneiodonium (DPI) can directly bind to gp91^{phox} to inhibit NOX2 activity. Ultralow doses of DPI show high specificity toward NOX2 and are capable of reducing microglial activation and their subsequent release of proinflammatory factors. Posttreatment with an ultralow dose of DPI is still capable of protecting dopamine neurons.

but not gp91^{phox}-deficient mice (Wang et al., 2014, 2015). Posttreatment with DPI at an ultralow dose inhibits neuroinflammation and protects dopamine neurons from toxic insults in vivo (Wang et al., 2015). The action of DPI on NOX2 was associated with preventing the translocation of the cytosolic subunit p47^{phox} to the membrane (Wang et al., 2014, 2015). Thus, subpicomolar DPI exhibits two striking features: (1) potent neuroprotection is still afforded in posttreatment regimens and (2) inhibition of microglial NOX2 was highly specific at ultralow concentrations. Subpicomolar-acting research is important as a new avenue for therapy, because the safety profile is greatly improved and there is a distinct advantage using minute amounts of drugs.

9. CONCLUDING REMARKS

Over the past decade, there has been convincing evidence that supports the role of inflammation in the pathogenesis of PD. Newly developed chronic inflammation-associated PD models provide useful tool for mechanistic studies on microglia–neuron interactions, gene–environmental interplay, premotor dysfunctions, and novel approaches to microglia-based therapeutic design will provide more insights into the etiology, the pathogenesis, and the treatment of PD.

ACKNOWLEDGMENTS

This work was supported by the National Natural Science Foundation of China (Nos. 31471006 and 21577004), National Basic Research Program of China Grants 2015BAI08B02, the national high technology research and development program of China (863 program; No. 2014AA021601), and the award to high-level innovative and entrepreneurial talents of Jiangsu Province of China to H-.M.G. and the Intramural Research Program of the NIH/NIEHS in the United States (ES090082) to J-.S.H.

Conflict of Interest: The authors declare no conflict of interest.

REFERENCES

Aldieri, E., Riganti, C., Polimeni, M., Gazzano, E., Lussiana, C., Campia, I., Ghigo, D., 2008. Classical inhibitors of NOX NAD(P)H oxidases are not specific. Curr. Drug Metab. 9, 686–696.

Aloisi, F., 1999. The role of microglia and astrocytes in CNS immune surveillance and immunopathology. Adv. Exp. Med. Biol. 468, 123–133.

Barron, K.D., 1995. The microglial cell. A historical review. J. Neurol. Sci. 134 (Suppl), 57–68.

Betarbet, R., Sherer, T.B., MacKenzie, G., Garcia-Osuna, M., Panov, A.V., Greenamyre, J.T., 2000. Chronic systemic pesticide exposure reproduces features of Parkinson's disease. Nat. Neurosci. 3, 1301–1306.

Block, M.L., Zecca, L., Hong, J.S., 2007. Microglia-mediated neurotoxicity: uncovering the molecular mechanisms. Nat. Rev. Neurosci. 8, 57–69.

Braak, H., Del Tredici, K., Rub, U., de Vos, R.A., Jansen Steur, E.N., Braak, E., 2003. Staging of brain pathology related to sporadic Parkinson's disease. Neurobiol. Aging 24, 197–211.

Brown, J.M., Gouty, S., Iyer, V., Rosenberger, J., Cox, B.M., 2006. Differential protection against MPTP or methamphetamine toxicity in dopamine neurons by deletion of ppN/OFQ expression. J. Neurochem. 98, 495–505.

Castano, J.P., Faught, W.J., Frawley, L.S., 1998. Multiple measurements of gene expression in single, living cells enable molecular analysis of endocrine cell heterogeneity. Ann. N. Y. Acad. Sci. 839, 336–340.

Chaudhuri, K.R., Healy, D.G., Schapira, A.H., 2006. Non-motor symptoms of Parkinson's disease: diagnosis and management. Lancet Neurol. 5, 235–245.

Chen, H., Zhang, S.M., Hernan, M.A., Schwarzschild, M.A., Willett, W.C., Colditz, G.A., Speizer, F.E., Ascherio, A., 2003. Nonsteroidal anti-inflammatory drugs and the risk of Parkinson disease. Arch. Neurol. 60, 1059–1064.

Chen, H., Jacobs, E., Schwarzschild, M.A., McCullough, M.L., Calle, E.E., Thun, M.J., Ascherio, A., 2005. Nonsteroidal antiinflammatory drug use and the risk for Parkinson's disease. Ann. Neurol. 58, 963–967.

Chen, H., O'Reilly, E.J., Schwarzschild, M.A., Ascherio, A., 2008. Peripheral inflammatory biomarkers and risk of Parkinson's disease. Am. J. Epidemiol. 167, 90–95.

Dauer, W., Przedborski, S., 2003. Parkinson's disease: mechanisms and models. Neuron 39, 889–909.

del Rio-Hortega, P., 1932. Cytology and Cellular Pathology of the Nervous System. Paul B. Hoeber, New York.

Dhillon, A.S., Tarbutton, G.L., Levin, J.L., Plotkin, G.M., Lowry, L.K., Nalbone, J.T., Shepherd, S., 2008. Pesticide/environmental exposures and Parkinson's disease in East Texas. J. Agromedicine 13, 37–48.

Di Monte, D.A., Lavasani, M., Manning-Bog, A.B., 2002. Environmental factors in Parkinson's disease. Neurotoxicology 23, 487–502.

Dick, F.D., De Palma, G., Ahmadi, A., Osborne, A., Scott, N.W., Prescott, G.J., Bennett, J., Semple, S., Dick, S., Mozzoni, P., Haites, N., Wettinger, S.B., Mutti, A., Otelea, M., Seaton, A., Soderkvist, P., Felice, A., Geoparkinson Study Group, 2007. Gene-environment interactions in parkinsonism and Parkinson's disease: the Geoparkinson study. Occup. Environ. Med. 64, 673–680.

Doty, R.L., Singh, A., Tetrud, J., Langston, J.W., 1992. Lack of major olfactory dysfunction in MPTP-induced parkinsonism. Ann. Neurol. 32, 97–100.

Duvoisin, R.C., Lobo-Antunes, J., Yahr, M.D., 1972. Response of patients with postencephalitic Parkinsonism to levodopa. J. Neurol. Neurosurg. Psychiatry 35, 487–495.

Elbaz, A., Clavel, J., Rathouz, P.J., Moisan, F., Galanaud, J.P., Delemotte, B., Alperovitch, A., Tzourio, C., 2009. Professional exposure to pesticides and Parkinson disease. Ann. Neurol. 66, 494–504.

Elizan, T.S., Casals, J., 1991. Astrogliosis in von Economo's and postencephalitic Parkinson's diseases supports probable viral etiology. J. Neurol. Sci. 105, 131–134.

Fan, S.T., Edgington, T.S., 1993. Integrin regulation of leukocyte inflammatory functions. CD11b/CD18 enhancement of the tumor necrosis factor-alpha responses of monocytes. J. Immunol. 150, 2972–2980.

Frank-Cannon, T.C., Tran, T., Ruhn, K.A., Martinez, T.N., Hong, J., Marvin, M., Hartley, M., Trevino, I., O'Brien, D.E., Casey, B., Goldberg, M.S., Tansey, M.G., 2008. Parkin deficiency increases vulnerability to inflammation-related nigral degeneration. J. Neurosci. 28, 10825–10834.

Gao, H.M., Hong, J.S., 2008. Why neurodegenerative diseases are progressive: uncontrolled inflammation drives disease progression. Trends Immunol. 29, 357–365.

Gao, H.M., Hong, J.S., 2011. Gene-environment interactions: key to unraveling the mystery of Parkinson's disease. Prog. Neurobiol. 94, 1–19.

Gao, H.M., Hong, J.S., Zhang, W., Liu, B., 2002. Distinct role for microglia in rotenone-induced degeneration of dopaminergic neurons. J. Neurosci. 22, 782–790.

Gao, H.M., Liu, B., Zhang, W., Hong, J.S., 2003a. Critical role of microglial NADPH oxidase-derived free radicals in the in vitro MPTP model of Parkinson's disease. FASEB J. 17, 1954–1956.

Gao, H.M., Liu, B., Zhang, W., Hong, J.S., 2003b. Novel anti-inflammatory therapy for Parkinson's disease. Trends Pharmacol. Sci. 24, 395–401.

Gao, H.M., Zhang, F., Zhou, H., Kam, W., Wilson, B., Hong, J.S., 2011a. Neuroinflammation and alpha-synuclein dysfunction potentiate each other, driving chronic progression of neurodegeneration in a mouse model of Parkinson's disease. Environ. Health Perspect. 119, 807–814.

Gao, H.M., Zhou, H., Zhang, F., Wilson, B.C., Kam, W., Hong, J.S., 2011b. HMGB1 acts on microglia Mac1 to mediate chronic neuroinflammation that drives progressive neurodegeneration. J. Neurosci. 31, 1081–1092.

Gao, H.M., Zhou, H., Hong, J.S., 2012. NADPH oxidases: novel therapeutic targets for neurodegenerative diseases. Trends Pharmacol. Sci. 33, 295–303.

Ghaemi, M., Rudolf, J., Schmulling, S., Bamborschke, S., Heiss, W.D., 2000. FDG- and Dopa-PET in postencephalitic parkinsonism. J. Neural. Transm. (Vienna) 107, 1289–1295.

Gorell, J.M., Johnson, C.C., Rybicki, B.A., Peterson, E.L., Richardson, R.J., 1998. The risk of Parkinson's disease with exposure to pesticides, farming, well water, and rural living. Neurology 50, 1346–1350.

Graeber, M.B., Streit, W.J., Kreutzberg, G.W., 1988. The microglial cytoskeleton: vimentin is localized within activated cells in situ. J. Neurocytol. 17, 573–580.

Guilarte, T.R., 2010. Manganese and Parkinson's disease: a critical review and new findings. Environ. Health Perspect. 118, 1071–1080.

Hamza, T.H., Zabetian, C.P., Tenesa, A., Laederach, A., Montimurro, J., Yearout, D., Kay, D.M., Doheny, K.F., Paschall, J., Pugh, E., Kusel, V.I., Collura, R., Roberts, J., Griffith, A., Samii, A., Scott, W.K., Nutt, J., Factor, S.A., Payami, H., 2010. Common genetic variation in the HLA region is associated with late-onset sporadic Parkinson's disease. Nat. Genet. 42, 781–785.

Holdorff, B., 2002. Friedrich Heinrich Lewy (1885–1950) and his work. J. Hist. Neurosci. 11, 19–28.

Jang, H., Boltz, D., Sturm-Ramirez, K., Shepherd, K.R., Jiao, Y., Webster, R., Smeyne, R.J., 2009. Highly pathogenic H5N1 influenza virus can enter the central nervous system and induce neuroinflammation and neurodegeneration. Proc. Natl. Acad. Sci. U.S.A. 106, 14063–14068.

Jang, H., Boltz, D., McClaren, J., Pani, A.K., Smeyne, M., Korff, A., Webster, R., Smeyne, R.J., 2012. Inflammatory effects of highly pathogenic H5N1 influenza virus infection in the CNS of mice. J. Neurosci. 32, 1545–1559.

Jellinger, K.A., 2001. The pathology of Parkinson's disease. Adv. Neurol. 86, 55–72.

Jenner, P., Marsden, C.D., 1986. The actions of 1-methyl-4-phenyl-1,2,3,6-tetrahydropyridine in animals as a model of Parkinson's disease. J. Neural Transm. Suppl. 20, 11–39.

Jiang, H., Wu, Y.C., Nakamura, M., Liang, Y., Tanaka, Y., Holmes, S., Dawson, V.L., Dawson, T.M., Ross, C.A., Smith, W.W., 2007. Parkinson's disease genetic mutations increase cell susceptibility to stress: mutant alpha-synuclein enhances H2O2- and Sin-1-induced cell death. Neurobiol. Aging 28, 1709–1717.

Kamel, F., Tanner, C., Umbach, D., Hoppin, J., Alavanja, M., Blair, A., Comyns, K., Goldman, S., Korell, M., Langston, J., Ross, G., Sandler, D., 2007. Pesticide exposure and self-reported Parkinson's disease in the agricultural health study. Am. J. Epidemiol. 165, 364–374.

Kelada, S.N., Checkoway, H., Kardia, S.L., Carlson, C.S., Costa-Mallen, P., Eaton, D.L., Firestone, J., Powers, K.M., Swanson, P.D., Franklin, G.M., Longstreth Jr., W.T., Weller, T.S., Afsharinejad, Z., Costa, L.G., 2006. 5′ and 3′ Region variability in the dopamine transporter gene (SLC6A3), pesticide exposure and Parkinson's disease risk: a hypothesis-generating study. Hum. Mol. Genet. 15, 3055–3062.

Kim, Y.S., Choi, D.H., Block, M.L., Lorenzl, S., Yang, L., Kim, Y.J., Sugama, S., Cho, B.P., Hwang, O., Browne, S.E., Kim, S.Y., Hong, J.S., Beal, M.F., Joh, T.H., 2007. A pivotal role of matrix metalloproteinase-3 activity in dopaminergic neuronal degeneration via microglial activation. FASEB J. 21, 179–187.

Kreutzberg, G.W., 1996. Microglia: a sensor for pathological events in the CNS. Trends Neurosci. 19, 312–318.

Kruger, R., Kuhn, W., Muller, T., Woitalla, D., Graeber, M., Kosel, S., Przuntek, H., Epplen, J.T., Schols, L., Riess, O., 1998. Ala30Pro mutation in the gene encoding alpha-synuclein in Parkinson's disease. Nat. Genet. 18, 106–108.

Langston, J.W., Ballard, P., Tetrud, J.W., Irwin, I., 1983. Chronic Parkinsonism in humans due to a product of meperidine-analog synthesis. Science 219, 979–980.

Levesque, S., Wilson, B., Gregoria, V., Thorpe, L.B., Dallas, S., Polikov, V.S., Hong, J.S., Block, M.L., 2010. Reactive microgliosis: extracellular micro-calpain and microglia-mediated dopaminergic neurotoxicity. Brain 133, 808–821.

Li, G., Cui, G., Tzeng, N.S., Wei, S.J., Wang, T., Block, M.L., Hong, J.S., 2005. Femtomolar concentrations of dextromethorphan protect mesencephalic dopaminergic neurons from inflammatory damage. FASEB J. 19, 489–496.

Lima, M.M., Martins, E.F., Delattre, A.M., Proenca, M.B., Mori, M.A., Carabelli, B., Ferraz, A.C., 2012. Motor and non-motor features of Parkinson's disease—a review of clinical and experimental studies. CNS Neurol. Disord. Drug Targets 11, 439–449.

Lindsay, R.M., 1994. Neurotrophic growth factors and neurodegenerative diseases: therapeutic potential of the neurotrophins and ciliary neurotrophic factor. Neurobiol. Aging 15, 249–251.

Ling, Z., Gayle, D.A., Ma, S.Y., Lipton, J.W., Tong, C.W., Hong, J.S., Carvey, P.M., 2002. In utero bacterial endotoxin exposure causes loss of tyrosine hydroxylase neurons in the postnatal rat midbrain. Mov. Disord. 17, 116–124.

Ling, Z., Zhu, Y., Tong, C., Snyder, J.A., Lipton, J.W., Carvey, P.M., 2006. Progressive dopamine neuron loss following supra-nigral lipopolysaccharide (LPS) infusion into rats exposed to LPS prenatally. Exp. Neurol. 199, 499–512.

Liu, B., Hong, J.S., 2003. Role of microglia in inflammation-mediated neurodegenerative diseases: mechanisms and strategies for therapeutic intervention. J. Pharmacol. Exp. Ther. 304, 1–7.

Liu, B., Du, L., Hong, J.S., 2000. Naloxone protects rat dopaminergic neurons against inflammatory damage through inhibition of microglia activation and superoxide generation. J. Pharmacol. Exp. Ther. 293, 607–617.

Liu, B., Gao, H.M., Hong, J.S., 2003. Parkinson's disease and exposure to infectious agents and pesticides and the occurrence of brain injuries: role of neuroinflammation. Environ. Health Perspect. 111, 1065–1073.

Liu, Y., Lo, Y.C., Qian, L., Crews, F.T., Wilson, B., Chen, H.L., Wu, H.M., Chen, S.H., Wei, K., Lu, R.B., Ali, S., Hong, J.S., 2011. Verapamil protects dopaminergic neuron damage through a novel anti-inflammatory mechanism by inhibition of microglial activation. Neuropharmacology 60, 373–380.

Mayadas, T.N., Cullere, X., 2005. Neutrophil beta2 integrins: moderators of life or death decisions. Trends Immunol. 26, 388–395.

McCulloch, C.C., Kay, D.M., Factor, S.A., Samii, A., Nutt, J.G., Higgins, D.S., Griffith, A., Roberts, J.W., Leis, B.C., Montimurro, J.S., Zabetian, C.P., Payami, H., 2008. Exploring gene-environment interactions in Parkinson's disease. Hum. Genet. 123, 257–265.

Mergler, D., Baldwin, M., 1997. Early manifestations of manganese neurotoxicity in humans: an update. Environ. Res. 73, 92–100.

Milligan, C.E., Cunningham, T.J., Levitt, P., 1991. Differential immunochemical markers reveal the normal distribution of brain macrophages and microglia in the developing rat brain. J. Comp. Neurol. 314, 125–135.

Mizuno, Y., Hattori, N., Kitada, T., Matsumine, H., Mori, H., Shimura, H., Kubo, S., Kobayashi, H., Asakawa, S., Minoshima, S., Shimizu, N., 2001. Familial Parkinson's disease. Alpha-synuclein and parkin. Adv. Neurol. 86, 13–21.

Oehmichen, W., Gencic, M., 1975. Experimental studies on kinetics and functions of monuclear phagozytes of the central nervous system. Acta Neuropathol. Suppl. 6, 285–290.

Pavlov, V.A., Wang, H., Czura, C.J., Friedman, S.G., Tracey, K.J., 2003. The cholinergic anti-inflammatory pathway: a missing link in neuroimmunomodulation. Mol. Med. 9, 125–134.

Pei, Z., Pang, H., Qian, L., Yang, S., Wang, T., Zhang, W., Wu, X., Dallas, S., Wilson, B., Reece, J.M., Miller, D.S., Hong, J.S., Block, M.L., 2007. MAC1 mediates LPS-induced production of superoxide by microglia: the role of pattern recognition receptors in dopaminergic neurotoxicity. Glia 55, 1362–1373.

Polymeropoulos, M.H., Lavedan, C., Leroy, E., Ide, S.E., Dehejia, A., Dutra, A., Pike, B., Root, H., Rubenstein, J., Boyer, R., Stenroos, E.S., Chandrasekharappa, S., Athanassiadou, A., Papapetropoulos, T., Johnson, W.G., Lazzarini, A.M., Duvoisin, R.C., Di Iorio, G., Golbe, L.I., Nussbaum, R.L., 1997. Mutation in the alpha-synuclein gene identified in families with Parkinson's disease. Science 276, 2045–2047.

Powers, K.M., Kay, D.M., Factor, S.A., Zabetian, C.P., Higgins, D.S., Samii, A., Nutt, J.G., Griffith, A., Leis, B., Roberts, J.W., Martinez, E.D., Montimurro, J.S., Checkoway, H., Payami, H., 2008. Combined effects of smoking, coffee, and NSAIDs on Parkinson's disease risk. Mov. Disord. 23, 88–95.

Qian, L., Xu, Z., Zhang, W., Wilson, B., Hong, J.S., Flood, P.M., 2007a. Sinomenine, a natural dextrorotatory morphinan analog, is anti-inflammatory and neuroprotective through inhibition of microglial NADPH oxidase. J. Neuroinflammation 4, 23.

Qian, L., Gao, X., Pei, Z., Wu, X., Block, M., Wilson, B., Hong, J.S., Flood, P.M., 2007b. NADPH oxidase inhibitor DPI is neuroprotective at femtomolar concentrations through inhibition of microglia over-activation. Parkinsonism Relat. Disord. 13 (Suppl. 3), S316–320.

Qian, L., Wei, S.J., Zhang, D., Hu, X., Xu, Z., Wilson, B., El-Benna, J., Hong, J.S., Flood, P.M., 2008. Potent anti-inflammatory and neuroprotective effects of TGF-beta1 are mediated through the inhibition of ERK and p47phox-Ser345 phosphorylation and translocation in microglia. J. Immunol. 181, 660–668.

Qin, L., Li, G., Qian, X., Liu, Y., Wu, X., Liu, B., Hong, J.S., Block, M.L., 2005a. Interactive role of the toll-like receptor 4 and reactive oxygen species in LPS-induced microglia activation. Glia 52, 78–84.

Qin, L., Block, M.L., Liu, Y., Bienstock, R.J., Pei, Z., Zhang, W., Wu, X., Wilson, B., Burka, T., Hong, J.S., 2005b. Microglial NADPH oxidase is a novel target for femtomolar neuroprotection against oxidative stress. FASEB J. 19, 550–557.

Qin, L., Wu, X., Block, M.L., Liu, Y., Breese, G.R., Hong, J.S., Knapp, D.J., Crews, F.T., 2007. Systemic LPS causes chronic neuroinflammation and progressive neurodegeneration. Glia 55, 453–462.

Qin, L., Liu, Y., Hong, J.S., Crews, F.T., 2013. NADPH oxidase and aging drive microglial activation, oxidative stress and dopaminergic neurodegeneration following systemic LPS administration. Glia 61, 855–868.

Ritz, B.R., Manthripragada, A.D., Costello, S., Lincoln, S.J., Farrer, M.J., Cockburn, M., Bronstein, J., 2009. Dopamine transporter genetic variants and pesticides in Parkinson's disease. Environ. Health Perspect. 117, 964–969.

Rock, R.B., Peterson, P.K., 2006. Microglia as a pharmacological target in infectious and inflammatory diseases of the brain. J. Neuroimmune Pharmacol. 1, 117–126.

Rommelfanger, K.S., Weinshenker, D., 2007. Norepinephrine: the redheaded stepchild of Parkinson's disease. Biochem. Pharmacol. 74, 177–190.

Ross, G.D., Vetvicka, V., 1993. CR3 (CD11b, CD18): a phagocyte and NK cell membrane receptor with multiple ligand specificities and functions. Clin. Exp. Immunol. 92, 181–184.

Schiller, F., 2000. Fritz Lewy and his bodies. J. Hist. Neurosci. 9, 148–151.

Semchuk, K.M., Love, E.J., Lee, R.G., 1992. Parkinson's disease and exposure to agricultural work and pesticide chemicals. Neurology 42, 1328–1335.

Streit, W.J., Graeber, M.B., Kreutzberg, G.W., 1988. Functional plasticity of microglia: a review. Glia 1, 301–307.

Streit, W.J., Walter, S.A., Pennell, N.A., 1999. Reactive microgliosis. Prog. Neurobiol. 57, 563–581.

Tacconi, M.T., 1998. Neuronal death: is there a role for astrocytes? Neurochem. Res. 23, 759–765.

Tanner, C.M., 2003. Is the cause of Parkinson's disease environmental or hereditary? Evidence from twin studies. Adv. Neurol. 91, 133–142.

Tomas-Camardiel, M., Rite, I., Herrera, A.J., de Pablos, R.M., Cano, J., Machado, A., Venero, J.L., 2004. Minocycline reduces the lipopolysaccharide-induced inflammatory reaction, peroxynitrite-mediated nitration of proteins, disruption of the blood-brain barrier, and damage in the nigral dopaminergic system. Neurobiol. Dis. 16, 190–201.

Ton, T.G., Heckbert, S.R., Longstreth Jr., W.T., Rossing, M.A., Kukull, W.A., Franklin, G.M., Swanson, P.D., Smith-Weller, T., Checkoway, H., 2006. Nonsteroidal anti-inflammatory drugs and risk of Parkinson's disease. Mov. Disord. 21, 964–969.

Vlajinac, H., Dzoljic, E., Maksimovic, J., Marinkovic, J., Sipetic, S., Kostic, V., 2013. Infections as a risk factor for Parkinson's disease: a case-control study. Int. J. Neurosci. 123, 329–332.

Wahner, A.D., Sinsheimer, J.S., Bronstein, J.M., Ritz, B., 2007. Inflammatory cytokine gene polymorphisms and increased risk of Parkinson disease. Arch. Neurol. 64, 836–840.

Wang, T., Zhang, W., Pei, Z., Block, M., Wilson, B., Reece, J.M., Miller, D.S., Hong, J.S., 2006. Reactive microgliosis participates in MPP+-induced dopaminergic neurodegeneration: role of 67 kDa laminin receptor. FASEB J. 20, 906–915.

Wang, Q., Zhou, H., Gao, H., Chen, S.H., Chu, C.H., Wilson, B., Hong, J.S., 2012. Naloxone inhibits immune cell function by suppressing superoxide production through a direct interaction with gp91^{phox} subunit of NADPH oxidase. J. Neuroinflammation 9, 32.

Wang, Q., Chu, C.H., Oyarzabal, E., Jiang, L., Chen, S.H., Wilson, B., Qian, L., Hong, J.S., 2014. Subpicomolar diphenyleneiodonium inhibits microglial NADPH oxidase with high specificity and shows great potential as a therapeutic agent for neurodegenerative diseases. Glia 62, 2034–2043.

Wang, Q., Qian, L., Chen, S.H., Chu, C.H., Wilson, B., Oyarzabal, E., Ali, S., Robinson, B., Rao, D., Hong, J.S., 2015. Post-treatment with an ultra-low dose of NADPH oxidase inhibitor diphenyleneiodonium attenuates disease progression in multiple Parkinson's disease models. Brain 138, 1247–1262.

Yang, S., Yang, J., Yang, Z., Chen, P., Fraser, A., Zhang, W., Pang, H., Gao, X., Wilson, B., Hong, J.S., Block, M.L., 2006. Pituitary adenylate cyclase-activating polypeptide (PACAP) 38 and PACAP4-6 are neuroprotective through inhibition of NADPH oxidase: potent regulators of microglia-mediated oxidative stress. J. Pharmacol. Exp. Ther. 319, 595–603.

Zarranz, J.J., Alegre, J., Gomez-Esteban, J.C., Lezcano, E., Ros, R., Ampuero, I., Vidal, L., Hoenicka, J., Rodriguez, O., Atares, B., Llorens, V., Gomez Tortosa, E., del Ser, T., Munoz, D.G., de Yebenes, J.G., 2004. The new mutation, E46K, of alpha-synuclein causes Parkinson and Lewy body dementia. Ann. Neurol. 55, 164–173.

Zhang, W., Wang, T., Pei, Z., Miller, D.S., Wu, X., Block, M.L., Wilson, B., Zhang, W., Zhou, Y., Hong, J.S., Zhang, J., 2005. Aggregated alpha-synuclein activates microglia: a process leading to disease progression in Parkinson's disease. FASEB J. 19, 533–542.

Zhang, W., Dallas, S., Zhang, D., Guo, J.P., Pang, H., Wilson, B., Miller, D.S., Chen, B., Zhang, W., McGeer, P.L., Hong, J.S., Zhang, J., 2007. Microglial PHOX and Mac-1 are essential to the enhanced dopaminergic neurodegeneration elicited by A30P and A53T mutant alpha-synuclein. Glia 55, 1178–1188.

Zhang, D., Hu, X., Wei, S.J., Liu, J., Gao, H., Qian, L., Wilson, B., Liu, G., Hong, J.S., 2008. Squamosamide derivative FLZ protects dopaminergic neurons against inflammation-mediated neurodegeneration through the inhibition of NADPH oxidase activity. J. Neuroinflammation 5, 21.

Zhang, F., Shi, J.S., Zhou, H., Wilson, B., Hong, J.S., Gao, H.M., 2010. Resveratrol protects dopamine neurons against lipopolysaccharide-induced neurotoxicity through its anti-inflammatory actions. Mol. Pharmacol. 78, 466–477.

Zhang, W., Phillips, K., Wielgus, A.R., Liu, J., Albertini, A., Zucca, F.A., Faust, R., Qian, S.Y., Miller, D.S., Chignell, C.F., Wilson, B., Jackson-Lewis, V., Przedborski, S., Joset, D., Loike, J., Hong, J.S., Sulzer, D., Zecca, L., 2011a. Neuromelanin activates microglia and induces degeneration of dopaminergic neurons: implications for progression of Parkinson's disease. Neurotox. Res. 19, 63–72.

Zhang, D., Hu, X., Qian, L., Chen, S.H., Zhou, H., Wilson, B., Miller, D.S., Hong, J.S., 2011b. Microglial MAC1 receptor and PI3K are essential in mediating beta-amyloid peptide-induced microglial activation and subsequent neurotoxicity. J. Neuroinflammation 8, 3.

Zheng, W., Fu, S.X., Dydak, U., Cowan, D.M., 2011. Biomarkers of manganese intoxication. Neurotoxicology 32, 1–8.

Zhou, H., Liao, J., Aloor, J., Nie, H., Wilson, B.C., Fessler, M.B., Gao, H.M., Hong, J.S., 2013. CD11b/CD18 (Mac-1) is a novel surface receptor for extracellular double-stranded RNA to mediate cellular inflammatory responses. J. Immunol. 190, 115–125.

Ziemssen, T., Reichmann, H., 2007. Non-motor dysfunction in Parkinson's disease. Parkinsonism Relat. Disord. 13, 323–332.

CHAPTER EIGHT

Mitochondrial Dynamics in Neurodegenerative Diseases

Jennifer Pinnell, Kim Tieu[1]
Florida International University, Miami, FL, United States
[1]Corresponding author: e-mail address: ktieu@fiu.edu

Contents

1. INTRODUCTION

Mitochondria are essential to the maintenance of normal neuronal function and viability through regulation of key processes such as energy production, calcium homeostasis, intracellular transport, neurotransmitter release, and reuptake (Hollenbeck, 2005; Ly and Verstreken, 2006). Studies in recent years have also highlighted that mitochondria are highly dynamic organelles. This dynamic feature is reflected in the term "mitochondrial dynamics," which refers to the processes of fission, fusion, and movement along microtubules that allow mitochondria to supply energy to cellular extensions, such as axons and dendrites (Chen and Chan, 2009). Through fusion, the effects of cellular stress can be mitigated by combining the contents of partially defective mitochondria to allow complementation of the undamaged components in the combined organelle. Fission is the process by which mitochondria divide and it can be beneficial to cell survival by eliminating destructive mitochondrial DNA mutations and components. Prior to fission, damaged DNA and proteins are segregated to one side of the mitochondrion such that only one of the "daughter" mitochondria contains damaged molecules and is targeted for mitophagy while the other daughter mitochondrion remains pristine (Youle and van der Bliek, 2012). However, fission also plays a role in the induction of cellular apoptosis and oxidative stress; thus, excessive mitochondrial fission can be detrimental (Youle and van der Bliek, 2012). Fission and fusion are controlled by specific proteins (Fig. 1). The outer mitochondrial mitofusin proteins

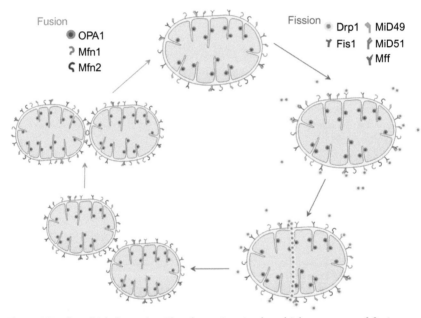

Fig. 1 Mitochondrial dynamics. The dynamic mitochondrial processes of fission and fusion work to maintain a healthy, functional population of mitochondria within cells. Fission is mediated by Drp1, a large GTPase which translocates from the cytosol to the mitochondrial outer membrane where it induces mitochondrial division by forming oligomeric chains which wrap around constriction points of dividing mitochondria. Fis1, Mff, MiD49, and MiD51 are all OMM-anchored fission factors to which Drp1 may be recruited. MiD49 and MiD51 are essential for the stabilization of multimeric Drp1 structures necessary for mitochondrial fission, and both may complex with Mff and Fis1. Fusion is mediated by proteins independent of the fission factors. OPA-1 controls mitochondrial inner membrane fusion, while Mfn1 and Mfn2 mediate fusion of the OMM. Together these proteins help to regulate the mitochondrial population within cells and to manage their response to metabolic stress.

(Mfn1 and Mfn2) and the inner mitochondrial protein optic atrophy 1 (OPA-1) are responsible for mitochondrial fusion. Fission requires the recruitment of dynamin-related protein 1 (Drp1) from the cytosol to the outer mitochondrial membrane (OMM) by mitochondrial fission factor (Mff), mitochondrial dynamic proteins of 49 and 51 kDA (MiD49 and MiD51, respectively), and Fis1 (Lee and Yoon, 2016). A balance of fusion and fission is crucial not only to mitochondrial morphology, but also to cell viability and synaptic function.

Accumulating evidence indicates that perturbed mitochondrial dynamics is a pathogenic mechanism in a number of neurodegenerative diseases. Although there are many types of neurodegenerative disease, the most prevalent and publicized ones are Alzheimer's disease (AD), Parkinson's disease (PD), Huntington's disease (HD), and amyotrophic lateral sclerosis (ALS). Over the past decades, major efforts have been made to understand the mechanisms of neurodegeneration in these diseases. From these intensive investigations, it is increasingly evident that despite the loss of differing cellular populations which distinguish the individual diseases from each other, there are common cell death mechanisms between them, such as mitochondrial dysfunction, impaired autophagic or proteasomal protein degradation, oxidative stress, and neuroinflammation. It is worth noting, however, that very often these seemingly different pathogenic mechanisms are not mutually exclusive (Fig. 2). The primary focus of this review is on impaired mitochondria. Overall, mitochondrial dysfunction is best documented in PD, but as will be discussed in this review, AD, HD, and ALS have also been demonstrated to be affected by this pathogenic mechanism.

2. PARKINSON'S DISEASE

PD is a hypokinetic disorder characterized by degeneration of the nigrostriatal dopaminergic pathway. It is the second most prevalent neurodegenerative disorder, affecting over 5 million individuals worldwide (Dorsey et al., 2007). As the principle neurotransmitter of the extrapyramidal system, dopamine (DA) plays a critical role in the regulation and control of motor nuclei and their associated pathways. Progressive loss of dopaminergic neurons in the substantia nigra (SN) presents as a clinically recognizable tetrad of Parkinsonian symptoms: rigidity, resting tremor, bradykinesia, and postural instability (Fahn and Prsedborski, 2010).

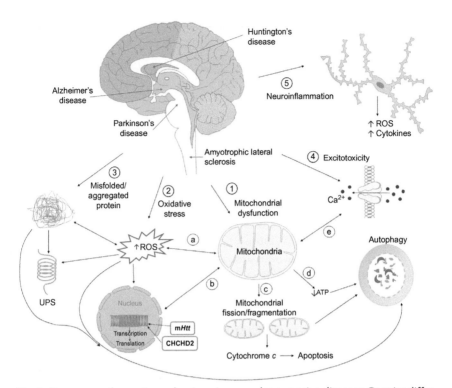

Fig. 2 Common pathogenic mechanisms in neurodegenerative diseases. Despite different genetic mutations and potential links to environmental toxicants giving rise to anatomical independence of the affected brain regions, AD, PD, HD, and ALS share the following neurotoxic mechanisms: (1) mitochondrial dysfunction, (2) increased oxidative stress, (3) presence of neurotoxic misfolded and aggregated proteins, (4) excitotoxicity, mediated by overactivation of the excitatory receptors, resulting in cellular calcium influx which then produces downstream defects such as calcium-induced mitochondrial depolarization, and (5) neuroinflammation induced by microglia, which release neurotoxic factors, such as cytokines, interleukins, and reactive oxygen species (ROS), leading to noncell autonomous neurotoxicity. Cross-talk between these five mechanisms, such as with mitochondrial dysfunction, stimulates further changes in neurodegeneration. (a) Mitochondrial dysfunction can result in increased generation of ROS, which has the capacity to promote the formation of toxic oligomers and protein aggregates, induce DNA damage, impair autophagic flux, and the ubiquitin proteasomal system (UPS), activate neuroinflammation, and damage mitochondria. (b) Mitochondrial dysfunction can also induce DNA damage, resulting in altered nuclear function and genomic instabilities, which may further aggravate disease progression. (c) Excessive mitochondrial fragmentation can result from dysfunction, resulting in apoptotic cell death mediated by cytochrome *c* release from the mitochondrial matrix. (d) ATP production is reduced when mitochondria fail to function normally, which can result in decreased autophagic clearance of damaged proteins and organelles, because autophagy and UPS are ATP-dependent processes. (e) Dysregulated cellular calcium is influenced by mitochondrial function, because mitochondria help to regulate intracellular calcium levels. When dysfunctional, mitochondria may release more calcium into the cytosol, thereby increasing cellular excitotoxicity. Of note, as discussed in the text, mutant Htt and CHCHD2 also directly impact nuclear function resulting in impaired transcription.

PD is a multifactorial disorder with complex etiology. Approximately 10% of PD cases are caused by monogenic, familial gene mutations, while the other 90% of cases are classified as idiopathic and sporadic (Cannon and Greenamyre, 2013). In reality, however, PD is most likely a polygenic disease with various genetic and environmental contributors cumulatively directing its pathological development.

2.1 Neurotoxicants and Impaired Mitochondria in PD

Prior to the first discovery of a genetic mutation in PD in 1997 (Polymeropoulos et al., 1997), environmental factors were the major driving force for the search of candidates responsible for PD pathogenesis. The rationale for this dominant environmental theory was largely based on observations such as postencephalitic infection, manganese, and MPTP exposure causing parkinsonism (Calne et al., 1994; Langston et al., 1983; Rail et al., 1981). Indeed, epidemiological studies and metaanalyses have implicated a number of environmental factors, including exposure to pesticides, herbicides, insecticides, and heavy metals, with PD risk (Kamel, 2013; Tanner et al., 2011; Wang et al., 2011a). Among these factors, pesticides have been the most widely studied and some are commonly utilized in laboratory models of PD.

The initial proposal of mitochondrial dysfunction as a pathogenic mechanism in PD and of exogenous chemicals as the culprit in causing PD originated from the discovery that 1-methyl-4-phenyl-1,2,3,6-tetrahydropyridine (MPTP) caused parkinsonism in humans (Langston et al., 1983). Subsequently, 1-methyl-4-phenylpyridinium ion (MPP^+), the active neurotoxic metabolite of MPTP, was identified as an inhibitor of complex I of the mitochondrial electron transport chain (Nicklas et al., 1985). This finding prompted many investigators to search for environmental contaminants with a similar structure or mechanism to MPP^+, as well as to evaluate mitochondrial defects in PD patients. Shortly thereafter, a reduction in complex I activity in platelets (Parker et al., 1989) and brain tissues, including the SN (Haas et al., 1995; Schapira et al., 1990), of PD patients was reported. Since its discovery, the $MPTP/MPP^+$ model has been used quite extensively to study PD. In recent years, MPP^+ has been shown to enhance mitochondrial fission, resulting in mitochondrial fragmentation and neurotoxicity in cell culture (Qi et al., 2013; Wang et al., 2011b) and animal models (Filichia et al., 2016; Rappold et al., 2014). In SH-SY5Y cells and primary dopaminergic neurons, MPP$^+$ increases the expression and Drp1 translocation from the cytosol to mitochondria, followed by mitochondrial fragmentation, loss of ATP levels, collapse of membrane potential, and subsequent cell death. Blocking Drp1 function genetically using Drp1–siRNA, the dominant negative Drp1–K38A, or the peptide (P110) to block the binding of Drp1 to Fis1, attenuated these neurotoxic effects of MPP$^+$ (Qi et al., 2013; Wang et al., 2011b). These results indicate that mitochondrial fragmentation is an upstream event of the neurotoxic cascade induced by MPP$^+$. Recent in vivo studies have confirmed that blocking mitochondrial fission in the MPTP mouse model is neuroprotective (Filichia et al., 2016; Rappold et al., 2014). We showed that Drp1 inhibition using gene-based and small-molecule approaches attenuates neurotoxicity and restores preexisting striatal dopamine release deficits in MPTP-treated mice (Rappold et al., 2014). A complementary approach using a peptide known as P110 to block the binding of Drp1 to Fis1 has also been shown to be protective in MPTP-treated mice (Filichia et al., 2016).

Rotenone, a pesticide associated with increased risk of developing PD (Kamel, 2013; Tanner et al., 2011), functions similarly to MPP$^+$ to induce mitochondrial dysfunction by direct, potent inhibition of complex I (Schuler and Casida, 2001). Rotenone was widely used as an organic pesticide and piscicide, as it is derived from plants of the *Leguminosa* family. The half-life of rotenone is relatively short (>3 days) (Hisata, 2002) so contamination of food is an unlikely exposure route; however, epidemiological studies have demonstrated an increased risk of ~1.5–3-fold of PD development in individuals who utilized rotenone in agriculture or lived in close proximity to its use (Tanner et al., 2011). Nevertheless, the role of rotenone in the induction of PD risk is most likely through low chronic exposure rather than acute toxicity. As a highly lipophilic molecule, rotenone freely crosses the blood–brain barrier and enters cells without the requirement of specific transporter action. The wide agricultural application of rotenone, combined with the observations that mitochondrial complex I activity is reduced in PD patients (Parker et al., 1989; Schapira et al., 1989), has partly led to an interest in using rotenone to model PD. Although the neuropathology is variable, rotenone treatment has been shown to induce motor impairment, neuroinflammation, nigrostriatal degeneration, and accumulation and phosphorylation of synuclein (Betarbet et al., 2000; Cannon et al., 2009) in rodents. Like MPP$^+$, rotenone also enhances mitochondrial fission. In primary cortical neurons, rotenone induces rapid mitochondrial fragmentation within 2 h of treatment, prior to any cellular changes indicative of cytotoxicity (Barsoum et al., 2006). Promoting mitochondrial fusion with Mfn2 and blocking Drp1 with Drp1–K38A prevents this neurotoxicity (Barsoum et al., 2006).

The widely used herbicide paraquat (PQ) was first hypothesized to be a potential environmental parkinsonian toxicant based on its similar structure to MPP$^+$ (Snyder and D'Amato, 1985). Supported by epidemiological studies, especially in recent years (Ritz et al., 2009; Tanner et al., 2011), there has been an interest in further investigating the mechanism by which this molecule induces dopaminergic neurodegeneration (Rappold et al., 2011; Richter et al., 2017). The herbicidal mechanism of PQ involves interference with photosynthesis and oxygen free radical production, resulting in damage of plant membranes (Tanner et al., 2011). When used as a toxicant to model PD, toxicity induced by PQ is primarily mediated by redox cycling with cellular diaphorases such as NADPH oxidase (NOX) and nitric oxide synthase (Day et al., 1999), leading to the generation of superoxide. Interestingly, despite its similar structure to MPP$^+$, PQ is not a complex I inhibitor per se (Richardson et al., 2005), although this is the site where it is reduced to form superoxide (Cochemé and Murphy, 2008). Regardless of whether reactive oxygen species (ROS) is generated from outside or inside of mitochondria, PQ impairs mitochondrial function and increases mitochondrial fission by reducing the levels of Mfn1 and Mfn2 (Tanaka et al., 2010). In rodent models, PQ has been shown to cause neuroinflammation, as well as α-synuclein upregulation and aggregation (Fernagut et al., 2007; Manning-Bog et al., 2002; Purisai et al., 2007; Wu et al., 2005). However, using PQ to model PD in wild-type mice produced variable results, with loss of nigral cell bodies but a lack of consistency in the loss of the corresponding striatal terminals (McCormack et al., 2002; Thiruchelvam et al., 2003). We previously reported that in mice with OCT3 (*Slc22a3*) deficiency, enhanced striatal damage was detected after PQ treatment; this increased sensitivity likely results from reduced buffering capacity by nondopaminergic cells, leading to increased availability of PQ for damaging dopaminergic terminals (Rappold et al., 2011).

Manganese (Mn) is widely distributed in the environment, partly due to industrial activity. Occupational overexposure of Mn causes manganism (Chen et al., 2016; Guilarte, 2013). Detailed information regarding the pharmacokinetics, metabolism, body distribution, and neurotoxic properties of Mn has been recently reviewed (Chen et al., 2016; Peres et al., 2016; Roth, 2014) and therefore will not be discussed here. In addition to manganism, Mn exposure may also increase the risk of developing PD because this heavy metal interacts with PD-linked gene products including α-synuclein, parkin, PINK1, DJ-1, LRRK2, ATP13A2, and VPS35 (Peres et al., 2016; Roth, 2014). Interestingly, Mn has been shown to suppress striatal DA release (Guilarte et al., 2006), a deficit that is commonly observed in animal models with PD gene mutations. It is possible that Mn interacts with these proteins at the presynaptic level to impair neurotransmitter release. Another potential mechanism is related to the fact that Mn induces mitochondrial fragmentation (Alaimo et al., 2014), which is consistent with its well-documented role in impairing mitochondrial function and the requirement of mitochondrial function for synaptic release. In this study, Alaimo and colleagues showed that Mn increased mitochondrial fission in cultured rat astrocytoma C6 cells by increasing the levels and translocation of Drp1 from the cytosol to mitochondria. Concomitantly, Mn impaired mitochondrial fusion by reducing OPA-1 levels. Overexpressing OPA-1 or blocking Drp1 function using Drp1–siRNA or a small-molecule inhibitor prevented these changes and the resultant neurotoxicity (Alaimo et al., 2014).

2.2 Genetic Mutations and Impaired Mitochondria in PD

2.2.1 Autosomal Dominant Genes

Tremendous work has been undertaken over the past 2 decades to identify genetic factors which may cause PD or confer increased risk of PD development. Multiple autosomal dominant and recessive genes linked to PD have been discovered (Hernandez et al., 2016; Przedborski, 2017) (Table 1), further highlighting the complexity of this disorder. The first mutation identified to cause PD was the autosomal dominant *SNCA* gene, which encodes α-synuclein (α-syn) protein (Polymeropoulos et al., 1997). Abundantly available in presynaptic terminals, α-syn is a small 14 kDa protein. Although the exact physiological function of α-syn is still incompletely understood, it has been found to play a role in regulating neurotransmitter release by interacting with synaptic vesicle lipids (Burré et al., 2010; Murphy et al., 2000) and SNARE proteins (Burré et al., 2010; Garcia-Reitböck et al., 2010). Missense mutations (A53T, A30P, E46K, H50Q, and G51D) or gene multiplication mutations leading to higher levels of wild-type α-syn cause autosomal dominant PD (Appel-Cresswell et al., 2013; Krüger et al., 1998; Lesage et al., 2013; Polymeropoulos et al., 1997; Singleton et al., 2003; Zarranz et al., 2004). The fact that duplications and triplications of *SNCA* can also cause PD (Singleton et al., 2003) is

Table 1 Monogenic PD-Linked Mutations and Associated Risk Genes

PARK Locus	Gene	Inheritance	Mutations	Clinical Phenotype	Neuropathology
PARK 1	SNCA	AD	A53T, A30P, E46K, G51D, H50Q, duplications, triplications	EOPD, rapid progression, L-DOPA responsive, common dementia	Diffuse LB and neurites, loss of SNpc and hippocampal neurons
PARK 2	Parkin	AR	>50 missense mutations, exon deletions, duplications, triplications	EOPD, slow progression, L-DOPA responsive	Loss of SNpc neurons, LB infrequently associated
PARK 6	PINK1	AR	>40 missense mutations, rare exon deletions	EOPD, L-DOPA responsive	Loss of SNpc neurons with LB
PARK 7	DJ-1	AR	>10 missense mutations, exon deletion (del [Ex1–5])	EOPD, slow progression, L-DOPA responsive	Unspecified
PARK 8	LRRK2	AD	>50 missense and nonsense mutations high risk variants, >15 of which are pathogenic	LOPD, slow progression, L-DOPA responsive	Loss of SNpc neurons, LB pathology in most cases
PARK 9	ATP13A2	AR	>10 missense mutations, some produce truncations due to splicing variants, frameshift mutations	EOPD variant with spasticity, supranuclear palsy, and dementia (Kufor–Rakeb syndrome), rapid progression, L-DOPA responsive	Unspecified
PARK 13	HTRA2/Omi	AD	G339S, A141S	LOPD, L-DOPA responsive	Unspecified

Continued

Table 1 Monogenic PD-Linked Mutations and Associated Risk Genes—cont'd

PARK Locus	Gene	Inheritance	Mutations	Clinical Phenotype	Neuropathology
PARK 14	PLA2G6	AR	R741Q, R747W, R635Q, D331Y	LOPD variant with dystonia, L-DOPA responsive	Unspecified
PARK 15	FBX07	AR	R378G, R498X, T22M	EOPD variant with parkinsonian–pyrimidal symptoms, L-DOPA	Unspecified
PARK 17	VPS35	AD	D620N	LOPD with predominant tremor, slow progression, L-DOPA responsive, MCI	Unspecified
PARK 18	EIF4G1	AD	G686C, S1164R, R1197W, R1205H	LOPD, slow progression, L-DOPA responsive	Unspecified
PARK 22	CHCHD2 (MNRR1)	AD	Missense mutations (Thr61Ile), rs10043, rs142444896	LOPD, L-DOPA responsive	Loss of SNpc neurons with LB pathology
PARK 23	VPS13C	PD	Loss of function mutation	EOPD, sporadic PD	Diffuse synuclein pathology
NA	CHCHD10	PD risk factor	>7 missense mutations, specifically P34S	Sporadic PD, also related to mitochondrial myopathy, ALS, FTD, LO-SMAJ, and cerebellar ataxia	Unspecified
NA	SMPD1	Risk factor	L302P mutation	Sporadic PD, causal gene in Niemann–Pick disease	Unspecified

NA	GCH1	Risk factor	>140 mutations, including N370S and R496H	Sporadic PD, L-DOPA responsive dystonia, tremors	Unspecified
NA	GBA	Risk factor	>10 missense mutations, L444P and N370S are best studied	Sporadic PD, homozygous mutations cause Gaucher's disease, also associated with DLB, cognitive impairments	Widespread Lewy body pathology
NA	MC1R	Risk factor	Homozygous R151C	Sporadic PD	Unspecified
NA	MAPT	Risk factor	H1 haplotype	Sporadic PD with cognitive decline	Unspecified
NA	GAK/DNAJC26	Risk factor	rs11248051C>T, rs1564282 variant	Sporadic PD	Alpha-synuclein pathology
NA	HLA-DRA	Risk factor	rs3129882G>A	Sporadic PD	Unspecified
NA	APOE	Risk factor	APOE-E4 allele rs7412-C, rs429358-C, APOE-E2 allele	Sporadic PD with dementia, rapid progression, DLB	Unspecified
NA	NOS2A	Risk factor	rs1060826T>C, rs2255929T>A	Sporadic PD	LB pathology
NA	ADH1C	Risk factor	G78X	Sporadic PD	Unspecified

Abbreviations: EOPD, early-onset PD (age at onset <50 years old); LOPD, late-onset PD (age at onset ≥50 years old); AD, autosomal dominant; AR, autosomal recessive; LBD, Lewy body disease; ALS, amyotrophic lateral sclerosis; FTD, frontotemporal dementia; LO-SMAJ, late-onset (age at onset ≥50 years old) spinal motor neuronopathy; DLB, dementia with Lewy bodies; MCI, mild-cognitive impairment; SNpc, substantia nigra pars compact; LB, Lewy body.

significant because it indicates that elevated wild-type α-syn alone is sufficient to cause disease. As discussed, environmental agents such as rotenone, PQ, and Mn have been reported to increase the levels and oligomerization of α-syn, and therefore may contribute to sporadic PD. Indeed, genome-wide association studies have identified *SNCA* as a major gene associated with sporadic PD (Hamza et al., 2010; Nalls et al., 2011; Satake et al., 2009). Identifying pathogenic mechanisms and effective therapies for α-syn neurotoxicity is thus relevant to familial PD, sporadic PD, and other synucleinopathies. α-syn is natively unfolded in solution (Weinreb et al., 1996), but it has a propensity to form aggregates under various pathological conditions. The aggregated and insoluble fibrillar form of α-syn constitutes a major component of the intracellular proteinaceous inclusions called Lewy bodies. Additionally, α-syn has been demonstrated to be capable of spreading from cell to cell (Alvarez-Erviti et al., 2011; Lee et al., 2012; Mao et al., 2016). Many interrelated pathogenic mechanisms of α-syn mutations have been proposed, including mitochondrial dysfunction (Dehay et al., 2015; Franco-Iborra et al., 2016). For example, a recent study from Greenamyre's laboratory reported that specific forms of posttranslationally modified α-syn bind with high affinity to the mitochondrial outer membrane protein TOM20 and inhibit mitochondrial protein import in rats injected with either rotenone or virally encoded α-syn, resulting in mitochondrial dysfunction, and the production of ROS (Di Maio et al., 2016). This study also demonstrated an increased interaction between TOM20 and alpha-synuclein in PD patient postmortem brains. Furthermore, α-syn has been reported in recent years to induce severe mitochondrial fragmentation both in vitro and in vivo due to increased fission, impaired fusion, and reduced connectivity between mitochondria and endoplasmic reticulum (Chen et al., 2015; Guardia-Laguarta et al., 2014; Gui et al., 2012; Kamp et al., 2010; Nakamura et al., 2011). Taken together, recent studies strongly support the negative impact of mutant α-syn on mitochondrial dynamics, leading to mitochondrial fragmentation and dysfunction.

A second autosomal dominant gene linked with late-onset PD is *leucine-rich repeat kinase 2* (*LRRK2*). Originally discovered in 2002 in a Japanese family, and subsequently independently confirmed in several other families from different countries, mutations in *LRRK2* are now recognized as the most common cause of late-onset familial PD and sporadic PD (Cookson, 2015). Although more than 100 LRRK2 mutations have been reported, only a few have been proven to cause PD (N1437H, R1441C, p.R1441G, R1441H, Y1699C, G2019S, and I2020T0) and two have been nominated as risk factors (R1628P and G2385R) (Dächsel and Farrer, 2010; Paisán-Ruiz et al., 2013), which might interact with environmental toxicants leading to the development of PD. Exactly how LRRK2 mutations cause PD is poorly understood but it appears that kinase activity is required for its toxicity. Overexpression of pathogenic LRRK2 mutants (R1441C and G20192) in cultured neuronal cells causes mitochondrial fragmentation and dysfunction through a Drp1-dependent mechanism (Niu et al., 2012; Wang et al., 2012). Abnormal mitochondrial morphology and alterations in mitochondrial fission/fusion proteins have also been reported in LRRK2 (G2019) knock-in mice (Yue et al., 2015).

A third gene associated with autosomal dominant PD is *vacuolar protein sorting 35* (*VPS35*) (Sharma et al., 2012; Vilariño-Güell et al., 2011; Zimprich et al., 2011). The most common mutation is D620N. VPS35 is a core subunit of a heteropentameric complex known as the retromer, which plays an important role in the transport of endosome to golgi and to plasma membrane, as well as in recycling of transmembrane protein cargo (Williams et al., 2017). VPS35 has also been shown in recent years to be involved in mediating the shuttling of cargo from mitochondria to peroxisomes and lysosomes through the formation of mitochondria-derived vesicles (Braschi et al., 2010; Sugiura et al., 2014). The pathogenic mechanism by which VPS35 induces PD is not clear but recent studies show that it impairs mitochondrial dynamics and function (Wang et al., 2016). This study demonstrated that overexpressing D620N induces mitochondrial fragmentation, dysfunction, and neurotoxicity in cultured neurons and in mouse nigral dopaminergic neurons in vivo. Mechanistically, D620N increases the interaction of VPS35 with Drp1, leading to enhanced turnover of the mitochondrial Drp1 complexes, via their mitochondria-derived vesicle-dependent transport to lysosomes for degradation. Another study shows that VPS35 deficiency or mutation induces mitochondrial fragmentation by impairing mitochondrial fusion (Tang et al., 2015). Using in vitro and in vivo models, these authors demonstrate that in the absence of VPS35 or the presence of D620N, mitochondrial E3 ubiquitin ligase 1 (MUL1) is upregulated, leading to increased ubiquitination and proteasomal degradation of MFN2. Together, these studies strongly support the role of VPS35 in imbalanced mitochondrial dynamics.

The latest autosomal dominant gene linked to late-onset PD is *CHCHD2* mutations (Funayama et al., 2015). Interestingly, the specific mutations reported so far are ethnicity specific, with some only detected in Asian populations or Caucasians (Zhou et al., 2016). The coiled-coil-helix-coiled-coil-helix (CHCHD2) protein contains a mitochondrial targeting sequence and has been detected in the intermembrane space where it binds to the complex IV cytochrome c oxidase (COX). This binding is required for COX activity. This critical role is consistent with the observation that loss of CHCHD2 function results in impaired electron transport chain flux, reduced mitochondrial membrane potential, and increased ROS levels (Aras et al., 2015). Although CHCHD2 predominantly localizes in mitochondria and directly binds to complex IV, CHCHD2 is also a transcription factor of the complex IV subunit 4 isoform COX4I2 under stress (Aras et al., 2015). Therefore, loss of CHCHD2 function negatively affects complex IV both functionally and structurally. Although the impact of this newly discovered mutation on mitochondrial dynamics has not been evaluated, with its dramatic influence of the electron transport chain, it is very likely that mitochondrial morphology and movement are also affected.

2.2.2 Autosomal-Recessive Genes

Several autosomal-recessive genes linked to PD have also been discovered (Table 1). Although these mutations have been shown to affect mitochondrial function either directly or indirectly, the genes that have received the most attention so far are *PINK1* and *Parkin*, both of which are responsible for most cases of recessive PD. PINK1 has a mitochondrial targeting signal at the N-terminus and a serine/threonine kinase domain at the C-terminus, facing the cytosol (Zhou et al., 2008). Although a portion of PINK1 is also present in the cytosol to promote dendritic outgrowth by enhancing anterograde transport of mitochondria to dendrites (Dagda et al., 2014), PINK1 is best known for its role in mitophagy. PINK1 is usually maintained at a low level in healthy mitochondria. However, in those with disrupted membrane potential, full-length PINK1 is rapidly accumulated at the OMM, where it recruits Parkin from the cytosol to mitochondria. Parkin is a multifunctional E3 ubiquitin ligase, capable of tagging ubiquitin to proteins at K48 for proteasomal degradation or at K63 for other biological functions. Parkin has been well reported to function downstream of PINK1. Once recruited, Parkin initiates the process of mitophagy to eliminate damaged mitochondria. The in vivo significance of mitophagy mediated by PINK1/Parkin has been controversial (Whitworth and Pallanck, 2017).

PINK1, and to a lesser extent parkin, also have an impact on mitochondrial fission and fusion, a role that has been extensively studied (Exner et al., 2012; Pilsl and Winklhofer, 2012). Briefly, loss of PINK1 function, or expression of human relevant PINK1 mutants, has been demonstrated to increase mitochondrial fission in most studies utilizing mammalian cell models (Exner et al., 2007). Parkin has been reported to ubiquitinate MFN2, leading to mitochondrial fission and thus facilitating mitophagy (Pallanck, 2013). However, parkin has also been reported by various laboratories to have mitochondrial profusion effects (Cui et al., 2010; Dagda et al., 2009; Kamp et al., 2010; Lutz et al., 2009). What is clear is that parkin is recruited by PINK1 to remove damaged mitochondria (Youle and Narendra, 2011).

3. ALZHEIMER'S DISEASE

AD is the most prevalent neurodegenerative disease; defined by progressive memory and cognition deficits with accompanying development of extracellular neuritic β-amyloid (Aβ) plaques and intracellular neurofibrillary tau tangles. Increasing age and genetic mutations are both risk factors for AD. Three autosomal dominant genes responsible for the early onset of AD have been identified: amyloid precursor protein (APP), and APP processing proteins presenilin-1 (*PS1*) and presenilin-2 (*PS2*) (Schellenberg and Montine, 2012). These mutations account for a fraction of AD cases. The remaining cases of AD are sporadic and late onset with unknown etiology. The characteristic neuropathology of AD includes a loss of cholinergic neurons in the nucleus basalis of Meynert, hippocampus, and cerebral cortex (Ferreira-Vieira et al., 2016; Whitehouse et al., 1982). Multiple investigations have demonstrated that energy deficiency is a fundamental characteristic feature of both brains and peripheral cells derived from AD patients (Beal, 2005; Gibson et al., 1998; Manczak et al., 2004).

Intraneuronal oligomeric Aβ has been reported to induce mitochondrial and synaptic deficiencies, disrupt calcium homeostasis, and induce apoptotic cell death (Calkins et al., 2011; Chen and Yan, 2006; Yan and Stern, 2005). Recent work (Richetin et al., 2017) demonstrated that mitochondrial function was impaired in adult-born neurons in an APPxPS1 mouse model of AD, and subsequent increasing of mitochondrial function in primary neurons stimulated dendritic growth and spine formation, supporting the importance of mitochondrial functionality in protecting against neuronal loss, and restoring neurogenesis capacity in AD. Hyperphosphorylated and truncated tau protein have been reported in a number of studies to alter mitochondrial structure and function in AD brains, and to exacerbate the neurotoxic effects of Aβ. Loss of mitochondrial membrane potential and oxidative stress levels in response to Aβ were significantly increased in neurons expressing caspase-cleaved tau in comparison with neurons that were expressing full-length tau (Quintanilla et al., 2012).

Several studies have reported impaired mitochondrial dynamics, involving the abnormal expression of mitochondrial fission and fusion proteins, in postmortem brains from AD patients, mouse, and cell models (Cho et al., 2009; Reddy et al., 2011; Wang et al., 2008, 2009). In human M17 neuronal cells, overexpression of mutant APP caused a drastic imbalance of mitochondrial fission/fusion proteins by reducing the levels of Drp1 and OPA1, and increasing Fis1 (Wang et al., 2008, 2009); this resulted in fragmented mitochondria, suggestive of excessive mitochondrial fission. Support work from this group reported that expression levels of Drp1, OPA1, Mfn1, and Mfn2 are significantly reduced, while Fis1 levels are increased in hippocampal tissue isolated from AD-afflicted individuals vs age-matched controls (Wang et al., 2009). With such a pattern of alterations, it is difficult to predict whether excessive fission or fusion occurred in these samples. In contrast to these studies, Manczak et al. found increased expression of Drp1 and Fis1, and decreased expression of Mfn1, Mfn2, Opa1, in AD frontal cortex and primary hippocampal neurons from mutant AβPP mice, suggestive of enhanced fission (Manczak et al., 2011). In their recent work, these investigators reported that crossing the $Drp1^{+/-}$ mice with either the transgenic AβPP mice (Tg2576) or with Tau P301L transgenic mice reduced toxic soluble proteins and improved mitochondrial function in these AD animal models (Kandimalla et al., 2016; Manczak et al., 2016). These studies strongly support blocking mitochondrial fission as a therapeutic strategy for AD.

The apolipoprotein E4 allele is a major genetic risk factor for AD; heterozygous carriers have four times increased risk of developing AD; and homozygous carriers have 15 times increased risk compared to APOE3 (Ashford and Mortimer, 2002; Raber et al., 2004). APOE4 has been linked to metabolic dysfunction (Huang, 2010; Mahley and Huang, 2012), alterations of synaptic function and development (Kim et al., 2014; Mahley and Huang, 2012); in Neuro-2A (N2A) neuroblastoma cells APOE4 expression reduced the levels of mitochondrial respiratory complexes I, IV, and V, with reduced complex IV activity also reported. Treatment with a small molecule which inhibits APOE4–domain interactions with mitochondria restored respiratory complex IV levels, indicating that the effect is specific to APOE4-mitochondrial contact (Chen et al., 2010). Altered energy metabolism has also been demonstrated in the posterior cingulate of young adult APOE4 carriers, indicating that metabolic dysregulation occurs decades before possible disease onset in this AD-risk population and supporting the contributing role of abnormal metabolism in AD (Perkins et al., 2016). Together, these results suggest that, through mitochondria, APOE4 may enhance the risk of developing AD when combined with environmental toxicants. Indeed, in a case–control study, APOE4 carriers were more susceptible to dichlorodiphenyldichloroethylene (DDE), the metabolite of the pesticide dichlorodiphenyltrichloroethane (DDT), and elevated serum levels of DDE were associated with increased risk of AD (Richardson et al., 2014).

4. HUNTINGTON'S DISEASE

HD is a dominantly inherited progressive hyperkinetic disease which results from the abnormal expansion of the glutamine/CAG repeat in exon one of the gene encoding Huntingtin protein (*Htt*) (Gusella et al., 1983; The Huntington's Disease Collaborative Research Group, 1993). In unaffected individuals, the *Htt* gene contains 6–35 CAG repeats. Once the repeat number exceeds 40 the likelihood of developing HD increases; typically, the more CAG repeats an individual displays, the younger the age of disease onset (Ashizawa et al., 1994; Gusella et al., 1983). Although HD is a genetic disorder, environmental factors also influence the age of disease onset, as demonstrated in animal models (Hockly et al., 2002; van Dellen et al., 2000) and HD patients (Wexlet et al., 2004). The neuropathology is characterized by selective neuronal degeneration, initially in the caudate and striatum, progressing to the cerebral cortex at later stages (Vonsattel and DiFiglia, 1998). The clinical progression is the result of a loss of striatal GABAergic medium spiny neurons in the striatum, and the presence of intracellular protein aggregation is a common feature in HD brains (DiFiglia et al., 1997). Currently, the pathogenic mechanism leading to neurodegeneration in HD has not been fully elucidated, however bioenergetic defects have been proposed to be instrumental.

Mitochondrial dysfunction in Huntington's was reported in 1974, following the discovery of defective succinate dehydrogenase (complex II) in postmortem HD brains (Stahl and Swanson, 1974). Subsequent studies showed that mutant Htt could impair complex II (Benchoua et al., 2006; Browne et al., 1997; Mann et al., 1990) and mitochondrial function (Milakovic and Johnson, 2005; Panov et al., 2002; Seong et al., 2005). Furthermore, inhibition of complex II by 3-nitropropionic acid (3-NP) in humans and in animal models produces striatal lesions and motor symptoms similar to those seen in HD patients (Brouillet et al., 2005). This neurotoxin was used to model HD in rodents and primates (Beal et al., 1993; Brouillet et al., 1995; Ferrante et al., 1991). However, because of its acute nature and the availability of transgenic HD models, the 3–NP model is no longer commonly used.

In recent years, mutant Htt has been reported to induce mitochondrial fragmentation and impair mitochondrial transport along neuronal processes (Jin et al., 2013; Kim et al., 2010; Shirendeb et al., 2011; Song et al., 2011). The excessive fission is shown to be a result of direct interaction between mutant Htt and Drp1, leading to increased Drp1 GTPase activity (Shirendeb et al., 2011; Song et al., 2011). In HD mouse models, Drp1 was found to be significantly increased and Mfn1 significantly decreased (Kim et al., 2010), further implicating mutant Htt interactions with mito-chondria, or proteins necessary for their dynamics, in the pathogenesis of HD. Together, these studies also suggest that mitochondrial fragmentation occurs prior to the onset of neurological deficits, neuronal cell death, and mutant Htt aggregate formation, supporting the theory that mitochondrial change is an early event, before the onset and progression of pathology in neurodegenerative diseases. Indeed, metabolic impairment occurs before any detected loss of striatal neurons in HD patients (Kuhl et al., 1982, 1984).

In addition to the electron transport chain, mutant Htt also impairs mito-chondrial function at the transcriptional level. Mutant Htt represses the expression of PGC-1α by interfering with TAF4/CREB signaling (Cui et al., 2006), thereby reducing cAMP-responsive element-binding protein (CREB)-binding protein (CBP)-dependent gene expression (Riley and Orr, 2006). Changes in mitochondrial biogenesis have been described in HD, with multiple reports of decreased PGC-1α expression in HD mouse models. PGC-1α is essential for mitochondrial biogenesis and the expression of numerous genes that detoxify ROS (Cui et al., 2006; Johri and Beal, 2012; St Pierre et al., 2006). Reduction of this protein will therefore not only impact mitochondrial biogenesis, but also reduce the antioxidant capacity of the cell to defend against the increased ROS associated with mutant Htt toxicity.

5. AMYOTROPHIC LATERAL SCLEROSIS

ALS is a progressive neurodegenerative disease with targeted loss of motor neurons in the brain stem, motor cortex, and spinal cord, resulting in muscle weakness, atrophy, and eventual death, as a result of paralysis of muscle function and consequent respiratory failure (Johri and Beal, 2012; Kawamata and Manfredi, 2010; Kiernan et al., 2011). Intracellular ubiquitin- and ALS-associated proteinaceous inclusions are visible prior to loss of these cells, suggestive of impaired protein clearance mechanisms as observed in other neurodegenerative disorders. Although about 90% of cases are sporadic, several genetic mutations (*SOD1*, *TARDBP*, *FUS*, *C9ORF72*, *OPTN*, *VCP*, *UBQLN*, and *PFN1*) have been identified which constitute the remaining 10% of ALS cases (Anderson and Al-Chalabi, 2011; Renton et al, 2014). Of these mutated genes, superoxide dismutase (*SOD1*), which represents approximately 20% of familial ALS cases, is best documented for its role in mitochondria.

Mitochondrial dysfunction due to impairment of the electron transport chain function has been well documented in cell and animal models with *SOD1* (*G931A*) mutation (Jiang et al., 2015; Johri and Beal, 2012). In more recent years, mitochondrial impairment due to excessive mitochondrial fission has been reported in experimental models of ALS. SOD1 (G931A) induces mitochondrial fragmentation and defects in anterograde and retrograde axonal movement in NSC34 motor neuronal cells (Magrané et al., 2009) and rat spinal cord neurons (Song et al., 2013). Blocking Drp1 rescues these defects and cell viability. In intact sciatic nerve of living mice with SOD1 (G93A) and TDP43, impaired mitochondrial dynamics occurred in a time-dependent manner, starting with mitochondrial fragmentation, impaired retrograde transport, anterograde transport, and then onset of symptoms (Magrané et al., 2014). By assessing these two models simultaneously, the authors demonstrated that impaired mitochondrial dynamics is a common mechanism of different genetic forms of ALS. In mice with selective expression of mitochondrial-targeted enhanced green fluorescent protein in motor neurons, axonal mitochondria of motor neurons are the primary in vivo targets for misfolded SOD1 (Vande Velde et al., 2011). In general, mutant SOD1 neurons have impaired axonal transport, smaller mitochondria, decreased mitochondrial density, increased production of free radical species, glutamate excitotoxicity, and protein aggregation (Pasinelli and Brown, 2006; Rothstein, 2009). These changes are consistent with those observed in motor neurons of ALS patients (Sasaki and Iwata, 2007). The appearance of mitochondrial abnormalities prior to the development of motor neuron degeneration and symptomatic onset in mutant SOD1 mouse models suggests that mitochondrial changes are actively involved in the disease pathogenesis and progression (Kawamata and Manfredi, 2010; Manfredi and Kawamata, 2016). Furthermore, defective mitochondria in ALS are not limited to neurons, as changes in muscle mitochondria may also contribute to weakness and atrophy during disease progression. Aside from its direct effects on mitochondria, mutant SOD1 in microglia increases ROS generation through NOX activity (Harraz et al., 2008; Marden et al., 2007), demonstrating that, similar to other neurodegenerative diseases, increased ROS in ALS is not limited to neuronal production alone.

As many cases of ALS are not the result of genetic changes, the role of the environmental factors may play a role in the pathogenesis of ALS. Recent investigations have studied the contribution of heavy metals (Vinceti et al., 2016), pesticides, solvents, and extremely low frequency magnetic fields (ELF-MFs) (Andrew et al., 2017; Koeman et al., 2017; Vinceti et al., 2017) to ALS risk. Reported results are conflicting, likely due to different investigative methods (biological sampling vs self-reported exposures) and sample sizes. While a number of questionnaire-based studies demonstrated increased ALS risk following exposure to pesticides, heavy metals, and ELF-MFs (Andrew et al., 2017; Koeman et al., 2017; Yu et al., 2014), further work utilizing CSF sampling failed to validate these results (Vinceti et al., 2016, 2017). More work is required before increased ALS risk can be attributed to any environmental toxicants, as current evidence for many factors remains scant.

6. THERAPEUTIC TARGETING OF MITOCHONDRIA

As discussed in this review, mitochondrial dysfunction has been implicated in multiple neurodegenerative diseases for decades. It seems logical therefore that efforts have been made to develop therapeutics targeting mitochondria. Over the years several strategies have been used. For example, Coenzyme Q10 (CoQ_{10}), an essential biological component of the mitochondrial transport chain and free radical scavenging antioxidant, has been shown to be neuroprotective in multiple models of neurodegeneration. However, the therapeutic efficacy of CoQ_{10} is variable and remains disputed. In two clinical trials for PD, CoQ_{10} was found to be well tolerated, but no change in the Unified Parkinson's Disease Rating Scale (UPDRS) motor score was reported (Shults et al., 1998; Strijks et al., 1997). MitoQ is a mitochondria-targeted antioxidant which may protect against oxidative damage (Kelso et al., 2001). It has been tested in various disease models with variable results (Gane et al., 2010; Smith and

Murphy, 2010), but it demonstrated no change in UPDRS score of PD patients following a 12-month clinical trial (Snow et al., 2010). Given that both mitochondria-oriented antioxidant therapies, CoQ_{10} and MitoQ, yielded no success it is worthwhile to consider other approaches to improve mitochondrial function.

6.1 L-Type Calcium Channel Targeting

L-type calcium (Ca^{2+}) channels are part of a family of voltage-dependent calcium channels (Alexander et al., 2015). Surmeier and colleagues demonstrated that the selective expression of this particular type of Ca^{2+} channel is responsible for higher cytosolic Ca^{2+} levels in nigral DA neurons, resulting in oxidative stress and mitochondrial damage (Chan et al., 2007; Surmeier et al., 2010). Subsequent work from Sulzer's group demonstrated that the combination of Ca^{2+}, α-synuclein, and DA in nigral DA neurons accounts for the selective vulnerability of nigral DA neurons (Mosharov et al., 2009). Targeting L-type calcium channels with inhibitory compounds has been demonstrated to be protective in nigral DA neurons against neurotoxic insult in multiple PD models (Surmeier and Schumacker, 2013). Isradipine, an L-type calcium channel blocker which specifically targets the $Ca_v1.3$ subunit, is currently in use as an antihypertensive medication. It has demonstrated efficacy in a mouse model of PD and is in phase III clinical trials as a potential therapeutic for PD (Chan et al., 2007; Ilijic et al., 2011; Parkinson Study Group, 2013).

6.2 Ketone Bodies

Ketone metabolism has been linked to the pathophysiology of brain disorders, such as AD (Hertz et al., 2015). White matter degeneration is a hallmark of AD and it has been proposed that catabolism of myelin lipids to produce ketones is an adaptive response to meet energy demands (Klosinski et al., 2015), thus providing an alternative ketone energy source may be protective in early AD and protect against white matter degeneration. Axona® is an oral agent which is digested and converted to ketone bodies by the liver. Axona® is a medium-chain triglyceride marketed as a medical food therapy for AD patients with mild-to-moderate symptoms (Roman, 2010). On the basis that AD brains may be ineffectively metabolizing glucose as a source of ATP, Axona® provides an alternative energy source, as ketone bodies can be metabolized to improve metabolic deficits

(Sharma et al., 2014). Clinical trials report a correlation between increased ketone body levels and improved cognition in AD patients, measured with the Alzheimer's Disease Assessment Score (ADAS-cog), compared with placebo controls (Henderson et al., 2009).

In addition to AD, ketone therapy has also been shown to be beneficial to PD. We demonstrated that the ketone body D-β-hydroxybutyrate (DβHB) conferred neuroprotection through a previously unidentified mechanism in the MPTP mouse model (Tieu et al., 2003). In general, electrons are fed into the mitochondrial transport chain at complexes I and II to generate ATP. In the MPTP model, complex I function is significantly reduced. DβHB fuels ATP production by supplying electrons to complex II. Subsequent to this study, a group of neurologists from St. Luke's–Roosevelt Hospital Center, New York, decided to adopt a similar approach by using ketogenic diet to treat a small group of patients with PD. The authors reported that this treatment significantly improved the symptoms of these patients (Vanitallie et al., 2005). DβHB is also protective in the R6/2 mouse model of HD (Lim et al., 2011). Together, ketone therapy appears to be promising for neurodegenerative diseases.

6.3 Drp1 and Mitochondrial Fission

The observation of mitochondrial fragmentation as a common abnormal feature in neurodegenerative disease has led investigators to develop strategies to either block mitochondrial fission or promote fusion. Because Drp1 can bind to multiple fission proteins (MFF, Fis1, MiD49, and MiD51) to sever mitochondria, directly targeting Drp1 is more effective to block mitochondrial fission, rather than targeting these receptor proteins individually. Furthermore, the translational potential of Drp1-based therapies is enhanced by the availability of small-molecule inhibitors.

Three small molecules have been described that inhibit the GTPase activity of Drp1, altering its functionality and reducing mitochondrial fission. The first described was Dynasore, a noncompetitive dynamin GTPase inhibitor of dynamin 1, dynamin 2, and Drp1 (Macia et al., 2006). In vitro reports have demonstrated the ability of Dynasore to completely and reversibly inhibit endocytosis in cultured hippocampal neurons (Newton et al., 2006). While Dynasore is not specific for just Drp1 and has been shown to alter other dynamin–independent cellular processes, such as lipid raft organization (Preta et al., 2015), its potential efficacy in neuronal restoration was recently described in a rat model of spinal cord injury. Dynasore ameliorates motor dysfunction, reduces Drp1 expression and apoptosis initiation in this model (Li et al., 2016). Despite the promising results from alternative applications of Dynasore, as of yet there are no studies assessing it's therapeutic potential in neurodegenerative diseases.

A second small-molecule inhibitor of Drp1, mitochondrial division inhibitor 1 (mdivi-1), was characterized in 2008 (Cassidy-Stone et al., 2008) following screening of approximately 23,000 compounds from several chemical libraries using a yeast two-hybrid system. Mdivi-1 inhibits the GTPase activity of Dnm1, the yeast homolog of the mammalian Drp1 (Cassidy-Stone et al., 2008). Subsequently, mdivi-1 was shown to inhibit recombinant mammalian Drp1 with an IC_{50} value of 13 µM (Numadate et al., 2014), although a recent study questioned this interaction (Bordt et al., 2017). To date, mdivi-1 has been demonstrated to be protective by blocking mitochondrial fission in cell culture models (Cassidy-Stone et al., 2008; Cui et al., 2010; Grohm et al., 2012; Zhao et al., 2014), as well as in mouse models of renal (Brooks et al., 2009), cardiac (Ong et al., 2010; Sharp et al., 2014), brain ischemic damage (Grohm et al., 2012; Zhao et al., 2014), rat neuropathic pain (Ferrari et al., 2011), epilepsy (Qiu et al., 2013; Xie et al., 2013), and PD (Rappold et al., 2014).

More recently a peptide inhibitor of Drp1 called P110 was described. This inhibitor was designed to prevent the binding of Drp1 to Fis1 (Qi et al., 2013). P110 has been shown to be protective in cultured cells treated with MPP^+ (Qi et al., 2013), mouse models of PD (Filichia et al., 2016), multiple sclerosis (Luo et al., 2017), and transgenic mouse models of HD (Guo et al., 2013).

Drp1 has also been targeted with genetic approaches, including strategies such as overexpression of Drp1 dominant negative mutants and knocking down Drp1 in various in vitro models of PD (Cui et al., 2010; Dagda et al., 2009; Lutz et al., 2009; Wang et al., 2011b). To evaluate the protective effect of blocking Drp1 in vivo, we performed supranigral injections of recombinant adeno-associated virus (rAAV2) to deliver Drp1–K38A (a dominant negative of Drp1) which attenuated neurotoxicity and restored preexisting striatal dopamine release deficits in $PINK1^{-/-}$ and MPTP mouse models (Rappold et al., 2014). Recent studies show that crossing $Drp1^{+/-}$ mice with either the transgenic AβPP mice (Tg2576) or with tau P301L transgenic mice reduced toxic soluble proteins and improved mitochondrial function in these animal models of AD (Kandimalla et al., 2016; Manczak et al., 2016). Altogether, Drp1 has emerged as a target with a tremendous therapeutic potential.

7. CONCLUSION

To understand the pathophysiology and find disease-modifying therapies for neurodegenerative diseases has been, and will continue to be, a major challenge. In this chapter, we discussed genetic mutations and the potential involvement of environmental exposure in the pathogenesis of AD, PD, HD, and ALS. From this review, it is evident that these disorders share common features: first, genetic mutations, environmental factors, and gene–environment interaction play a role in pathogenesis, disease onset, and progression. Although multiple mutations linked to PD, AD, and ALS have been discovered, they still represent less than 10% of cases, further implicating the role of the environment in these diseases. Despite the rarity of these mutations, over the past 2 decades, these genes have provided tremendous insights into the mechanisms of neuronal dysfunction and degeneration. Such mechanisms have then been subsequently investigated and confirmed to also be mediated by neurotoxicants such as Mn, rotenone, and paraquat. Together, as illustrated in Fig. 2, these studies indicate that regardless of the diverse causes of neurodegenerative diseases, there exists common pathogenic mechanisms such as mitochondrial dysfunction, oxidative stress, neuroinflammation, and impairment in protein degradation pathways. This chapter highlights the emerging view that perturbed mitochondrial dynamics may represent a converging and common mechanism that underlies both genetic and toxicant-induced neurodegeneration. However, this knowledge does not address the intriguing question of how a common pathogenic mechanism may give rise to such diverse forms of neurodegenerative diseases. That is, if mitochondrial dysfunction is a common pathogenic feature in AD, PD, HD, and ALS, then it is not clear how this abnormality would lead to specific cell loss in particular brain regions, not only within, but also between these disorders. The answer to this fundamental question may hold the key to developing ground-breaking disease-modifying therapies for neurodegenerative diseases.

ACKNOWLEDGMENTS

We thank Peter Hodges for the illustrations. This work was supported in part by National Institute of Environmental Health Services (R01-ES022274) and Stempel College of Environmental & Occupational Health, Florida International University.

REFERENCES

Alaimo, A., Gorojod, R.M., Beauquis, J., Muñoz, M.J., Saravia, F., Kotler, M.L., 2014. Deregulation of mitochondria-shaping proteins opa-1 and drp-1 in manganese-induced apoptosis. PLoS One 9, e91848.

Alexander, S., Catterall, W.A., Kelly, E., Marrion, N., Peters, J., Benson, H., et al., 2015. The concise guide to pharmacology 2015/16: voltage-gated ion channels. Br. J. Pharmacol. 172, 5904–5941.

Alvarez-Erviti, L., Seow, Y., Schapira, A.H., Gardiner, C., Sargent, I.L., Wood, M.J., et al., 2011. Lysosomal dysfunction increases exosome-mediated alpha-synuclein release and transmission. Neurobiol. Dis. 42, 360–367.

Anderson, P.M., Al-Chalabi, A., 2011. Clinical genetics of amyotrophic lateral sclerosis: what do we really know? Nat. Rev. Neurol. 7, 603–615.

Andrew, A.S., Caller, T.A., Tandan, R., Duell, E.J., Henegan, P.L., Field, N.C., et al., 2017. Environmental and occupational exposures and amyotrophic lateral sclerosis in New England. Neurodegener. Dis. 17, 110–116.

Appel-Cresswell, S., Vilarino-Guell, C., Encarnacion, M., Sherman, H., Yu, I., Shah, B., et al., 2013. Alpha-synuclein p.H50q, a novel pathogenic mutation for Parkinson's disease. Mov. Disord. 28, 811–813.

Aras, S., Bai, M., Lee, I., Springett, R., Huttemann, M., Grossman, L., 2015. Mnrr1 (formerly chchd2) is a bi-organellar regulator of mitochondrial metabolism. Mitochondrion 20, 43–51.

Ashford, J., Mortimer, J., 2002. Non-familial Alzheimer's disease is mainly due to genetic factors. J. Alzheimers Dis. 4, 169–177.

Ashizawa, T., Wong, L., Richards, C., Caskey, C., Jankovic, J., 1994. Cag repeat size and clinical presentation in Huntington's disease. Neurology 44, 1137–1143.

Barsoum, M.J., Yuan, H., Gerencser, A.A., Liot, G., Kushnareva, Y., Gräber, S., et al., 2006. Nitric oxide-induced mitochondrial fission is regulated by dynamin-related GTPases in neurons. EMBO J. 25, 3900–3911.

Beal, M., 2005. Mitochondria take center stage in aging and neurodegeneration. Ann. Neurol. 58, 495–505.

Beal, M., Brouillet, E., Jenkins, B., Ferrante, R., Kowall, N., Miller, J., et al., 1993. Neurochemical and histologic characterization of striatal excitotoxic lesions produced by the mitochondrial toxin 3-nitropropionic acid. J. Neurosci. 13, 4181–4192.

Benchoua, A., Trioulier, Y., Zala, D., Gaillard, M.C., Lefort, N., Dufour, N., et al., 2006. Involvement of mitochondrial complex II defects in neuronal death produced by n-terminus fragment of mutated huntingtin. Mol. Biol. Cell 17, 1652–1663.

Betarbet, R., Sherer, T., MacKenzie, G., Garcia-Osuna, M., Panov, A., Greenamyre, J., 2000. Chronic systemic pesticide exposure reproduces features of Parkinson's disease. Nat. Neurosci. 3, 1301–1306.

Bordt, E.A., Clerc, P., Roelofs, B.A., Saladino, A.J., Tretter, L., Adam-Vizi, V., et al., 2017. The putative drp1 inhibitor mdivi-1 is a reversible mitochondrial complex I inhibitor that modulates reactive oxygen species. Dev. Cell 40, 583–594, e586.

Braschi, E., Goyon, V., Zunino, R., Mohanty, A., Xu, L., McBride, H.M., 2010. Vps35 mediates vesicle transport between the mitochondria and peroxisomes. Curr. Biol. 20, 1310–1315.

Brooks, C., Wei, Q., Cho, S.G., Dong, Z., 2009. Regulation of mitochondrial dynamics in acute kidney injury in cell culture and rodent models. J. Clin. Invest. 119, 1275–1285.

Brouillet, E., Hantraye, P., Ferrante, R., Dolan, R., Leroy-Willig, A., Kowall, N., et al., 1995. Chronic mitochondrial energy impairment produces selective striatal degeneration and abnormal choreiform movements in primates. Proc. Natl. Acad. Sci. U.S.A. 92, 7105–7109.

Brouillet, E., Jacquard, C., Bizat, N., Blum, D., 2005. 3-Nitropropionic acid: a mitochondrial toxin to uncover physiopathological mechanisms underlying striatal degeneration in Huntington's disease. J. Neurochem. 95, 1521–1540.

Browne, S.E., Bowling, A.C., MacGarvey, U., Baik, M.J., Berger, S.C., Muqit, M.M., et al., 1997. Oxidative damage and metabolic dysfunction in Huntington's disease: selective vulnerability of the basal ganglia. Ann. Neurol. 41, 646–653.

Burré, J., Sharma, M., Tsetsenis, T., Buchman, V., Etherton, M.R., Südhof, T.C., 2010. Alpha-synuclein promotes snare-complex assembly in vivo and in vitro. Science 329, 1663–1667.

Calkins, M., Manczak, M., Mao, P., Shirendeb, U., Reddy, P., 2011. Impaired mitochondrial biogenesis, defective axonal transport or mitochondria, abnormal mitochondrial dynamics and synaptic degeneration in a mouse model of Alzheimer's disease. Hum. Mol. Genet. 20, 4515–4529.

Calne, D.B., Chu, N.S., Huang, C.C., CS, L., Olanow, W., 1994. Manganism and idiopathic Parkinsonism: similarities and differences. Neurology 44, 1583–1586.

Cannon, J.R., Greenamyre, J.T., 2013. Gene–environment interactions in Parkinson's disease: specific evidence in humans and mammalian models. Neurobiol. Dis. 57, 38–46.

Cannon, J.R., Tapias, V.M., Na, H.M., Honick, A.S., Drolet, R.E., Greenamyre, J.T., 2009. A highly reproducible rotenone model of Parkinson's disease. Neurobiol. Dis. 34, 279–290.

Cassidy-Stone, A., Chipuk, J.E., Ingerman, E., Song, C., Yoo, C., Kuwana, T., et al., 2008. Chemical inhibition of the mitochondrial division dynamin reveals its role in bax/bak-dependent mitochondrial outer membrane permeabilization. Dev. Cell 14, 193–204.

Chan, C., Guzman, J., Ilijic, E., Mercer, J., Rick, C., Tkatch, T., et al., 2007. "Rejuvenation" protects neurons in mouse models of Parkinson's disease. Nature 447, 1081–1086.

Chen, H., Chan, D., 2009. Mitochondrial dynamics—fission, fusion, movement and mitophagy—in neurodegenerative diseases. Hum. Mol. Genet. 18, 169–176.

Chen, X., Yan, S., 2006. Mitochondrial Abeta: a potential cause of metabolic dysfunction in Alzheimer's disease. IUBMB Life 58, 686–694.

Chen, H.-K., Ji, Z.-S., Dodson, S.E., Miranda, R.D., Rosenblum, C.I., Reynolds, I.J., et al., 2010. Apolipoprotein e4 domain interaction mediates detrimental effects on mitochondria and is a potential therapeutic target for Alzheimer's disease. J. Biol. Chem. 286, 5215–5221.

Chen, L., Xie, Z., Turkson, S., Zhuang, X., 2015. A53t human alpha-synuclein overexpression in transgenic mice induces pervasive mitochondria macroautophagy defects preceding dopamine neuron degeneration. J. Neurosci. 35, 890–905.

Chen, P., Culbreth, M., Aschner, M., 2016. Exposure, epidemiology, and mechanism of the environmental toxicant manganese. Environ. Sci. Pollut. Res. Int. 23, 13802–13810.

Cho, D.-H., Nakamura, T., Fang, J., Cieplak, P., Godzik, A., Gu, Z., et al., 2009. S-nitrosylation of drp1 mediates β-amyloid-related mitochondrial fission and neuronal injury. Science 324, 102.

Cochemé, H.M., Murphy, M.P., 2008. Complex I is the major site of mitochondrial superoxide production by paraquat. J. Biol. Chem. 283, 1786–1798.

Cookson, M.R., 2015. Lrrk2 pathways leading to neurodegeneration. Curr. Neurol. Neurosci. Rep. 15, 42.

Cui, L., Jeong, H., Borovecki, F., Parkhurst, C., Tanese, N., Krainc, D., 2006. Transcriptional repression of pgc-1alpha by mutant huntingtin leads to mitochondrial dysfunction and neurodegeneration. Cell 127, 59–69.

Cui, M., Tang, X., Christian, W.V., Yoon, Y., Tieu, K., 2010. Perturbations in mitochondrial dynamics induced by human mutant pink1 can be rescued by the mitochondrial division inhibitor mdivi-1. J. Biol. Chem. 285, 11740–11752.

Dächsel, J.C., Farrer, M.J., 2010. LRRK2 and Parkinson disease. Arch. Neurol. 67, 542–547.

Dagda, R.K., Cherra, S.J., Kulich, S.M., Tandon, A., Park, D., Chu, C.T., 2009. Loss of pink1 function promotes mitophagy through effects on oxidative stress and mitochondrial fission. J. Biol. Chem. 284, 13843–13855.

Dagda, R.K., Pien, I., Wang, R., Zhu, J., Wang, K.Z., Callio, J., et al., 2014. Beyond the mitochondrion: cytosolic PINK1 remodels dendrites through protein kinase A. J. Neurochem. 128, 864–877.

Day, B.J., Patel, M., Calavetta, L., Stamler, J.S., 1999. A mechanism of paraquat toxicity involving nitric oxide synthase. Proc. Natl. Acad. Sci. U.S.A. 96, 12760–12765.

Dehay, B., Bourdenx, M., Gorry, P., Przedborski, S., Vila, M., Hunot, S., et al., 2015. Targeting α-synuclein for treatment of Parkinson's disease: mechanistic and therapeutic considerations. Lancet Neurol. 14, 855–866.

Di Maio, R., Barrett, P.J., Hoffman, E.K., Barrett, C.W., Zharikov, A., Borah, A., et al., 2016. A-synuclein binds to TOM20 and inhibits mitochondrial protein import in Parkinson's disease. Sci. Transl. Med. 8.

DiFiglia, M., Sapp, E., Chase, K., Davies, S., Bates, G., Vonsattel, J., et al., 1997. Aggregation of huntingtin in neuronal intracellular inclusions and dystrophic neurites in brain. Science 277, 1990–1993.

Dorsey, E.R., Constantinescu, R., Thompson, J.P., Biglan, K.M., Holloway, R.G., Kieburtz, K., et al., 2007. Projected number of people with Parkinson disease in the most populous nations, 2005 through 2030. Neurology 68, 384–386.

Exner, N., Treske, B., Paquet, D., Holmström, K., Schiesling, C., Gispert, S., et al., 2007. Loss-of-function of human pink1 results in mitochondrial pathology and can be rescued by parkin. J. Neurosci. 27, 12413–12418.

Exner, N., Lutz, A.K., Haass, C., Winklhofer, K.F., 2012. Mitochondrial dysfunction in Parkinson's disease: molecular mechanisms and pathophysiological consequences. EMBO J. 31, 3038–3062.

Fahn, S., Prsedborski, S., 2010. Parkinsonism, 10th ed. Lippincott Williams and Wilkins, New York, NY.

Fernagut, P.O., Hutson, C.B., Fleming, S.M., Tetreaut, N.A., Salcedo, J., Masliah, E., et al., 2007. Behavioral and histopathological consequences of paraquat intoxication in mice: effects of alpha-synuclein over-expression. Synapse 61, 991–1001.

Ferrante, R., Kowall, N., Richardson, E.J., 1991. Proliferative and degenerative changes in striatal spiny neurons in Huntington's disease: a combined study using section-Golgi method and calbindin d28k immunocytochemistry. J. Neurosci. 11, 3877–3887.

Ferrari, L.F., Chum, A., Bogen, O., Reichling, D.B., Levine, J.D., 2011. Role of drp1, a key mitochondrial fission protein, in neuropathic pain. J. Neurosci. 31, 11404–11410.

Ferreira-Vieira, T.H., Guimaraes, I.M., Silva, F.R., Ribeiro, F.M., 2016. Alzheimer's disease: targeting the cholinergic system. Curr. Neuropharmacol. 14, 101–115.

Filichia, E., Hoffer, B., Qi, X., Luo, Y., 2016. Inhibition of drp1 mitochondrial translocation provides neural protection in dopaminergic system in a Parkinson's disease model induced by MPTP. Sci. Rep. 6, 32656.

Franco-Iborra, S., Vila, M., Perier, C., 2016. The Parkinson disease mitochondrial hypothesis: where are we at? Neuroscientist 22, 266–277.

Funayama, M., Ohe, K., Amo, T., Furuya, N., Yamaguchi, J., Saiki, S., et al., 2015. Chchd2 mutations in autosomal dominant late-onset Parkinson's disease: a genome-wide linkage and sequencing study. Lancet Neurol. 14, 274–282.

Gane, E., Weilert, F., Orr, D., Keogh, G., Gibson, M., Lockhart, M., et al., 2010. The mitochondria-targeted anti-oxidant mitoquinone decreases liver damage in a phase II study of hepatitis c patients. Liver Int. 30, 1019–1026.

Garcia-Reitböck, P., Anichtchik, O., Bellucci, A., Iovino, M., Ballini, C., Fineberg, E., et al., 2010. Snare protein redistribution and synaptic failure in a transgenic mouse model of Parkinson's disease. Brain 133, 2032–2044.

Gibson, G., Sheu, K., Blass, J., 1998. Abnormalities of mitochondrial enzymes in Alzheimer's disease. J. Neural Transm. 105, 855–870.

Grohm, J., Kim, S.W., Mamrak, U., Tobaben, S., Cassidy-Stone, A., Nunnari, J., et al., 2012. Inhibition of drp1 provides neuroprotection in vitro and in vivo. Cell Death Differ. 19, 1446–1458.

Group THsDCR, 1993. A novel gene containing a trinucleotide repeat that is expanded and unstable on Huntington's disease chromosomes. The Huntington's disease collaborative research group. Cell 72, 971–983.

Guardia-Laguarta, C., Area-Gomez, E., Rüb, C., Liu, Y., Magrané, J., Becker, D., et al., 2014. A-synuclein is localized to mitochondria-associated ER membranes. J. Neurosci. 34, 249–259.

Gui, Y.-X., Wang, X.-Y., Kang, W.-Y., Zhang, Y.-J., Zhang, Y., Zhou, Y., et al., 2012. Extracellular signal-related kinase is involved in alpha-synuclein-induced mitochondrial dynamic disorders by regulating dynamin-like protein 1. Neurobiol. Aging 33, 2841–2854.

Guilarte, T.R., 2013. Manganese neurotoxicity: new perspectives from behavioral, neuroimaging, and neuropathological studies in humans and non-human primates. Front. Aging Neurosci. 5, 23.

Guilarte, T.R., Chen, M.K., McGlothan, J.L., Verina, T., Wong, D.F., Zhou, Y., et al., 2006. Nigrostriatal dopamine system dysfunction and subtle motor deficits in manganese-exposed non-human primates. Exp. Neurol. 202, 381–390.

Guo, X., Disatnik, M.H., Monbureau, M., Shamloo, M., Mochly-Rosen, D., Qi, X., 2013. Inhibition of mitochondrial fragmentation diminishes Huntington's disease-associated neurodegeneration. J. Clin. Invest. 123, 5371–5388.

Gusella, J., MacDonald, M., Ambrose, C., Duyao, M., 1983. Molecular genetics of Huntington's disease. Arch. Neurol. 50, 1157–1163.

Haas, R.H., Nasirian, F., Nakano, K., Ward, D., Pay, M., Hill, R., et al., 1995. Low platelet mitochondrial complex I and complex II/III activity in early untreated Parkinson's disease. Ann. Neurol. 37, 714–722.

Hamza, T.H., Zabetian, C.P., Tenesa, A., Laederach, A., Montimurro, J., Yearout, D., et al., 2010. Common genetic variation in the HLA region is associated with late-onset sporadic Parkinson's disease. Nat. Genet. 42, 781–785.

Harraz, M.M., Marden, J.J., Zhou, W., Zhang, Y., Williams, A., Sharov, V.S., et al., 2008. Sod1 mutations disrupt redox-sensitive Rac regulation of NADPH oxidase in a familial ALS model. J. Clin. Invest. 118, 659–670.

Henderson, S., Vogel, J., Barr, L., Garvin, F., Jones, J., Constantini, L., 2009. Study of the ketogenic agent ac-1202 in mild to moderate Alzheimer's disease: a randomized, double-blind, placebo-controlled, multicenter trial. Nutr. Metab. 6, 31.

Hernandez, D.G., Reed, X., Singleton, A.B., 2016. Genetics in Parkinson disease: Mendelian versus non-Mendelian inheritance. J. Neurochem. 139 (Suppl 1), 59–74.

Hertz, L., Chen, Y., Waagepeterson, H., 2015. Effects of ketone bodies in Alzheimer's disease in relation to neural hypometabolism, b-amyloid toxicity and astrocyte function. J. Neurochem. 134, 7–20.

Hisata, J., 2002. Final Supplemental Environmental Impact Statement. Lake and Stream Rehabilitation: Rotenone Use and Health Risks. Washington State Department of Fish and Wildlife, North Olympia, WA.

Hockly, E., Cordery, P.M., Woodman, B., Mahal, A., van Dellen, A., Blakemore, C., et al., 2002. Environmental enrichment slows disease progression in r6/2 Huntington's disease mice. Ann. Neurol. 51, 235–242.

Hollenbeck, P., 2005. Mitochondria and neurotransmission: evacuating the synapse. Neuron 47, 331–333.

Huang, Y., 2010. Abeta-independent roles of apolipoprotein e4 in the pathogenesis of Alzheimer's disease. Trends Mol. Med. 16, 287–294.

Ilijic, E., Guzman, J., Surmeier, D., 2011. The L-type calcium channel antagonist isradipine is neuroprotective in a mouse model of Parkinson's disease. Neurobiol. Dis. 43, 364–371.

Jiang, Z., Wang, W., Perry, G., Zhu, X., Wang, X., 2015. Mitochondrial dynamic abnormalities in amyotrophic lateral sclerosis. Transl. Neurodegener. 4, 14.

Jin, Y.N., YV, Y., Gundemir, S., Jo, C., Cui, M., Tieu, K., et al., 2013. Impaired mitochondrial dynamics and Nrf2 signaling contribute to compromised responses to oxidative stress in striatal cells expressing full-length mutant huntingtin. PLoS One 8, e57932.

Johri, A., Beal, M.F., 2012. Mitochondrial dysfunction in neurodegenerative diseases. J. Pharmacol. Exp. Ther. 342, 619–630.

Kamel, F., 2013. Paths from pesticides to Parkinson's. Science 341, 722–723.

Kamp, F., Exner, N., Lutz, A.K., Wender, N., Hegermann, J., Brunner, B., et al., 2010. Inhibition of mitochondrial fusion by a-synuclein is rescued by PINK1, Parkin and DJ-1. EMBO J. 29, 3571–3589.

Kandimalla, R., Manczak, M., Fry, D., Suneetha, Y., Sesaki, H., Reddy, P.H., 2016. Reduced dynamin-related protein 1 protects against phosphorylated tau-induced mitochondrial dysfunction and synaptic damage in Alzheimer's disease. Hum. Mol. Genet. 25, 4881–4897.

Kawamata, H., Manfredi, G., 2010. Mitochondrial dysfunction and intracellular calcium dysregulation in ALS. Mech. Ageing Dev. 131, 517–526.

Kelso, G., Porteous, C., Coulter, C., Hughers, G., Porteous, W., Ledgerwood, E., et al., 2001. Selective targeting of a redox-active ubiquinone to mitochondria within cells: antioxidant and antiapoptotic properties. J. Biol. Chem. 276, 4588–4596.

Kiernan, M., Vucic, S., Cheah, B., Turner, M., Eisen, A., Hardiman, O., et al., 2011. Amyotrophic lateral sclerosis. Lancet 377, 942–955.

Kim, J., Moody, J., Edgerly, C., Bordiuk, O., Cormier, K., Smith, K., et al., 2010. Mitochondrial loss, dysfunction and altered dynamics in Huntington's disease. Hum. Mol. Genet. 19, 3919–3935.

Kim, J., Yoon, H., Basak, J., Kim, J., 2014. Apolipoprotein e in synaptic plasticity and Alzheimer's disease: potential cellular and molecular mechanisms. Mol. Cells 37, 767–776.

Klosinski, L., Yao, J., Yin, F., Fonteh, A., Harrington, M., Christensen, T., et al., 2015. White matter lipids as a ketogenic fuel supply in aging female brain: implications for Alzheimer's disease. EBioMedicine 2, 1888–1904.

Koeman, T., Slottje, P., Schouten, L.J., Peters, S., Huss, A., Veldink, J.H., et al., 2017. Occupational exposure and amyotrophic lateral sclerosis in a prospective cohort. Occup. Environ. Med. 74 (8), 578–585.

Krüger, R., Kuhn, W., Müller, T., Woitalla, D., Graeber, M., Kösel, S., et al., 1998. Ala30pro mutation in the gene encoding alpha-synuclein in Parkinson's disease. Nat. Genet. 18, 106–108.

Kuhl, D.E., Phelps, M.E., Markham, C.H., Metter, E.J., Riege, W.H., Winter, J., 1982. Cerebral metabolism and atrophy in Huntington's disease determined by 18fdg and computed tomographic scan. Ann. Neurol. 12, 425–434.

Kuhl, D.E., Metter, E.J., Riege, W.H., Markham, C.H., 1984. Patterns of cerebral glucose utilization in Parkinson's disease and Huntington's disease. Ann. Neurol. 15 (Suppl), S119–125.

Langston, J., Ballard, P., Tetrad, J., Irwin, I., 1983. Chronic Parkinsonism in humans due to a product of meperidine–analog synthesis. Science 25, 979–980.

Lee, H., Yoon, Y., 2016. Mitochondrial fission and fusion. Biochem. Soc. Trans. 44, 1725–1735.

Lee, S.-J., Desplats, P., Lee, H.-J., Spencer, B., Masliah, E., 2012. Cell-to-cell transmission of α-synuclein aggregates. Methods Mol. Biol. 849, 347–359.

Lesage, S., Anheim, M., Letournel, F., Bousset, L., Honoré, A., Rozas, N., et al., 2013. G51d α-synuclein mutation causes a novel parkinsonian–pyramidal syndrome. Ann. Neurol. 73, 459–471.

Li, G., Shen, F., Fan, Z., Wang, Y., Kong, X., Yu, D., et al., 2016. Dynasore improves motor function recovery via inhibition of neuronal apoptosis and astrocytic proliferation after spinal cord injury in rats. Mol. Biol. 10, 241.

Lim, S., Chesser, A.S., Grima, J.C., Rappold, P.M., Blum, D., Przedborski, S., et al., 2011. D-β-hydroxybutyrate is protective in mouse models of Huntington's disease. PLoS One 6, e24620.

Luo, F., Herrup, K., Qi, X., Yang, Y., 2017. Inhibition of Drp1 hyper-activation is protective in animal models of experimental multiple sclerosis. Exp. Neurol. 292, 21–34.

Lutz, A.K., Exner, N., Fett, M.E., Schlehe, J.S., Kloos, K., Lämmermann, K., et al., 2009. Loss of parkin or PINK1 function increases Drp1-dependent mitochondrial fragmentation. J. Biol. Chem. 284, 22938–22951.

Ly, C., Verstreken, P., 2006. Mitochondria at the synapse. Neuroscientist 12, 291–299.

Macia, E., Erlich, M., Massol, R., Boucrot, E., Brunner, C., Kirchhausen, T., 2006. Dynasore, a cell-permeable inhibitor of dynamin. Dev. Cell 10, 839–850.

Magrané, J., Hervias, I., Henning, M.S., Damiano, M., Kawamata, H., Manfredi, G., 2009. Mutant SOD1 in neuronal mitochondria causes toxicity and mitochondrial dynamics abnormalities. Hum. Mol. Genet. 18, 4552–4564.

Magrané, J., Cortez, C., Gan, W.B., Manfredi, G., 2014. Abnormal mitochondrial transport and morphology are common pathological denominators in SOD1 and TDP43 ALS mouse models. Hum. Mol. Genet. 23, 1413–1424.

Mahley, R., Huang, Y., 2012. Apolipoprotein e sets the stage: response to injury triggers neuropathology. Neuron 76, 871–885.

Manczak, M., Park, B., Jung, Y., Reddy, P., 2004. Differential expression of oxidative phosphorylation genes in patients with Alzheimer's disease: implications for early mitochondrial dysfunction and oxidative damage. Neuromolecular Med. 5, 147–162.

Manczak, M., Calkins, M., Reddy, P., 2011. Impaired mitochondrial dynamics and abnormal interaction of amyloid beta with mitochondrial protein Drp1 in neurons from patients with Alzheimer's disease: implications for neuronal damage. Hum. Mol. Genet. 20, 2495–2509.

Manczak, M., Kandimalla, R., Fry, D., Sesaki, H., Reddy, P.H., 2016. Protective effects of reduced dynamin-related protein 1 against amyloid beta-induced mitochondrial dysfunction and synaptic damage in Alzheimer's disease. Hum. Mol. Genet. 25, 5148–5166.

Manfredi, G., Kawamata, H., 2016. Mitochondrial and endoplasmic reticulum crosstalk in amyotrophic lateral sclerosis. Neurobiol. Dis. 90, 35–42.

Mann, V.M., Cooper, J.M., Javoy-Agid, F., Agid, Y., Jenner, P., Schapira, A.H., 1990. Mitochondrial function and parental sex effect in Huntington's disease. Lancet 336, 749.

Manning-Bog, A.B., McCormack, A.L., Li, J., Uversky, V.N., Fink, A.L., Di Monte, D.A., 2002. The herbicide paraquat causes up-regulation and aggregation of a-synuclein in mice. J. Biol. Chem. 277, 1641–1644.

Mao, X., Ou, M.T., Karuppagounder, S.S., Kam, T.-I., Tin, X., Xiong, Y., et al., 2016. Pathological a-synuclein transmission initiated by binding lymphocyte-activated gene 3. Science 353, 1513.

Marden, J.J., Harraz, M.M., Williams, A.J., Nelson, K., Luo, M., Paulson, H., et al., 2007. Redox modifier genes in amyotrophic lateral sclerosis in mice. J. Clin. Invest. 117, 2913–2919.

McCormack, A.L., Thiruchelvam, M., Manning-Bog, A.B., Thiffault, C., Langston, W.J., Cory-Slechta, D.A., et al., 2002. Environmental risk factors and Parkinson's disease: selective degeneration of nigral dopaminergic neurons caused by the herbicide paraquat. Neurobiol. Dis. 10, 119–127.

Milakovic, T., Johnson, G.V., 2005. Mitochondrial respiration and ATP production are significantly impaired in striatal cells expressing mutant huntingtin. J. Biol. Chem. 280, 30773–30782.

Mosharov, E., Larsen, K., Kanter, E., Phillips, K., Wilson, K., Schmitz, Y., et al., 2009. Interplay between cytosolic dopamine, calcium and a-synuclein causes selective cell death of substantia nigra neurons. Neuron 62, 218–229.

Murphy, D.D., Rueter, S.M., Trojanowski, J.Q., Lee, V.M., 2000. Synucleins are developmentally expressed, and alpha-synuclein regulates the size of the presynaptic vesicular pool in primary hippocampal neurons. J. Neurosci. 20, 3214–3220.

Nakamura, K., Nemani, V., Azarbal, F., Skibinski, G., Levy, J., Egami, K., et al., 2011. Direct membrane association drives mitochondrial fission by the Parkinson disease-associated protein alpha-synuclein. J. Biol. Chem. 286, 20701–20726.

Nalls, M.A., Plagnol, V., Hernandez, D.G., Sharma, M., Sheerin, U.M., Saad, M., et al., 2011. Imputation of sequence variants for identification of genetic risks for Parkinson's disease: a meta-analysis of genome-wide association studies. Lancet 377, 641–649.

Newton, A., Kirchhausen, T., Murthy, V., 2006. Inhibition of dynamin completely blocks compensatory synaptic vesicle endocytosis. Proc. Natl. Acad. Sci. U.S.A. 103, 17955–17960.

Nicklas, W.J., Vyas, I., Heikkila, R.E., 1985. Inhibition of NADH-linked oxidation in brain mitochondria by 1-methyl-4-phenyl-pyridine, a metabolite of the neurotoxin, 1-methyl-4-phenyl-1,2,5,6-tetrahydropyridine. Life Sci. 36, 2503–2508.

Niu, J., Yu, M., Wang, C., Xu, Z., 2012. Leucine-rich repeat kinase 2 disturbs mitochondrial dynamics via dynamin-like protein. J. Neurochem. 122, 650–658.

Numadate, A., Mita, Y., Matsumoto, Y., Fujii, S., Hashimoto, Y., 2014. Development of 2-thioxoquinazoline-4-one derivatives as dual and selective inhibitors of dynamin-related protein 1 (drp1) and puromycin-sensitive aminopeptidase (PSA). Chem. Pharm. Bull. (Tokyo) 62, 979–988.

Ong, S.B., Subrayan, S., Lim, S.Y., Yellon, D.M., Davidson, S.M., Hausenloy, D.J., 2010. Inhibiting mitochondrial fission protects the heart against ischemia/reperfusion injury. Circulation 121, 2012–2022.

Paisán-Ruiz, C., Lewis, P.A., Singleton, A.B., 2013. Lrrk2: cause, risk, and mechanism. J. Parkinsons Dis. 3, 85–103.

Pallanck, L., 2013. Mitophagy: mitofusin recruits a mitochondrial killer. Curr. Biol. 23, R570–572.

Panov, A., Gutekunst, C., Leavitt, B., Hayden, M., Burke, J., Strittmatter, W., et al., 2002. Early mitochondrial calcium defects in huntington's disease are a direct effect of polyglutamines. Nat. Neurosci. 5, 731–736.

Parker, W.J., Boyson, S., Parks, J., 1989. Abnormalities of the electron transport chain in idiopathic Parkinson's disease. Ann. Neurol. 26, 719–723.

Parkinson Study Group, 2013. Phase II safety, tolerability, and dose selection study of isradipine as a potential disease-modifying intervention in early Parkinson's disease (STEADY-PD). Mov. Disord. 28, 1823–1831.

Pasinelli, P., Brown, R., 2006. Molecular biology of amyotrophic lateral sclerosis: insights from genetics. Nat. Rev. Neurosci. 7, 710–723.

Peres, T.V., Parmalee, N.L., Martinez-Finley, E.J., Aschner, M., 2016. Untangling the manganese-α-synuclein web. Front. Neurosci. 10, 364.

Perkins, M., Wolf, A., B C, S.D., Meckel, J., Leung, L., et al., 2016. Altered energy metabolism pathways in the posterior cingulate in young adult apolipoprotein e ε4 carriers. J. Alzheimers Dis. 53, 95–106.

Pilsl, A., Winklhofer, K.F., 2012. Parkin, PINK1 and mitochondrial integrity: emerging concepts of mitochondrial dysfunction in Parkinson's disease. Acta Neuropathol. 123, 173–188.

Polymeropoulos, M.H., Lavedan, C., Leroy, E., Ide, S.E., Dehejia, A., Dutra, A., et al., 1997. Mutation in the alpha-synuclein gene identified in families with Parkinson's disease. Science 276, 2045–2047.

Preta, G., Cronin, J.G., Sheldon, I.M., 2015. Dynasore—not just a dynamin inhibitor. Cell Commun. Signal. 13, 24.

Przedborski, S., 2017. The two-century journey of Parkinson disease research. Nat. Rev. Neurosci. 18, 251–259.

Purisai, M.G., McCormack, A.L., Cumine, S., Li, J., Isla, M.Z., Di Monte, D.A., 2007. Microglial activation as a priming event leading to paraquat-induced dopaminergic cell degeneration. Neurobiol. Dis. 25, 392–400.

Qi, X., Qvit, N., Y-C, S., Mochly-Rosen, D., 2013. A novel drp1 inhibitor diminishes aberrant mitochondrial fission and neurotoxicity. J. Cell Sci. 126, 789–802.

Qiu, X., Cao, L., Yang, X., Zhao, X., Liu, X., Han, Y., et al., 2013. Role of mitochondrial fission in neuronal injury in pilocarpine-induced epileptic rats. Neuroscience 245, 157–165.

Quintanilla, R.A., Dolan, P.J., Jin, Y.N., Johnson, G.V., 2012. Truncated tau and Aβ cooperatively impair mitochondria in primary neurons. Neurobiol. Aging 33, 619.e625–619. e635.

Raber, J., Huang, Y., Ashford, J., 2004. ApoE genotype accounts for the vast majority of ad risk and ad pathology. Neurobiol. Aging 25, 641–650.

Rail, D., Scholtz, C., Swash, M., 1981. Post-encephalitic Parkinsonism: current experience. J. Neurol. Neurosurg. Psychiatry 44, 670–676.

Rappold, P.M., Cui, M., Chesser, A.S., Tibbett, J., Grima, J.C., Duan, L., et al., 2011. Paraquat neurotoxicity is mediated by the dopamine transporter and organic cation transporter-3. Proc. Natl. Acad. Sci. U.S.A. 108, 20766–20771.

Rappold, P., Cui, M., Grima, J., Fan, R., de Mesy-Bentley, K., Chen, L., et al., 2014. Drp1 inhibition attenuates neurotoxicity and dopamine release deficits in vivo. Nat. Commun. 5, 5244.

Reddy, P., Reddy, T., Manczak, M., Calkins, M., Shirendeb, U., Mao, P., 2011. Dynamin-related protein 1 and mitochondrial fragmentation in neurodegenerative diseases. Brain Res. Rev. 67, 103–118.

Renton, A.E., Chìo, A., Traynor, B.J., 2014. State of play in amyotrophic lateral sclerosis genetics. Nat. Neurosci. 17, 17–23.

Richardson, J.R., Quan, Y., Sherer, T.B., Greenamyre, J.T., Miller, G.W., 2005. Paraquat neurotoxicity is distinct from that of MPTP and rotenone. Toxicol. Sci. 88, 193–201.

Richardson, J.R., Roy, A., Shalat, S.L., von Stein, R.T., Hossain, M.M., Buckley, B., et al., 2014. Elevated serum pesticide levels and risk for Alzheimer's disease. JAMA Neurol. 71, 284–290.

Richetin, K., Moulis, M., Millet, A., Arrazola, M.S., Andraini, T., Hua, J., et al., 2017. Amplifying mitochondrial function rescues adult neurogenesis in a mouse model of Alzheimer's disease. Neurobiol. Dis. 102, 113–124.

Richter, F., Gabby, L., McDowell, K.A., Mulligan, C.K., De La Rosa, K., Sioshansi, P.C., et al., 2017. Effects of decreased dopamine transporter levels on nigrostriatal neurons and paraquat/maneb toxicity in mice. Neurobiol. Aging 51, 54–66.

Riley, B., Orr, H., 2006. Polyglutamine neurodegenerative diseases and regulation of transcription: assembling the puzzle. Genes Dev. 20, 2183–2192.

Ritz, B.R., Manthripragada, A.D., Costello, S., Lincoln, S.J., Farrer, M.J., Cockburn, M., et al., 2009. Dopamine transporter genetic variants and pesticides in Parkinson's disease. Environ. Health Perspect. 117, 964–969.

Roman, M., 2010. Axona (Accera, Inc): a new medical food therapy for persons with Alzheimer's disease. Issues Ment. Health Nurs. 31, 435–436.

Roth, J.A., 2014. Correlation between the biochemical pathways altered by mutated Parkinson-related genes and chronic exposure to manganese. Neurotoxicology 44, 314–325.

Rothstein, J., 2009. Current hypotheses for the underlying biology of amyotrophic lateral sclerosis. Ann. Neurol. 65, S3–9.

Sasaki, S., Iwata, M., 2007. Mitochondrial alterations in the spinal cord of patients with sporadic amyotrophic lateral sclerosis. J. Neuropathol. Exp. Neurol. 66, 10–16.

Satake, W., Nakabayashi, Y., Mizuta, I., Hirota, Y., Ito, C., Kubo, M., et al., 2009. Genome-wide association study identifies common variants at four loci as genetic risk factors for Parkinson's disease. Nat. Genet. 41, 1303–1307.

Schapira, A.H., Cooper, J.M., Dexter, D., Jenner, P., Clark, J.B., Marsden, C.D., 1989. Mitochondrial complex I deficiency in Parkinson's disease. Lancet 1, 1269.

Schapira, A., Cooper, J., Dexter, D., Clack, J., Jenner, P., Marsden, C., 1990. Mitochondrial complex I deficiency in Parkinson's disease. J. Neurochem. 54, 823–827.

Schellenberg, G.D., Montine, T.J., 2012. The genetics and neuropathology of Alzheimer's disease. Acta Neuropathol. 124, 305–323.

Schuler, F., Casida, J.E., 2001. Functional coupling of PSST and ND1 subunits in NADH: ubiquinone oxidoreductase established by photoaffinity labeling. Biochim. Biophys. Acta 1506, 79–87.

Seong, I.S., Ivanova, E., Lee, J.M., Choo, Y.S., Fossale, E., Anderson, M., et al., 2005. HD CAG repeat implicates a dominant property of huntingtin in mitochondrial energy metabolism. Hum. Mol. Genet. 14, 2871–2880.

Sharma, M., Ioannidis, J.P., Aasly, J.O., Annesi, G., Brice, A., Bertram, L., et al., 2012. A multi-centre clinico-genetic analysis of the vps35 gene in Parkinson disease indicates reduced penetrance for disease-associated variants. J. Med. Genet. 49, 721–726.

Sharma, A., Bemis, M., Desilets, A., 2014. Role of medium chain triglycerides (Axona®) in the treatment of mild to moderate Alzheimer's disease. Am. J. Alzheimers Dis. Other Demen. 29, 409–414.

Sharp, W.W., Fang, Y.H., Han, M., Zhang, H.J., Hong, Z., Banathy, A., et al., 2014. Dynamin-related protein 1 (Drp1)-mediated diastolic dysfunction in myocardial ischemia-reperfusion injury: therapeutic benefits of Drp1 inhibition to reduce mitochondrial fission. FASEB J. 28, 316–326.

Shirendeb, U., Reddy, A.P., Manczak, M., Calkins, M.J., Mao, P., Tagle, D.A., et al., 2011. Abnormal mitochondrial dynamics, mitochondrial loss and mutant huntingtin oligomers in Huntington's disease: implications for selective neuronal damage. Hum. Mol. Genet. 20, 1438–1455.

Shults, C., Beal, M., Fontaine, D., Nakano, K., Haas, R., 1998. Absorption, tolerability, and effects on mitochondrial activity of oral coenzyme q10 in Parkinsonian patients. Neurology 50, 793–795.

Singleton, A.B., Farrer, M., Johnson, J., Singleton, A., Hague, S., Kachergus, J., et al., 2003. Alpha-synuclein locus triplication causes Parkinson's disease. Science 302, 841.

Smith, R., Murphy, M., 2010. Animal and human studies with the mitochondria-targeted antioxidant MitoQ. Ann. N.Y. Acad. Sci. 1201, 96–103.

Snow, B., Rolfe, F., Lockhart, M., Frampton, C., O'Sullivan, J., Fung, V., et al., 2010. A double-blind, placebo-controlled study to assess the mitochondria-targeted antioxidant MitoQ as a disease-modifying therapy in Parkinson's disease. Mov. Disord. 25, 1670–1674.

Snyder, S.H., D'Amato, R.J., 1985. Predicting Parkinson's disease. Nature 317, 198–199.

Song, W., Chen, J., Petrilli, A., Liot, G., Klinglmayr, E., Zhou, Y., et al., 2011. Mutant huntingtin binds the mitochondrial fission GTPase drp1 and increases its enzymatic activity. Nat. Med. 17, 377–382.

Song, W., Song, Y., Kincaid, B., Bossy, B., Bossy-Wetzel, E., 2013. Mutant sod1g93a triggers mitochondrial fragmentation in spinal cord motor neurons: neuroprotection by sirt3 and pgc-1α. Neurobiol. Dis. 51, 72–81.

St Pierre, J., Drori, S., Uldry, M., Silvaggi, J., Rhee, J., Jager, S., et al., 2006. Suppression of reactive oxygen species and neurodegeneration by the PGC1 transcriptional coactivators. Cell 127, 397–408.

Stahl, W.L., Swanson, P.D., 1974. Biochemical abnormalities in Huntington's chorea brains. Neurology 24, 813–819.

Strijks, E., Kremer, H., Horstink, M., 1997. Q10 therapy in patients with idiopathic Parkinson's disease. Mol. Aspects Med. 18, S237–240.

Sugiura, A., McLelland, G.L., Fon, E.A., McBride, H.M., 2014. A new pathway for mitochondrial quality control: mitochondrial-derived vesicles. EMBO J. 33, 2142–2156.

Surmeier, D., Schumacker, P., 2013. Calcium, bioenergetics, and neuronal vulnerability in Parkinson's disease. J. Biol. Chem. 288, 10736–10741.

Surmeier, D.J., Guzman, J.N., Sanchez-Padilla, J., Goldberg, J.A., 2010. What causes the death of dopaminergic neurons in Parkinson's disease? Prog. Brain Res. 183, 59–77.

Tanaka, A., Cleland, M.M., Xu, S., Narendra, D.P., Suen, D.F., Karbowski, M., et al., 2010. Proteasome and p97 mediate mitophagy and degradation of mitofusins induced by Parkin. J. Cell Biol. 191, 1367–1380.

Tang, F.L., Liu, W., JX, H., Erion, J.R., Ye, J., Mei, L., et al., 2015. Vps35 deficiency or mutation causes dopaminergic neuronal loss by impairing mitochondrial fusion and function. Cell Rep. 12, 1631–1643.

Tanner, C., Kamel, F., Ross, G., Hoppin, J., Goldman, S., Korell, M., et al., 2011. Rotenone, paraquat, and Parkinson's disease. Environ. Health Perspect. 119, 866–872.

Thiruchelvam, M., McCormack, A., Richfield, E.K., Baggs, R.B., Tank, A.W., Di Monte, D.A., et al., 2003. Age-related irreversible progressive nigrostriatal dopaminergic neurotoxicity in the paraquat and maneb model of the Parkinson's disease phenotype. Eur. J. Neurosci. 18, 589–600.

Tieu, K., Perier, C., Caspersen, C., Teisman, P., C-C, W., Yan, S.-D., et al., 2003. D-b-hydroxybutyrate rescues mitochondrial respiration and mitigates features of Parkinson disease. J. Clin. Investig. 112, 892–901.

van Dellen, A., Blakemore, C., Deacon, R., York, D., Hannan, A.J., 2000. Delaying the onset of Huntington's in mice. Nature 404, 721–722.

Vande Velde, C., McDonald, K., Boukhedimi, Y., McAlonis-Downes, M., Lobsiger, C., Bel Hadj, S., et al., 2011. Misfolded sod1 associated with motor neuron mitochondria alters mitochondrial shape and distribution prior to clinical onset. PLoS One 6, e22031.

Vanitallie, T.B., Nonas, C., Di Rocco, A., Boyar, K., Hyams, K., Heymsfield, S.B., 2005. Treatment of Parkinson disease with diet-induced hyperketonemia: a feasibility study. Neurology 64, 728–730.

Vilariño-Güell, C., Wider, C., Ross, O.A., Dachsel, J.C., Kachergus, J.M., Lincoln, S.J., et al., 2011. Vps35 mutations in Parkinson disease. Am. J. Hum. Genet. 89, 162–167.

Vinceti, M., Filippini, T., Mandrioli, J., Violi, F., Bargellini, A., Weuve, J., et al., 2016. Lead, cadmium and mercury in cerebrospinal fluid and risk of amyotrophic lateral sclerosis: a case–control study. J. Trace Elem. Med. Biol. 43, 121–125.

Vinceti, M., Violi, F., Tzatzarakis, M., Mandrioli, J., Malagoli, C., Hatch, E.E., et al., 2017. Pesticides, polychlorinated biphenyls and polycyclic aromatic hydrocarbons in cerebrospinal fluid of amyotrophic lateral sclerosis patients: a case–control study. Environ. Res. 155, 261–267.

Vonsattel, J., DiFiglia, M., 1998. Huntington disease. J. Neuropathol. Exp. Neurol. 57, 369–384.

Wang, X., Su, B., Siedlak, S., Moreira, P., Fujioka, H., Wang, Y., et al., 2008. Amyloid-beta overproduction causes abnormal mitochondrial dynamics via differential modulation of mitochondrial fission/fusion proteins. Proc. Natl. Acad. Sci. U.S.A. 105, 19318–19323.

Wang, X., Su, B., Lee, H., Li, X., Perry, G., Smith, M., et al., 2009. Impaired balance of mitochondrial fission and fusion in Alzheimer's disease. J. Neurosci. 29, 9090–9103.

Wang, A., Costello, S., Cockburn, M., Zhang, X., Bronstein, J., Ritz, B., 2011a. Parkinson's disease risk from ambient exposure to pesticides. Eur. J. Epidemiol. 26, 547–555.

Wang, X., Su, B., Liu, W., He, X., Gao, Y., Castellani, R.J., et al., 2011b. Dlp1-dependent mitochondrial fragmentation mediates 1-methyl-4-phenyl-1,2,3,6-tetrahydropyridinium toxicity in neurons: implications for Parkinson's disease. Aging Cell 10, 807–823.

Wang, X., Yan, M.H., Fujioka, H., Liu, J., Wilson-Delfosse, A., Chen, S.G., et al., 2012. LRRK2 regulates mitochondrial dynamics and function through direct interaction with DLP1. Hum. Mol. Genet. 21, 1931–1944.

Wang, W., Wang, X., Fujioka, H., Hoppel, C., Whone, A.L., Caldwell, M.A., et al., 2016. Parkinson's disease-associated mutant vps35 causes mitochondrial dysfunction by recycling DLP1 complexes. Nat. Med. 22, 54–63.

Weinreb, P.H., Zhen, W., Poon, A.W., Conway, K.A., Lansbury, P.T., 1996. NACP, a protein implicated in Alzheimer's disease and learning, is natively unfolded. Biochemistry 35, 13709–13715.

Wexlet, N.S., Lorimer, J., Porter, J., Gomez, F., Moskowitz, C., Shackell, E., et al., 2004. Venezuelan kindreds reveal that genetic and environmental factors modulate Huntington's disease age of onset. Proc. Natl. Acad. Sci. U.S.A. 101, 3498–3503.

Whitehouse, P.J., Price, D.L., Struble, R.G., Clark, A.W., Coyle, J.T., Delon, M.R., 1982. Alzheimer's disease and senile dementia: loss of neurons in the basal forebrain. Science 215, 1237–1239.

Whitworth, A.J., Pallanck, L.J., 2017. PINK1/Parkin mitophagy and neurodegeneration— what do we really know in vivo? Curr. Opin. Genet. Dev. 44, 47–53.

Williams, E.T., Chen, X., Moore, D.J., 2017. VPS35, the retromer complex and Parkinson's disease. J. Parkinsons Dis. 7, 219–233.

Wu, X.F., Block, M.L., Zhang, W., Qin, L., Wilson, B., Zhang, W.Q., et al., 2005. The role of microglia in paraquat-induced dopaminergic neurotoxicity. Antioxid. Redox Signal. 7, 654–661.

Xie, N., Wang, C., Lian, Y., Zhang, H., Wu, C., Zhang, Q., 2013. A selective inhibitor of Drp1, mdivi-1, protects against cell death of hippocampal neurons in pilocarpine-induced seizures in rats. Neurosci. Lett. 545, 64–68.

Yan, S.D., Stern, D.M., 2005. Mitochondrial dysfunction and Alzheimer's disease: role of amyloid-beta peptide alcohol dehydrogenase (ABAD). Int. J. Exp. Pathol. 86, 161–171.

Youle, R., Narendra, D., 2011. Mechanisms of mitophagy. Nat. Rev. Mol. Cell Biol. 12, 9–14.

Youle, R.J., van der Bliek, A.M., 2012. Mitochondrial fission, fusion and stress. Science 337, 1062–1065.

Yu, Y., FC, S., Callaghan, B.C., Goutman, S.A., Batterman, S.A., Feldman, E.L., 2014. Environmental risk factors and amyotrophic lateral sclerosis (ALS): a case–control study of ALS in Michigan. PLoS One 9, e101186.

Yue, M., Hinkle, K.M., Davies, P., Trushina, E., Fiesel, F.C., Christenson, T.A., et al., 2015. Progressive dopaminergic alterations and mitochondrial abnormalities in LRRK2 G2019S knock-in mice. Neurobiol. Dis. 78, 172–195.

Zarranz, J.J., Alegre, J., Gómez-Esteban, J.C., Lezcano, E., Ros, R., Ampuero, I., et al., 2004. The new mutation, E46k, of alpha-synuclein causes Parkinson and Lewy body dementia. Ann. Neurol. 55, 164–173.

Zhao, Y.-X., Cui, M., Chen, S.-F., Dong, Q., Liu, X.-Y., 2014. Amelioration of ischemic mitochondrial injury and Bax-dependent outer membrane permeabilization by Mdivi-1. CNS Neurosci. Ther, 20, 1–11.

Zhou, C., Huang, Y., Shao, Y., May, J., Prou, D., Perier, C., et al., 2008. The kinase domain of mitochondrial PINK1 faces the cytoplasm. Proc. Natl. Acad. Sci. U.S.A. 105, 12022–12027.

Zhou, Z.D., Saw, W.T., Tan, E.K., 2016. Mitochondrial CHCHD-containing proteins: Physiologic functions and link with neurodegenerative diseases. Mol. Neurobiol. 54, 5534–5546.

Zimprich, A., Benet-Pagès, A., Struhal, W., Graf, E., Eck, S.H., Offman, M.N., et al., 2011. A mutation in VPS35, encoding a subunit of the retromer complex, causes late-onset Parkinson disease. Am. J. Hum. Genet. 89, 168–175.

FURTHER READING

Cui, M., Aras, R., Christian, W.V., Rappold, P.M., Hatwar, M., Jo, P., et al., 2009. The organic cation transporter 3 is a pivotal modulator of neurodegeneration in the nigrostriatal dopaminergic pathway. Proc. Natl. Acad. Sci. U.S.A. 106, 8043–8048.

CHAPTER NINE

Food Plant Chemicals Linked With Neurological and Neurodegenerative Disease

Peter S. Spencer[1], Valerie S. Palmer
Oregon Health & Science University, Portland, OR, United States
[1]Corresponding author: e-mail address: spencer@ohsu.edu

Contents

1. INTRODUCTION

Recent emphasis on the human neurotoxic potential of industrial chemicals and environmental pollutants runs the risk of deemphasizing the many naturally occurring substances that can also compromise the human nervous system. It bears repeating that chemicals with toxic potential elaborated by plants are often designed to stop predatory animals, and humans using their products for food are predators in this regard.

This chapter addresses advances in understanding the mechanisms and effects of a diverse collection of plant-derived substances, including cyanobacterial beta-methylamino–L-alanine (L-BMAA), that have direct or indirect actions on the human nervous system. There is no attempt to provide a comprehensive overview; instead, a few examples are chosen

Advances in Neurotoxicology, Volume 1
ISSN 2468-7480
http://dx.doi.org/10.1016/bs.ant.2017.07.009

for purposes of illustration and to emphasize common principles associated with neurological disease induced by plant-derived biological toxins.

2. FOOD PLANTS WITH NEUROTOXIC POTENTIAL

Although humans regularly employ potentially toxic plants for food, adverse effects on the nervous system usually do not result because intake is below the threshold that must be crossed for toxicity to surface. Thus, the cyanogenic plant cassava (*Manihot esculenta*) is a staple for an estimated 500 million people in the tropics and subtropics, but only those who subsist on incompletely detoxified plant material develop neurological disease (cassavism). The same is true for grasspea (*Lathyrus sativus*), long banned from sale in India because of its potential chronic motorsystem effects (lathyrism) but recently restored as a legal food product. Ingestion of fruit of the Soapberry family (Sapindaceae), such as lychee (*Litchi chinensis*) and ackee (*Blighia sapida*), can trigger a fatal hypoglycemic encephalopathy but is also enjoyed without adverse effect by millions worldwide. Poverty, nutritional state, hunger, food availability, and extent of detoxication and amount of plant consumed are the critical threads that link food plants and neurotoxic illness. Drought or water logging that results in survival only of those species that withstand environmental extremes causes increased food dependence on resilient plants. For example, seed of grasspea, roots of cassava, and seed and sago of cycad (*Cycas* spp.) are eaten in normal times, but they also serve as "famine foods" when intake of plant chemicals may exceed their toxic thresholds. Toxicity may also result from climate-related increases in chemical concentration within the plant, ingestion of unripe and acutely toxic fruit, or failure to carry out postharvest detoxication of otherwise edible plant components.

The threshold separating harmless and harmful effects of potentially toxic plants varies critically as a function of individual susceptibility: this is sometimes age- or gender-related but most commonly associated with hunger and resultant undernutrition. Nutritional state, as well as dose and duration of exposure to plant products with toxic potential, is inseparate from any consideration of dietary phytotoxins and induction of neurological disease. Dietary exposure to plant products has been linked with acute (ackee, lychee), chronic (grasspea, cassava), and long-latency neurological disease, the latter particularly associated with food and medicinal use of plants in the families Annonaceae and Cycadaceae. Photographs of these plants and the chemical formulas of the toxins they harbor are readily found on the World Wide Web.

2.1 Sapindaceae Juss.

2.1.1 Ackee: Hypoglycemic Encephalopathy

Among the ~140 genera of the Soapberry family of flowering plants (Sapindaceae) are commercially important species, such as the lychee (*L. chinensis*), longan (*Dimocarpus longan*), rambutan (*Nephelium lappaceum*), and ackee (*B. sapida*). While the mature pulp (aril) of many of these species is sweet and aromatic, it harbors unusual L-amino acids with potential in susceptible subjects (notably undernourished, hungry children) rapidly to induce hypoglycemic encephalopathy. Known from Asia (lychee), Africa and the Caribbean (ackee), this rapidly evolving acute illness is especially seen among those who eat unripe fruit or seed that contain higher concentrations of the potentially toxic hypoglycemic amino acids. These include hypoglycin A (2-methylene*cyclo*propanealanine) and methylene*cyclo*propylglycine (MCPG). While neurotoxic effects of ackee were first noted in 1875 and documented in the early 1900s (Hill, 1952; Scott, 1916), more than 100 years passed before a similar illness in southern Asia was correctly attributed to lychee fruit ingestion (John and Das, 2014; Shrivastava et al., 2017; Spencer et al., 2015a). The following historical account of ackee poisoning serves to introduce the recent reports of lychee toxicity in southern Asia.

Ackee (from the *twi* language of Ghana) is an evergreen tree native to West Africa (known as *ankye* or *ishin*), including Benin, Cameroon, Côte d'Ivoire, Ghana, Guinea, Liberia, Nigeria, Senegal, and Togo (Ekué, 2011; Morton, 1987). It is one of 270 plant species with hypoglycemic potential, many of which in low-income countries are used for the treatment of diabetes mellitus (Oloyede et al., 2014; Reisman, 2015). Ackee was introduced into Jamaica in 1778 and was subsequently adopted as the country's national fruit. The plant may also be found elsewhere in the Caribbean Basin (Bahamas, Haiti, Trinidad.), Central America (Costa Rica, Guatemala, and Panama), and South America (Brazil, Colombia, Ecuador, French Guyana, Suriname, and Venezuela), as well as in the United States (Florida). The roots, bark, leaves, capsules, and seed of ackee have been used in traditional medicine in West Africa, with indications for use in 22 diseases in Benin and for the treatment of diabetes mellitus in Nigeria. Ripe arils (fresh, dried, fried, roasted) and sometimes young leaves are eaten in West Africa, and parboiled arils with onions, tomato, and saltfish with bread or rice form an important national dish in Jamaica. The nutritional components of ackee aril, flour, and oil are detailed elsewhere (Dossou, 2014). Canned, frozen, and other processed ackee products are distributed from Belize, Haiti, and Jamaica to Canada, United Kingdom, and the United States, where the FDA limit for content of hypoglycin A is 100 ppm (US FDA, 2014).

Suspected and confirmed cases of ackee-induced toxic hypoglycemic encephalopathy in Jamaica (U.S. CDC, 1992), known locally as Jamaican Vomiting Sickness, are recorded by the Ministry of Health National Surveillance Unit (a Class 1 Notifiable Event from 2011). Risk factors include chronic malnutrition, ingestion of unripe or forcibly opened ackee fruit, and reuse of water used to cook ackee aril (Figeroa and Nembhard, 1992; Henry et al., 1998). Most cases are reported between November and April, increasing in weeks 48–51 with peak occurrence in weeks 2–5 of the following year. Cold weather resulting in delayed maturation (opening) of ackee fruit has been implicated in annual illness spikes. Unripe ackee arils contain as much as 0.1% dry weight of hypoglycin, 10 times the concentration when ripe (Hassal and Reyle, 1955; Hassal et al., 1954). Children over age 5 were once most commonly affected by ackee poisoning but, in recent years, the age range of affected subjects has expanded. Whereas, in January–March 1991, three-quarters of 28 patients were under 15 years of age, in 2011, most of the 320 cases of ackee poisoning and deaths (68%, 72%, respectively) occurred among 15–54-year-old subjects, with a small number of cases affecting 17–70-year-old subjects in 2013. Most affected in the recent past have been impoverished rural communities of northeastern Jamaica, including St. Ann (54 cases in 2011), Trelawny (14 in 2012), and Portland (13 in 2012 and 10 in 2013) (Barrett, 2016; Blake et al., 2004). While adults are considered less sensitive to ackee intoxication, the toxic–metabolic effects are essentially the same (Golden et al., 1984).

In addition to poverty, ackee intoxication has occurred in the setting of postdrought food shortages in Côte d'Ivoire in 1984 (Moyal, 2014) and in Burkina Faso in 1998 (Meda et al., 1999), in Haiti in 2001 (Joskow et al., 2006), the border of Suriname and French Guyana between 1998 and 2001 (Gaillard et al., 2011), and, in 2015, in Kwara state, Nigeria (Katibi et al., 2015). Cases are also known from other parts of Africa, including Togo and Benin (Goldson, 2005; Quere et al., 1999). In Nigeria, 8 siblings aged 4–10 years were poisoned after eating 2–6 roasted ackee seed with or without the aril, while, in Burkina Faso between January and May 1988, 29 children aged 2–6 years (55% girls, 31-847/100,000 population) died within 2–48 h of the first signs of poisoning. A case–control study showed the only significant factor was the presence of ackee trees within 100 m of the household. Most villagers were unaware of the dangers of eating unripe ackee fruit (Meda et al., 1999). Clinical manifestations of the resulting hypoglycemic encephalopathy included vomiting, hypotonia, convulsions, and coma coupled with very high levels (4–200 × normal) of certain urinary dicarboxylic acids (ethylmalonic, glutaric, and adipic acids), which serve as biomarkers of ackee poisoning (Tanaka, 1972; Tanaka et al., 1976). Since these three dicarboxylic acids accumulate individually in rare human genetic disorders (*ETHE1* mutation, *GCDH* mutation, *ABCD1* mutation, respectively), all of which themselves are associated with neurological deficits (Mohamed et al., 2015; Pigeon et al., 2009; Rocchiccioli et al., 1986), their toxic effects plausibly may be related to the persistent neurological deficits seen in some children who survive acute hypoglycemic encephalopathy syndrome.

The acute toxicity of hypoglycin A from ingestion of unripe ackee arils arises from a liver metabolite, methylene*cyclo*propylacetic acid. Hypoglycin A is transaminated to methylene*cyclo*propyl-alanine (MCPA) and subsequently undergoes oxidative decarboxylation to form MCPA-CoA, which accumulates in the matrix of liver mitochondria of fasted rats (Melde et al., 1989, 1991) and inhibits the β-oxidation of long-chain fatty acids at the butyryl-CoA dehydrogenase stage (Tanaka, 1972; Tanaka and Ikeda, 1990; von Holt et al., 1964, 1966). Oxidation of the carbon chains of valine and isoleucine is also impaired. Inhibition of fatty acid oxidation increases dependence on glucose as a source of energy, but gluconeogenesis is severely inhibited because the replacement of glucose is dependent on products of fatty acid oxidation (acetyl-CoA, NADH, and ATP) (Osmundsen et al., 1978). Hypoglycin A thus induces rapid carbohydrate utilization while simultaneously inhibiting carbohydrate synthesis, the outcome of which is a fall in serum glucose (Nunn et al., 2010). Dramatic reduction of blood glucose levels (e.g., from 5 to 0.5 mM) disturbs brain function and portends the graphic signs of hypoglycemic encephalopathy.

The human response to ackee poisoning is similar to Reye syndrome (acute noninflammatory encephalopathy and fatty degenerative liver failure). In Jamaican Vomiting Sickness, generalized epigastric discomfort begins 2–6 h after ackee ingestion and is followed by sudden onset of vomiting. This is followed by diaphoresis, tachypnea, tachycardia, headache, generalized weakness, paresthesias, disturbed mental state, and prostration. The mainstay of treatment in ackee and lychee fruit poisoning is to maintain a normal blood glucose level with 10% intravenous dextrose (Holson, 2015; Shah and John, 2014). Dehydration, acidosis, and electrolyte disturbances may also require attention (Sherratt and Turnbull, 1999). Unless treatment is given, a second round of vomiting is followed by tonic–clonic convulsions, coma, and death.

Amino acid analysis of the hypoglycin A content of unripe ackee fruit components revealed 939, 711, and 41.6 mg/100 g in seed, aril, and husk components, respectively. As the fruit matures, the aril concentration decreases apparently by transfer to the seed, where the compound is stored as γ-glutamylhypoglycin (hypoglycin B). In the ripe fruit, the seed concentration decreased to 269 mg/100 g, remained unchanged in the husk, and decreased below the detection limit of 1.2 mg/100 g in the edible aril (Chase et al., 1990). The maximum tolerated dose of hypoglycin A in Sprague Dawley rats, determined over a 30-day feeding period, was 1.50 + 0.07 mg hypoglycin/kg/body weight/day. The acute toxic dose in feed was 215–230 mg hypoglycin A/kg body weight, with males slightly more sensitive than female rats. A lower oral dose of hypoglycin A (*ca.* 100 mg/kg body weight) is sufficient to induce the acute toxic response (Blake et al., 2006). There was no change in insulin levels in Wistar rats given hypoglycin A (Mills et al., 1987).

2.1.2 Lychee: Hypoglycemic Encephalopathy

Cousins of ackee include a number of delicious tropical fruits of southern Asia, including the lychee (*L. chinensis*), longan (*D. longan*), and mamoncillo (*Melicoccus bijugatus*): these harbor the same hypoglycemic amino acids that interrupt gluconeogenesis and β-oxidation of amino acids (Gray and Fowden, 1962; Isenberg et al., 2016). Other widely eaten members of the Sapindaceae, including *N. lappaceum* (rambutan) and *N. mutabile* (pulasan), would also be expected to harbor these potentially toxic compounds. As with ackee, the levels of hypoglycin A and MCPG diminish as lychee fruit ripens (Bowen-Forbes and Minott, 2011; Brown et al., 1991; Das et al., 2015). Seed of lychee contains the highest levels of hypoglycin A and MCPG (Das et al., 2015; Gray and Fowden, 1962; Melde et al., 1991).

To date, only lychee among these tropical fruits has been documented to be associated with acute hypoglycemic encephalopathy, and recognition of the true phytotoxic cause of this illness is very recent, essentially a century after description of ackee toxicity. Just as the clinical picture of ackee poisoning may be mistaken for nonfebrile cerebral malaria in West Africa (Quere et al., 1999), lychee intoxication has been misdiagnosed as heat stress in northern India, pesticide intoxication in northwest Bangladesh, and Japanese B encephalitis or an unknown neuroviral disorder in northeast Vietnam (Spencer et al., 2015b). The explosive growth of the lychee industry in these regions, driven by the commercial value and export potential of lychee fruit, is a plausible explanation for why unexplained local disease outbreaks affecting multiple children have been only recently recognized in these countries (Spencer and Palmer, 2017).

Affected children in all three countries reside in areas of commercial lychee fruit production, and outbreaks of "acute seasonal encephalitis" (AES) are confined to lychee harvest periods (May–July). The children typically are poorly nourished, hungry from lack of daytime food, and consume lychee fruit, including unripe fruit that has fallen to the ground. After sleeping overnight, the child awakens, develops convulsions, coma, and may die within 36 hours, usually in association with markedly depressed serum glucose values. In Vietnam's Bắc Giang Province, where 25–50 affected children have died annually between 2004 and 2009, a team of Vietnamese and French investigators showed a spatial–temporal association between outbreaks of AES (known locally as *Ac Mộng*, meaning bad nightmares) and the lychee harvest season, the annual AES risk increasing with the area under lychee cultivation (Paireau et al., 2012). AME outbreaks occurred earlier in communes ($n = 230$) harvesting lychee fruit in May–June versus those harvesting in June–July, and there was a negative association with animals, rainfall, mean temperature, and relative humidity. AME cases occurred in boys > girls (1:2:1:0) of median age 5 (range 2–7.5 years) with disease onset May 1 to July 31. Brain imaging ($n = 50$) showed edema (18%), focal signs (14%), or no abnormality (58%). PCR-based viral screens were negative for Japanese B encephalitis, dengue 1–4, West Nile, Chikungunya, *Herpes simplex* 1 and 2, *Herpes zoster*, Rift Valley, Crimean Congo hemorrhagic fever; most Enteroviruses and most Alphaviruses. While the *Ac Mộng*-causing agent could not be identified, Paireau and colleagues postulated that mosquitoes fed on fruit-eating bats and transmitted an unknown viral pathogen to children. The present authors challenged this interpretation and proposed that seasonal AES was correctly interpreted as immature lychee fruit intoxication because of its clinical similarities to the toxic effects of unripe ackee fruit (Spencer et al., 2015b).

A seasonal AES-like illness also affects the northerly lychee-growing areas of northwest Bangladesh and India, where the illness has been variously attributed to an unknown virus, heat stroke, pesticide contamination, or chemical-induced coloration of fruit (Dinesh et al., 2013; Islam, 2012; Islam et al., 2017). In Bangladesh, in May–June 2012, there was an outbreak of unexplained fatal illness of young children (median age: 4.2 years) living near the lychee orchards of Dinajpur and Thakurgaon Districts. Males were more often affected than females in the ratio of 9:5. Clinical manifestations included fatigue, altered mental status, convulsions, coma, and death. The median time from onset of illness to death was 20 h, with a range of 6–130 h (Biswas, 2012; Islam, 2012). A similar illness occurs seasonally among young children in lychee-producing areas of Bihar and West Bengal, India, the latter geographically close to northwest Bangladesh (Bandyopadhyay et al., 2014; John and Das, 2014; Shrivastava et al., 2015), where the extremely rapid evolution of illness is also atypical for viral disorders (Spencer and Palmer, 2017). A recent hospital-based comprehensive case–control study also failed to identify a neurotropic virus and indicted lychee amino acids as probably culpable (Shrivastava et al., 2017). Contemporary methods for quantification of hypoglycin A in blood and of the glycine adducts of hypoglycin A and MCPG in urine are available (Carlier et al., 2015; Isenberg et al., 2015, 2016; Sander et al., 2017).

MCPG is metabolized to the toxic CoA ester methylene*cyclo*propyl formyl-CoA (Isenberg et al., 2015; Li et al., 1999). The metabolic effects of lychee-derived MCPG were long ago explored in young adult male Sprague Dawley rats fasted for 24 h before intraperitoneal administration of a single dose (100 mg MCPG/kg body weight) (Melde et al., 1991). Glucose was decreased by 75% at 4 h posttreatment, accompanied by small increases in lactate and pyruvate, and a decrease in the lactate/pyruvate ratio (20:9). Acetoacetate and 3-hydroxybuyrate concentrations were lowered to 0.01 mM. Plasma concentrations of most amino acids were increased, the largest in branched-chain amino acids (500%–550%) and basic amino acids (200%–260%). There was little effect on the citric acid cycle, the respiratory chain, or oxidative phosphorylation. Major targets of the toxic metabolites of MCPG in the β-oxidation pathway of liver, namely 3-oxoacyl-CoA and acetoacetyl-CoA thiolases, were depressed to 10% and 20% of controls, representing a strong inhibition of mitochondrial β-oxidation. This decreases gluconeogenesis by reducing the supply of NADH, required for glucose synthesis from many of its precursors, and of acetyl-CoA, which allosterically activates pyruvate carboxylase necessary for glucose synthesis from pyruvate. Administration of MCPG to fasted rats also caused a marked organic aciduria, with accumulation of multiple long-chain monocarboxylic and dicarboxylic fatty acids (Melde et al., 1991).

2.2 Annonaceae

Several members of the Annonaceae family of largely tropical plants contain neurotoxic long-chain fatty acids known as acetogenins, including annonacin from the soursop (*Annona muricata*), a potent lipophilic, mitochondrial complex I inhibitor acting via inhibition of NADH ubiqui-none oxidoreductase.

Soursop is widely cultivated and popular in parts of Latin America, the Caribbean, Africa, Southeast Asia, and Pacific Islands, and its derivative products are consumed across the world. Closely related species are also used for food and beverages, including: *Annona reticulata* (custard apple), a hybrid of the sugar apple (*A. squamosa*) cultivated and consumed in and exported from Australia, and *A. cherimola* (cherimoya), widely cultivated in South America, Spain, and Portugal. The temperate annonaceous plant *Asimina triloba* (pawpaw) is native to the Eastern United States, and its products are available online. All contain acetogenins, including annonacin (Le Ven et al., 2014; Potts et al., 2012; Yuan et al., 2003).

Soursop is used in food, beverages, and other preparations. Unripe fruit, seeds, leaves, and roots are used as pesticides, insecticides, and insect repellents (Moghadamtousi et al, 2015). The ripe fruit is eaten raw or processed into ice cream, smoothies, jelly, nectar and fruit juice, or eaten as a vegetable. Tea is prepared from dried twigs and leaves, which also contain annonacin. Biomedical interest in Annonaceae has also focused on possible therapeutic uses in cancer. Over 200 bioactive compounds have been isolated from *A. muricata*, mostly six types of more than 210 acetogenins, many alkaloids (mostly isoquinolines, aporphines, and protoberberines, notably reticuline and coreximine), phenols (e.g., quercetin and gallic acids), and other compounds, including sesquiterpene derivatives. There are at least 50 reports of pharmacological studies, including two-thirds in vitro and one-third in vivo studies with mouse models. Recent comprehensive reviews of the foregoing are available (Coria-Tellez et al., 2017; Kedari and Khan, 2014) such that present coverage of the extensive literature is brief and confined to soursop neurotoxicity and its possible relationship to neurodegenerative disease.

Consumption of soursop fruit and leaf-derived tea has been widespread on the island of Guadeloupe, where French biomedical scientists linked the practice with a high incidence of atypical parkinsonism (Caparros-Lefebvre and Elbaz, 1999). One study of 160 parkinsonian patients in Guadeloupe used neuropsychological tests and magnetic resonance imaging of the brain to classify approximately one-third with idiopathic Parkinson's disease (PD), another one-third with progressive supranuclear palsy (PSP), and the balance with parkinsonism–dementia complex (PDC). The PSP patients were resistant to levodopa, had early postural instability and supranuclear oculomotor dysfunction, but differed from the classical disease by the frequency of tremor (>50%), dysautonomia (50%), and the occurrence of hallucinations

(59%). The PDC patients had hallucinations, no oculomotor dysfunction, and levodopa-resistant parkinsonism associated with frontosubcortical dementia. Consumption of soursop was significantly greater in both PSP-like and PDC patients than in controls and PD patients (Lannuzel et al., 2007).

Controlled studies with dopaminergic neurons in vitro have shown that certain acetogenins (annonacin, annonacinone, isoannonacinone, and solamin) and alkaloids (reticuline and coreximine) are toxic to dopaminergic cells by impairing energy production (Lannuzel et al., 2002, 2003, 2006, 2008). Annonacin is about 1000 times more toxic for neuronal cell cultures than reticuline and showed 700 times greater neurotoxic potency than rotenone (Champy et al., 2005), which has been linked to parkinsonism (Betarbet et al., 2000). Annonacin induced ATP depletion, retrograde transport of mitochondria to the cell soma, changes in the intracellular distribution of microtubule-associated tau protein (MAPT) (Escobar-Khondiker et al., 2007; Höllerhage et al., 2009), and reduced degradation, increased phosphorylation and redistribution of neuronal tau in a mouse genetic model of tauopathy (Yamada et al., 2014). Chronic consumption of *A. muricata* juice triggered and aggravated cerebral tau phosphorylation in wild-type and MAPT-transgenic mice (Rottscholl et al., 2016). Annonacin entered the brain, decreased ATP levels, and induced neurodegeneration in the basal ganglia but without changes in behavior or locomotion (Lannuzel et al., 2006). In rats treated with drugs intravenously, annonacin was 100 times more potent than 1-methyl-4-phenylpyridinium, the active metabolite of 1-methyl-4-phenyl-1,2,3,6-tetrahydropyridine, a well-established cause of parkinsonism in humans and animal models (Langston et al., 1984).

While experimental data available in 2010 were considered by the French food safety agency (Agence française de sécurité sanitaire des aliments) to be insufficient to conclude that human consumption of soursop caused atypical parkinsonism in Guadeloupe, the agency called for further study on potential risks to human health (AVIS, 2010). Given the widespread consumption of *Annona* species, it seems appropriate to call for such studies to be extended worldwide, especially in the tropics. An average soursop fruit has been estimated to contain ∼15 mg of annonacin, a can of commercial nectar 36 mg, and a cup of infusion or decoction 140 µg, such that an adult who consumes one fruit or can of nectar a day is estimated to ingest over 1 year the amount of annonacin that induced brain lesions in rats receiving purified annonacin by intravenous infusion (Champy et al., 2005).

2.3 Grasspea (*Lathyrus sativus*)

The seed of this legume harbors an excitotoxic amino acid (β-*N*-oxalylamino-L-alanine) held responsible for induction of spastic paraparesis (lathyrism) in those with heavy and prolonged dependence on grasspea as a dietary staple. Other factors including malnutrition and physical exertion probably increase susceptibility to the neurotoxin. While lathyrism in the past century has affected poor people in Bangladesh (Rajshahi, Kushtia), Ethiopia (Dembia, Fogera), China (Gansu), central and northern India, and southern Europe (Greece, Spain), the disease has disappeared from Europe and is presently restricted to the Horn of Africa and Indian subcontinent (Giménez-Roldán and Spencer, 2016; Mishra et al., 2014; Woldemanuel et al., 2012). Historically, outbreaks of lathyrism in India have occurred in midyear following drought conditions in which dietary intake of grasspea has steadily increased as other food plants have died.

Lathyrism more frequently affects adults than children, males than females, and male adults usually develop more severe forms of spastic paraparesis. Cases form a pyramid in which the base comprises a majority who ambulates without walking aids, fewer use one, and even fewer two crutches to walk, and the apex is made up of a minority of individuals with complete spastic paraplegia. While the clinical course and neurophysiology of established lathyrism are well documented (Hugon et al., 1993; Ludolph et al., 1987), the neuropathology of the disease is poorly understood, and a complete animal model has yet to be generated (Reis and Spencer, 2015).

The culpable neurotoxic agent in grasspea is the nonprotein amino acid L-BOAA (Nunn et al., 2010), also known as L-β-*N*-oxalyl-α,β-diaminopropionic acid (β-ODAP) and dencichine. Ross et al. (1989) demonstrated L-BOAA to be a potent agonist of the murine α-amino-3-hydroxy-5-methyl-4-isoxazolepropionic acid (AMPA) neuronal receptor (Spencer, 1999), an observation confirmed by others (Krogsgaard-Larsen and Hansen, 1992; Kusama-Eguchi et al., 2014; van Moorhem et al., 2010, 2011). AMPA receptor activation by L-BOAA results in a prolonged rise of intracellular calcium levels (Kusama-Eguchi et al., 2014; van Moorhem et al., 2010, 2011); however, unlike AMPA, which utilizes L-type Ca^{2+} channels to passage calcium ions, L-BOAA-stimulated calcium entry uniquely involves transient receptor potential channels in association with the metabotropic mGluR 1 receptor (Kusama-Eguchi et al., 2014). Free intracellular calcium ions are able to activate calcium-dependent proteases able to destroy cellular integrity.

Continuous infusion of AMPA into the rat spinal cord induced anterior horn degeneration and rear limb paralysis by 7 days, a phenomenon attenuated by mitochondrial energy substrates (pyruvate and β-hydroxybutyrate) but not by antioxidants (ascorbate and glutathione ethyl ester) (Netzahualcoyotzi and Tapia, 2015). Incubation of mouse brain slices with subpicomolar concentrations of L-BOAA reportedly resulted in dose- and time-dependent inhibition of mitochondrial NADH-dehydrogenase (NADH-DH) followed by lactate dehydrogenase leakage, pathophysiological events in vitro that were blocked by an AMPA receptor antagonist and by glutathione (Pai and Ravinbdranath, 1993). However, Sabri et al. (1995) found that millimolar concentrations of L-BOAA failed to inhibit NADH-DH activity in mouse brain homogenate and isolated brain mitochondria.

Despite evidence that L-BOAA induces AMPA receptor-mediated motor neuronal degeneration in cell culture (Nunn et al., 1987; Ross et al., 1987), human lathyrism primarily involves degeneration of the projections of cortical motor neurons to the distal spinal cord. There is sparse evidence of Betz cell loss and no evidence of spinal anterior horn cell degeneration in cases of recent onset (Giménez-Roldán et al., 2017). Three Ethiopian patients with a history of grasspea ingestion and spastic paraparesis thought to be lathyrism showed normal cortical activation by functional magnetic resonance imaging and no detectable loss of descending motor tracts in the pons upon diffusion tensor imaging, suggesting a spinal cord location of pathology (Bick et al., 2016). While distal damage to corticospinal spinal tracts might result from repetitive sublethal L-BOAA attacks on cortical motor neurons, a phenomenon characterized as central distal axonopathy, this does not explain the totality of spinal cord pathology in lathyrism. Given that subcutaneous L-BOAA induces transient hemorrhage and lower vascular endothelial growth factor levels in the rat lumbosacral spinal cord (Kusama-Eguchi et al., 2010), the grasspea amino acid might also perturb small blood vessels and their blood–brain barrier function (Giménez-Roldán et al., 2017). Additionally, since β-ODAP also impacts astrocytes in culture, there is the possibility that astrocytes participate early in pathogenesis as well as forming a glial scar in areas of nerve fiber loss (Bridges et al., 1991; Miller et al., 1993). A recent toxicogenomics study showed that L-BOAA upregulates β-integrin, which is proposed to activate the focal adhesion pathway and participate in neuronal degeneration (Tan et al., 2017).

Lathyrus sativus continues to attract research attention because the plant tolerates extreme environmental conditions (drought, water-logging), fixes soil nitrogen, grows well without inputs, and the edible seed has a high protein content (Enneking, 2011; Smartt et al., 1994). The plant is therefore seen as a potentially viable source of food and feed for a rapidly expanding human population (especially in Africa) on a planet undergoing climate change (Kumar et al., 2011; Reis et al., 2017). Coordinated multidisciplinary efforts dating from the 1980s have sought to develop safe, low-toxin strains of grasspea for human consumption (Spencer, 1989). In India, native strains of grasspea have continued to be eaten in Maharashtra (Khandare et al., 2014; Singh and Rao, 2013), and a 55-year-old ban on the sale of the legume across India was recommended for removal in 2016 in part for economic reasons (Fernandes, 2016). Similar interest in the cultivation of grasspea has been expressed in China and the West Balkans (Bosnia, Herzegovina, Serbia) (Mikić et al., 2011; Yan et al., 2006). Stable low-BOAA strains of grasspea that have been tested for neurotoxic properties in a suitable animal species (the horse is most susceptible) are needed to prevent future outbreaks of lathyrism (Yan et al., 2006).

2.4 Cassava (*Manihot esculenta*)

The starchy tuberous root of cassava is used for food carbohydrates for an estimated half billion people (FAO, 2004), with production centers in South America, southern Asia, and Africa. The tropical plant was introduced to the African continent in the 16th century and has subsequently spread across sub-Saharan countries. Since the plant is resistant to drought, grows in marginal soils, and is a reliable and cheap source of energy, it is used both as a staple, famine-prevention food, and during other times of food insecurity (Chabwine et al., 2011; Cliff et al., 2011; Lim, 2016; Oluwole, 2015). Impoverished, poorly nourished people in Africa and India (Kerala) are at risk for chronic neurological disease because of their heavy dependence on bitter cassava root and leaves, both of which harbor two cyanogenic glucosides: linamarin and lotaustralin (in a ratio of 97:3) that are hydrolyzed by endogenous plant or gut enzyme linamarase to acetone cyanohydrin and cyanide (CN), which probably reacts with plasma cysteine to form the seizuregenic compound 2-imino-4-thiazolidine carboxylic acid (Nunn et al., 2010). Other cyanogenic metabolites (cyanate, thiocyanate) have received most attention in regard to the neurotoxic potential of a cassava-based staple diet.

While cyanogens occur in >2500 plant taxa, including 26 economically important crops (Abraham et al., 2016; Jones, 1998; Vetter, 2000), the level of human consumption of numerous other cyanogenic plants (e.g., sorghum grass, lima beans) presumably exposes the vast majority of individuals to subthreshold, harmless doses of these toxic elements. However, the results of a controlled cyanogen bioavailability crossover study suggest this explanation may be too simplistic. This was based on 12 healthy human adults who consumed apricot kernel paste (persipan) (68 mg/kg), linseed (220 mg/kg), bitter apricot kernels (3250 mg/kg), and fresh cassava roots (76–150 mg/kg). Mean levels of CN at different time points were highest after consumption of fresh cassava roots (15.4 µM, after 37.5 min) and bitter apricot kernels (14.3 µM, after 20 min), followed by linseed (5.7 µM, after 40 min) and 100 g persipan (1.3 µM, after 105 min) (Abraham et al., 2016). The apparently high bioavailability of CN from cassava might be a factor in its human toxic potential. CN is mostly metabolized by the mitochondrial liver enzyme thiosulfate sulfurtransferase (TST, rhodanese, EC 2.8.1.1) and by other sulfur transferases, including mercaptopyruvate sulfurtransferase [EC 2.8.1.2]. These enzymes catalyze the transfer of sulfur from a donor to CN to form the less acutely toxic thiocyanate molecule (SCN), which is readily excreted in urine. Several TST polymorphisms are known, some associated with diminished rhodanese activity but with little effect on CN detoxication (Billaut-Laden et al., 2006). Sulfur-containing donor molecules are the rate-limiting factor in the detoxication of CN (JECFA, 2012; WHO, 2004). Since cassava roots contain almost no protein, those with daily dietary dependence on this plant gradually become deficient in endogenous sulfur amino acids required to convert CN to SCN. Rodent studies show that KCN-treated animals on a diet deficient in sulfur amino acids can metabolize CN to cyanate (OCN) (Tor-Agbidye et al., 1999). While both SCN and OCN are less acutely toxic than CN, either or both may hold the keys to understanding the chronic neurotoxic effects of cassava dependency. Cyanate is an established primate and rodent motorsystem neurotoxin, but macaques and rats differ in their ability to detoxify CN (Kimani et al., 2014a). Carbamoylation of myelin-associated proteins may mediate cyanate neurotoxicity (Kimani et al., 2013). Thiocyanate increases the binding of glutamate to the AMPA receptor in vitro, which raises the possibility of a mechanistic relationship between the induction of upper motor neuron disease by grasspea (vide supra) and cassava, but there are no studies to evaluate whether chronic exposure to a high level of SCN has chronic neurotoxic potential in primates. In fact, there are no satisfactory models of the motorsystem disorder of cassavism in any species.

Understanding of the human neurotoxic response to dietary dependency on cassava is based on a number of epidemiological studies. Three distinct neurodegenerative disorders have been described: (a) a form of subacute onset spastic paraparesis in central, eastern, and southern Africa that predominantly affects children and women of childbearing age (Oluwole, 2015; Tshala-Katumbay et al., 2013); (b) a slowly developing ataxic myeloneuropathy, which affects one-quarter of subjects aged 60–69 years in Nigeria (male:female prevalence 1:3) (Oluwole and Oludiran, 2013; Oluwole et al., 2003); and (c) a poorly described cerebellar-parkinsonism–dementia syndrome among elderly Nigerians (Akinyemi, 2012). This pattern is suggestive of high, moderate, and low rates and durations of exposure to cassava cyanogens that impacts different components of the nervous system as a function of subject age. Thiamine deficiency has also been proposed as a unifying hypothesis for the first two cassava-associated disorders (Adamolekun, 2011), but overt vitamin deficiency or malabsorption syndrome was not present in cassava-associated ataxic neuropathy in Kerala, India (Madhusudanan et al., 2008). A recent study of CN- or CNO-treated rats found changes in working memory thought to be related to the cognition deficits found in *konzo* (Boivin et al., 2013; Kimani et al., 2014b), the local name in the Democratic Republic of the Congo for cassava-associated spastic paraparesis. Children with *konzo* had low levels of selenium, copper, and zinc relative to controls, but elemental deficiencies were unrelated to poor cognition (Bumoko et al., 2015). Cognitive deficits greater in females than males and increasing with age may reflect long-term effects of malnutrition and/or cyanogenic exposure (Bumoko et al., 2014). The characteristic *konzo*-associated elevation of urinary SCN, an established thyroid toxin that can induce hypothyroidism (Gaitan, 1989), was not associated with poor cognition (Bumoko et al., 2014). In areas endemic for goiter, the deterioration in thyroid function during and after weaning is linked to persistent iodine deficiency accompanied by an increase in SCN overload. This conclusion was reached in a study of 200 neonates and 347 children exposed to dietary goitrogenic factors (iodine deficiency and SCN overload). A high serum SCN concentration at birth ($129 \pm 5\,\mu mol/L$), presumably from a mother consuming cyanogenic plants, decreased during the suckling period to normal values between 3 and 12 months of age (suggesting mother's breast milk lacks cyanogens) and increased again during and after weaning (1 to 3 years of age) to reach a value of 138 μmol/L, a value comparable to that observed in adults in the same area (Vanderpas et al., 1984).

In summary, the irreversible neurological effects of food dependency on cassava represent a significant contribution to the global burden of disease (Gibb et al., 2015). Unofficial estimates of up to 100,000 cases of cassavism may have occurred across the African continent in year-2000 (Nzwalo and Cliff, 2011). Disease prevention requires improved nutrition, utilization of stable, low-toxin strains of cassava, and improved food processing

(Nambisan, 2011). Additionally, given the synergistic effects of SCN and iodine deficiency in relation to thyroid toxicity, the importance of thyroid function in brain development and maturation, renewed attention should be given to earlier controlled human and experimental animal (rat, pig) studies of this relationship (Bourdoux et al., 1978; Gaitan, 1989). Human studies showed that whereas severe iodine deficiency alone triggers adaptive mechanisms that maintain nearly normal thyroid hormone levels, the addition of cassava-derived SCN compromised thyroid adaptation and diminished capacity of thyroid secretion. Both in humans and rats, chronic SCN overload markedly depletes thyroid iodine stores, and thus the yield of thyroid hormonal synthesis, in spite of markedly increased uptake of radiolabeled iodine by the thyroid. Although chronic hypothyroidism can precipitate peripheral neuropathy, a possible relationship with the motor and cognitive features of the various forms of cassavism is virtually unexplored.

2.5 Cycad (*Cycas* spp.) and BMAA

The nonprotein amino acid L-α-amino-β-methylaminopropionic acid, renamed L-β-methylamino-L-alanine (L-BMAA) by Spencer et al. (1987a),was originally identified in cycad plants (*Cycas circinalis,* renamed *C. micronesica*) (Bell et al., 1967; Vega and Bell, 1967; Vega et al., 1968), and subsequently as a product of a diverse range of cyanobacteria, diatoms, and dinoflagellates in both aquatic and terrestrial ecosystems and in temporal and tropical climes globally (Cox et al., 2005; Jiang and Ilag, 2014; Jiang et al., 2014; Lage et al., 2014). Whereas, among Chamorro people on the island of Guam, the disappearing traditional use of washed cycad seed for food was historically associated with a high incidence of amyotrophic lateral sclerosis and parkinsonism–dementia (ALS–PDC) (Borenstein et al., 2007), L-BMAA exposure routes for humans globally are through direct contact with phytoplankton-infested waters, especially during algal blooms, and consumption of fish and invertebrate filter feeders in which the amino acid has accumulated. There is significant current interest in the possible role of L-BMAA exposure in human neurodegenerative disorders beyond Guam, including ALS and Alzheimer disease (AD). Interest was stimulated by demonstration that macaques given daily oral doses of L-BMAA developed a neuropathic motorsystem disorder reminiscent of ALS–PDC (Spencer et al., 1987b) and, more recently, induction of subclinical neurofibrillary pathology in the brains of vervets fed a diet containing L-BMAA (Cox et al., 2016a,b; Spencer et al., 2016a).

While the current high level of research interest in L-BMAA is based on the hypothesis the amino acid is responsible for western Pacific ALS–PDC, the assignment by Cox et al. (2017) that L-BMAA is causal of this prototypical neurodegenerative disease is premature. Cycads contain multiple nonprotein amino acids (Pan et al., 1997), virtually none of which has been tested toxicologically either alone in combination. Additionally, the major toxic factor in cycad is not L-BMAA but cycasin, the aglycone of which (methylazoxymethanol) is a potent genotoxin, carcinogen, and developmental neurotoxin that activates system biology pathways in rodent brain linked to human neurodegenerative disease (Kisby et al., 2011a,b). The complex relationship between cycad-associated L-BMAA, MAM, cancer, and ALS–PDC in Guam, Japan, and New Guinea is reviewed elsewhere (Spencer et al., 2012a, 2016b). In sum, while there are sound reasons to pursue the possibility of an etiological relationship between L-BMAA and human neurodegenerative disease, the hypothesis is unproven in our opinion and that of others (Chernoff et al., 2017).

Unfortunately, the large majority of experimental studies has failed to include a negative chemical control that would allow assessment of the specificity of measured responses to L-BMAA (or D-BMAA). The obvious control is L-BOAA (and D-BOAA) because of its well-established pharmacological and toxicological properties (vide supra), including a distinct primate response and strong relationship with a clinically and neuropathologically distinct human neurological disease (lathyrism). Among a number of biochemical mechanisms attributed to L-BMAA (Nunn and Ponnusamy, 2009), most experimental studies have addressed two: the excitotoxic properties of the amino acid and its proposed misincorporation into the backbone of brain protein (Spencer et al., 2012b). This body of research has been based on the L-enantiomer found in cycads and microorganisms although, surprisingly, D-BMAA was isolated from the cerebrospinal fluid of L-BMAA-treated vervets (Metcalf et al., 2017). Although this report found D-BMAA acts via AMPA receptors in vitro, D-BMAA was previously identified as a probable stereospecific modulator of N-methyl-D-aspartate (NMDA) receptor function by acting as an agonist at the strychnine-insensitive glycine modulatory site of the NMDA receptor (Allen et al., 1995). Transport experiments show that both L- and D-enantiomers are taken up by human-derived SH-SY5Y neuroblastoma cells (Andersson et al., 2017). In SY5Y cells, L-BMAA altered alanine, aspartate, and glutamate metabolism, as well as various neurotransmitters/neuromodulators, such as γ-aminobutyric acid and taurine (Engskog et al., 2017; Matsuoka et al., 1993). Notably, L-BMAA acts via glutamatergic ionotropic and metabotropic glutamate receptors, causes oxidative stress, inducing endoplasmic reticulum stress and accumulation of TAR DNA-binding protein 43 (TDP-43) fragments, increases activity of the proapoptotic enzyme caspase-3, and activates mechanisms that would promote tau pathology (Cucchiaroni et al., 2010; Goto et al., 2012; Lobner, 2009; Muñoz-Saez et al., 2013; Shen et al., 2016). Except for the latter, how these observations relate to the neurotoxic properties of L-BMAA in primates is unclear since neuronal pathology in both macaques and vervets bears none of the hallmarks of excitotoxic or apoptotic cell degeneration.

The acute and often reversible actions of L-BMAA on various pharmacological and metabolic mechanisms in cell systems and laboratory animals also do not address the long latent period of years or decades that intervenes between exposure to cycad materials and the appearance of progressive neurodegenerative disease in humans. Neurological disease of this type is quite distinct from the self-limiting disorder lathyrism, which is mediated by a non-protein excitotoxic amino acid (L-BOAA) with acute neurotoxic properties distinct from those of L-BMAA (Spencer et al., 1987b). More plausible theories to explain how L-BMAA might induce irreversibly progressive neurodegeneration might propose perturbations of the DNA or protein structure of neurons that activate a cascade of effects culminating in slowly evolving neuronal demise. L-BMAA substituted for alanine and serine in a cell-free protein synthesis system and for serine in human-derived cell proteins and mouse brain (Dunlop et al., 2013; Glover et al., 2014; Xie et al., 2013), but misincorporation of L-BMAA into proteins was not found in bacteria, PC12 cells, mussels, or macaque brain (Rosén et al., 2016; Spencer et al., 2016a; van Onselen et al., 2015, 2017). As with endogenous amino acids, L-BMAA may undergo N-nitrosation to form toxic alkylating agents that can cause DNA strand breaks in vitro and SH-SY5Y cell toxicity under conditions in which L-BMAA was minimally toxic (Potjewyd et al., 2017).

Intraperitoneal injection of L-BMAA into neonatal and young adult mice results in systemic distribution, with higher levels detected in the brains and livers of adult males than females (Al-Sammak et al., 2015; Karlsson et al., 2015b). L-BMAA is taken up by the hippocampus and striatum of neonatal rodents, which exhibit long-term learning and memory deficits, as well as regionally restricted changes in protein expression, neuronal degeneration and mineralization in the hippocampal CA1 region (Karlsson et al., 2009, 2012, 2015a). Whereas prolonged intrathecal infusion of L-BMAA induces degeneration of anterior horn cells (Yin et al., 2014), rats treated systemically with L-BMAA selectively develop cerebellar damage (Muñoz-Sáez et al.,

2015; Seawright et al., 1990), a pattern of neuropathology different from that seen in ALS and western Pacific ALS–PDC. Macaques dosed orally with L-BMAA develop a L-DOPA-sensitive neurological disorder characterized by corticomotoneuronal dysfunction, parkinsonian features, and behavioral anomalies, with chromatolytic and degenerative changes of motor neurons in cerebral cortex and spinal cord, and clusters of paired helical filaments (Spencer, 1987; Spencer et al., 1987b, 2016a). Vervets with chronic dietary exposure to L-BMAA remain clinically unremarkable but develop neurofibrillary tangles and sparse β-amyloid plaque-like deposits in the brain (Cox et al., 2016a,b). While further studies are needed to determine, in the absence of continued L-BMAA dosing, whether these disorders can show long-term progression akin to ALS–PDC, the vervet and macaque responses to chronic oral L-BMAA, respectively, resemble the subclinical and clinical expressions of cycad-associated western Pacific ALS–PDC, respectively.

Several studies have sought to assess geospatial relationships of high-incidence neurodegenerative disease and proximity to bodies of water rich in cyanobacteria potentially generating L-BMAA. Satellite remote sensing of thousands of lakes across northern New England contaminated with harmful algal blooms concluded proximity posed a 48% increase in average risk for ALS (Torbick et al., 2017). The incidence of sporadic ALS near Lake Mascoma in Enfield, New Hampshire, was found to be 10–25 times higher than expected, with L-BMAA detectable in fish samples and filtered aerosolized samples (Banack et al., 2015). Among a few case reports, three patients who developed sporadic ALS in Annapolis, Maryland, reported frequent consumption of blue crab found to contain L-BMAA (Field et al., 2013). There is also a high incidence of ALS proximate to the most important area of shellfish production and consumption along the French Mediterranean coast (Masseret et al., 2013; Réveillon et al., 2015). L-BMAA was also detected in the cerebrospinal fluid of Swedes living near the Baltic Sea, but there was no preference for ALS versus normal subjects (Berntzon et al., 2015). One group reported L-BMAA in postmortem brain tissue from Guam Chamorros with ALS–PDC, in Canadians who died of neurodegenerative disease, and in sporadic ALS and AD subjects in the continental United States (Murch et al., 2004; Pablo et al., 2009). However, other groups have failed to find evidence of free or protein-associated BMAA in autopsy samples of frontal or temporal cortex from Chamorros with PDC and continental Americans with AD (Montine et al., 2005; Snyder et al., 2009). Additionally, no trace of L-BMAA was detected in 20 subjects with pathologically confirmed AD or in 20 healthy controls (Meneely et al., 2016).

A recent critical review sought to assess evidence for the postulated role of ʟ-BMAA in human neurodegenerative diseases (Chernoff et al., 2017). The authors point out correctly that:

- ʟ-BMAA has an unproven causal relationship with western Pacific ALS–PDC;
- analytical challenges in ʟ-BMAA identification and quantitation cast doubt on results of some studies;
- doses of ʟ-BMAA used to create motorsystem disease in primates and rodents have been high relative to human intake via cycad-derived food, the major exposure route for ʟ-BMAA on Guam;
- conflicting data on the presence of ʟ-BMAA in human brain tissue in normal and disease states;
- weak evidence for the misincorporation of ʟ-BMAA in protein and neuroprotein.

These points notwithstanding, it is clear that western Pacific ALS–PDC is a slowly disappearing disorder in all three high-incidence disease foci (Guam and Rota; Kii Peninsula, Honshu Island, Japan; West Papua, Indonesia). This argues strongly against a hereditable etiology and for declining exposures to one or more environmental factors peculiar to the disease-affected populations. Declining use of neurotoxic cycad components for medicine and/or food in all three affected populations is clearly the most plausible candidate for the etiology of ALS–PDC (Spencer et al., 2015a). The chemical etiology is unknown but, among Chamorros of Guam, the flour content of cycasin, but not of the 10-fold lower concentrations of ʟ-BMAA, was very strongly correlated with the historical incidence of ALS–PDC among males and females (Kisby et al., 1992; Zhang et al., 1996).

2.6 Summary

There is strong evidence that plants used by humans for food include species with toxic potential resulting in acute (ackee, lychee) and chronic, self-limiting (grasspea, cassava) neurological disease. There is also evidence that other plant species (cycad) induce neuromuscular disease in animals, and there is ongoing research into the possibility that humans respond to cycad and soursop toxins with progressive neurodegenerative disease. The neurotoxic properties of naturally occurring chemicals present in plants used for food is a subject of immense importance for human health, especially among those who must rely on such plants as a major component of their nutrition. Research on this subject has provided important insights into the possible causes of neurological disorders.

REFERENCES

Abraham, K., Buhrke, T., Lampen, A., 2016. Bioavailability of cyanide after consumption of a single meal of foods containing high levels of cyanogenic glycosides: a crossover study in humans. Arch. Toxicol. 90, 559–574.

Adamolekun, B., 2011. Neurological disorders associated with cassava diet: a review of putative etiological mechanisms. Metab. Brain Dis. 26, 79–85.

Akinyemi, R.O., 2012. Epidemiology of parkinsonism and Parkinson's disease in sub-Sahara Africa: Nigerian profile. J. Neurosci. Rural Pract. 3, 233–234.

Allen, C.N., Omelchenko, I., Ross, S.M., Spencer, P., 1995. The neurotoxin, β-N-methylamino-L-alanine (BMAA) interacts with the strychnine-insensitive glycine modulatory site of the N-methyl-D-aspartate receptor. Neuropharmacology 34, 651–658.

Al-Sammak, M.A., Rogers, D.G., Hoagland, K.D., 2015. Acute β-N-methylamino-L-alanine toxicity in a mouse model. J. Toxicol. 2015, 739746. http://dx.doi.org/10.1155/2015/739746.

Andersson, M., Ersson, L., Brandt, I., Bergström, U., 2017. Potential transfer of neurotoxic amino acid β–N-methylamino-alanine (BMAA) from mother to infant during breast-feeding: predictions from human cell lines, Toxicol. Appl. Pharmacol. 320, 40–50. http://dx.doi.org/10.1016/j.taap.2017.02.004. Epub 2017 Feb 4.

AVIS, 2010. de l'Agence française de sécurité sanitaire des ailments relative aux risques liés à la consummation de corossol et de ses. Agence Française de Sécurité Sanitairte des Aliments, Maison-Alfor, 28 April.

Banack, S.A., Caller, T., Henegan, P., Haney, J., Murby, A., Metcalf, J.S., Powell, J., Cox, P.A., Stommel, E., 2015. Detection of cyanotoxins, β-N-methylamino-L-alanine and microcystins, from a lake surrounded by cases of amyotrophic lateral sclerosis. Toxins (Basel) 7, 322–336.

Bandyopadhyay, B., Chakraborty, D., Ghosh, S., Mishra, R., Rahman, M., Bhattacharya, N., Alam, S., Mandal, A., Das, A., Mishra, A., Mishra, A.K., Kumar, A., Haldar, S., Pathak, T., Mahapatra, N., Mondal, D.K., Maji, D., Basu, N., 2014. Epidemiological investigation of an outbreak of acute encephalitis syndrome (AES) in Malda District of West Bengal, India, Clin. Microbiol. 4, 1000181.

Barrett, M.D.P., 2016. A Review: Ackee Poisoning in Jamaica. Jamaica Observer. https://www.jamaicaobserver.com/NEWS/A-review–Ackee-poisoning-in-Jamaica_60767 (accessed June 2017).

Bell, E.A., Vega, A., Nunn, P.B., 1967. A new amino acid from seeds of *Cycas circinalis*. In: Whiting, M.G. (Ed.), Fifth Conference on Cycad Toxicity. Third World Medical Research Foundation, New York and London, pp. XI-1–XI-4.

Berntzon, L., Ronnevi, L.O., Bergman, B., Eriksson, J., 2015. Detection of BMAA in the human central nervous system. Neuroscience 292, 137–147.

Betarbet, R., Sherer, T.B., MacKenzie, G., Garcia-Osuna, M., Panov, A.V., Greenamyre, J.T., 2000. Chronic systemic pesticide exposure reproduces features of Parkinson's disease. Nat. Neurosci. 3, 1301–1306.

Bick, A.S., Meiner, Z., Gotkine, M., Levin, N., 2016. Using advanced imaging methods to study neurolathyrism. Isr. Med. Assoc. J. 18, 341–345.

Billaut-Laden, I., Allorge, D., Crunelle-Thibaut, A., Rat, E., Cauffiez, C., Chevalier, D., Houdret, N., Lo-Guidice, J.M., Broly, F., 2006. Evidence for a functional genetic polymorphism of the human thiosulfate sulfurtransferase (Rhodanese), a cyanide and H$_2$S detoxification enzyme. Toxicology 225, 1–11.

Biswas, S.K., 2012. Outbreak of illness and deaths among children living near lychee orchards in northern Bangladesh. ICDDRB Health Sci. 10, 15–22.

Blake, O.A., Jackson, J.C., Jackson, M.A., Gordon, L.A., 2004. Assessment of dietary exposure to the natural toxin hypoglycin in ackee (*Blighia sapida*) by Jamaican consumers. Food Res. Int. 37, 833–838.

Blake, O.A., Bennink, M.R., Jackson, J.C., 2006. Ackee (*Blighia sapida*) hypoglycin A toxicity: dose response assessment in laboratory rats. Food Chem. Toxicol. 44, 207–213.

Boivin, M.J., Okitundu, D., Makila-Mabe Bumoko, G., Sombo, M.T., Mumba, D., Tylleskar, T., Page, C.F., Tamfum Muyembe, J.J., Tshala-Katumbay, D., 2013. Neuropsychological effects of konzo: a neuromotor disease associated with poorly processed cassava. Pediatrics 131, e1231–1239.

Borenstein, A.R., Mortimer, J.A., Schofield, E., Wu, Y., Salmon, D.P., Gamst, A., Olichney, J., Thal, L.J., Silbert, L., Kaye, J., Craig, U.L., Schellenberg, G.D., Galasko, D.R., 2007. Cycad exposure and risk of dementia, MCI, and PDC in the Chamorro population of Guam. Neurology 68, 1764–1771.

Bourdoux, P., Delange, F., Gerard, M., Mafuta, M., Hanson, A., Ermans, A.M., 1978. Evidence that cassava ingestion increases thiocyanate formation: a possible etiologic factor in endemic goiter. J. Clin. Endocrinol. Metab. 46, 613–621.

Bowen-Forbes, C.S., Minott, D.A., 2011. Tracking hypoglycin A and B over different maturity stages: implications for detoxification of ackee (*Blighia sapida*, K.D. Koenig) fruits. J. Agric. Food Chem. 59, 3869–3875.

Bridges, R.J., Hatalski, C., Shim, S.N., Nunn, P.B., 1991. Gliotoxic properties of the *Lathyrus* excitotoxin β-N-oxalyl-L-a,β-diaminopropionic acid (β-L-ODAP). Brain Res. 561, 262–268.

Brown, M., Bates, R.P., McGowan, C., Cornell, J.A., 1991. Influence of fruit maturity on the hypoglycin A level in ackee (*Blighia sapida*). J. Food Saf. 12, 167–177.

Bumoko, G.M., Sombo, M.T., Okitundu, L.D., Mumba, D.N., Kazadi, K.T., Tamfum-Muyembe, J.J., Lasarev, M.R., Boivin, M.J., Banea, J.P., Tshala-Katumbay, D.D., 2014. Determinants of cognitive performance in children relying on cyanogenic cassava as staple food. Metab. Brain Dis. 29, 359–366.

Bumoko, G.M., Sadiki, N.H., Rwatambuga, A., Kayembe, K.P., Okitundu, D.L., Mumba Ngoyi, D., Muyembe, J.J., Banea, J.P., Boivin, M.J., Tshala-Katumbay, D., 2015. Lower serum levels of selenium, copper, and zinc are related to neuromotor impairments in children with konzo. J. Neurol. Sci. 349, 149–153.

Caparros-Lefebvre, D., Elbaz, A., 1999. Possible relation of atypical parkinsonism in the French West Indies with consumption of tropical plants: a case-control study. Caribbean Parkinsonism Study Group. Lancet 354, 281–286.

Carlier, J., Guitton, J., Moreau, C., Boyer, B., Bévalot, F., Fanton, L., Habyarimana, J., Gault, G., Gaillard, Y., 2015. A validated method for quantifying hypoglycin A in whole blood by UHPLC-HRMS/MS. J. Chromatogr. B Analyt. Technol. Biomed. Life Sci. 978–979: 70–77.

Chabwine, J.N., Masheka, C., Balol'ebwami, Z., Maheshe, B., Balegamire, S., Rutega, B., Wa Lola, M., Mutendela, K., Bonnet, M.J., Shangalume, O., Balegamire, J.M., Nemery, B., 2011. Appearance of konzo in South-Kivu, a wartorn area in the Democratic Republic of Congo. Food Chem. Toxicol. 49, 644–649.

Champy, P., Melot, A., Guérineau Eng, V., Gleye, C., Fall, D., Höglinger, G.U., Ruberg, M., Lannuzel, A., Laprévote, O., Laurens, A., Hocquemiller, R., 2005. Quantification of acetogenins in *Annona muricata* linked to atypical parkinsonism in Guadeloupe. Mov. Disord. 20, 1629–1633.

Chase Jr., G.W., Landen Jr., W.O., Soliman, A.G., 1990. Hypoglycin A content in the aril, seeds, and husks of ackee fruit at various stages of ripeness. J. Assoc. Off. Anal. Chem. 73, 318–319.

Chernoff, N., Hill, D.J., Diggs, D.L., Faison, B.D., Francis, B.M., Lang, J.R., Larue, M.M., Le, T.T., Loftin, K.A., Lugo, J.N., Schmid, J.E., Winnik, W.M., 2017. A critical review of the postulated role of the non-essential amino acid, β-N-methylamino-L-alanine, in neurodegenerative disease in humans. J. Toxicol. Environ. Health B Crit. Rev. 20 (4), 1–47.

Cliff, J., Muquingue, H., Nhassico, D., Nzwalo, H., Bradbury, J.H., 2011. Konzo and continuing cyanide intoxication from cassava in Mozambique. Food Chem. Toxicol. 49, 631–635.

Coria-Tellez, A.V., Montalvo-Gonzalez, E., Yahi, E.M., Obledo-Vázquez, E.N., 2017. *Annona muricata:* a comprehensive review on its traditional medicinal uses, phytochemicals, pharmacological activities, mechanisms of action and toxicity. Arab. J. Chem. in press. http://dx.doi.org/10.1016/j.arabjc.2016.01.004 (accessed June 2016).

Cox, P.A., Banack, S.A., Murch, S.J., Rasmussen, U., Tien, G., Bidigare, R.R., Metcalf, J.S., Morrison, L.F., Codd, G.A., Bergman, B., 2005. Diverse taxa of cyanobacteria produce β-N-methylamino-L-alanine, a neurotoxic amino acid. Proc. Natl. Acad. Sci. U.S.A. 102, 5074–5078.

Cox, P.A., Davis, D.A., Mash, D.C., Metcalf, J.S., Banack, S.A., 2016a. Dietary exposure to an environmental toxin triggers neurofibrillary tangles and amyloid deposits in the brain. Proc. Biol. Sci. 283, 1823.

Cox, P.A., Davis, D.A., Mash, D.C., Metcalf, J.S., Banack, S.A., 2016b. Do vervets and macaques respond differently to BMAA? Neurotoxicology 57, 310–311.

Cox, P.A., Kostrzewa, R.M., Guillemin, G.J., 2017. BMAA and neurodegenerative illness. Neurotox. Res. http://dx.doi.org/10.1007/s12640-017-9753-6. [Epub ahead of print].

Cucchiaroni, M.L., Viscomi, M.T., Berebardi, G., Molinari, M., Guatteo, E., Mercuri, N.B., 2010. Metabotropic glutamate receptor 1 mediates the electrophysiological and toxic actions of the cycad derivative β-N-methylamino-L-alanine on substantia nigra pars compacta DAergic neurons. J. Neurosci. 30, 5176–5188.

Das, M., Asthana, S., Singh, S.P., Dixit, S., Tripathi, A., John, T.J., 2015. Litchi fruit contains methylene cyclopropyl-glycine. Curr. Sci. 109, 2195–2197.

Dinesh, D.S., Pandey, K., Das, V.N.R., Topno, R.K., Kesari, S., Kumar, V., Ranjan, A., Sinha, P.K., Das, P., 2013. Possible factors causing acute encephalitis syndrome outbreak in Bihar, India, Int. J. Curr. Microbiol. App. Sci. 2, 531–538.

Dossou, V.M., 2014. *Physicochemical and Functional Properties of Different Ackee* (Blighia sapida) *Aril Flours.* MSc thesis, Department of Food Science and Technology, College of Science, Kwame Nkrumah University of Science and Technology, 86 pages.

Dunlop, R.A., Cox, P.A., Banack, S.A., Rodgers, K.J., 2013. The non-protein amino acid BMAA is misincorporated into human proteins in place of L-serine causing protein misfolding and aggregation. PLoS One 8, e75376.

Ekué, M.R.M., 2011. *Blighia sapida, Ackee: Conservation and Sustainable Uses of Genetic Resources of Priority Food Tree Species in Sub-Sahara Africa.* Bioversity International, Rome, Italy.

Engskog, M.K., Ersson, L., Haglöf, J., Arvidsson, T., Pettersson, C., Brittebo, E., 2017. β-N-Methylamino-L-alanine (BMAA) perturbs alanine, aspartate and glutamate metabolism pathways in human neuroblastoma cells as determined by metabolic profiling. Amino Acids 49, 905–919.

Enneking, D., 2011. The nutritive value of grasspea (*Lathyrus sativus*) and allied species, their toxicity to animals and the role of malnutrition in neurolathyrism. Food Chem. Toxicol. 49, 694–709.

Escobar-Khondiker, M., Höllerhage, M., Muriel, M.P., Champy, P., Bach, A., Depienne, C., Respondek, G., Yamada, E.S., Lannuzel, A., Yagi, T., Hirsch, E.C., Oertel, W.H., Jacob, R., Michel, P.P., Ruberg, M., Höglinger, G.U., 2007. Annonacin, a natural mitochondrial complex I inhibitor, causes tau pathology in cultured neurons. J. Neurosci. 27, 7827–7837.

FAO, Food and Agriculture Organization, 2004. *Global Cassava Market Study: Business Opportunities for the Use of Cassava*, sixth ed. Rome, Italy: FAO.

Fernandes, V. 2016 As Prices Pulses Soar, Can *Khesari Dal* be an Alternative? https://www.thequint.com/india/2016/01/09/as-prices-of-pulses-soar-can-khesari-dal-be-an-alternative (accessed June 2017).

Field, N.C., Metcalf, J.S., Caller, T.A., Banack, S.A., Cox, P.A., Stommel, E.W., 2013. Linking β-methylamino-L-alanine exposure to sporadic amyotrophic lateral sclerosis in Annapolis, MD. Toxicon 70, 179–183.

Figeroa, P., Nembhard, O., 1992. Toxic hypoglycemic syndrome—Jamaica 1989–1991. Morb. Mortal. Wkly. Rep. 41, 53–55.

Gaillard, Y., Carlier, J., Berscht, M., Mazoyer, C., Bevalot, F., Guitton, J., Fanton, L., 2011. Fatal intoxication due to ackee (Blighia sapida) in Suriname and French Guyana. GC–MS detection and quantification of hypoglycin-A. Forensic Sci. Int. 206, 103–e107.

Gaitan, E., 1989. Environmental Goitrogenesis. CRC Press. Boca Raton, Fl.

Gibb, H., Devleesschauwer, B., Bolger, P.M., Wu, F., Ezendam, J., Cliff, J., Zeilmaker, M., Verger, P., Pitt, J., Baines, J., Adegoke, G., Afshari, R., Liu, Y., Bokkers, B., van Loveren, H., Mengelers, M., Brandon, E., Havelaar, A.H., Bellinger, D., 2015. World Health Organization estimates of the global and regional disease burden of four foodborne chemical toxins, 2010: a data synthesis, F1000Res 4, 1393.

Giménez-Roldán, S., Spencer, P.S., 2016. Azañón's disease. A 19th century epidemic of neurolathyrism in Spain. Rev. Neurol. (Paris). 172, 748–755.

Giménez-Roldán S., Morales-Asin F., Ferrer C.P., Spencer P.S., 2017. New insights in the neuropathology of lathyrism: a 1944 report by Oliveras de la Riva, submitted, 2017.

Glover, W.B., Mash, D.C., Murch, S.J., 2014. The natural non-protein amino acid N-β-methylamino-L-alanine (BMAA) is incorporated into protein during synthesis. Amino Acids 46, 2553–2559.

Golden, K.D., Kean, E., Terry, S.I., 1984. Jamaican vomiting sickness: a study of two adult cases. Clin. Chim. Acta 142, 293–298.

Goldson, A., 2005. The ackee fruit (Blighia sapida) and its associated toxic effects. Sci. Creat. Quart. https://www.scq.ubc.ca/the-ackee-fruit-blighia-sapida-and-its-associated-toxic-effects/(accessed June 2017).

Goto, J.J., Koenig, J.H., Ikeda, K., 2012. The physiological effect of ingested β-N-methylamino-L-alanine on a glutamatergic synapse in an in vivo preparation. Comp. Biochem. Physiol. C Toxicol. Pharmacol. 156, 171–177.

Gray, D.O., Fowden, L., 1962. α-(Methylenecyclopropyl)glycine from Litchi seeds. Biochem. J. 82, 385–389.

Hassal, C.H., Reyle, K., 1955. Hypoglycin A and B: biologically active polypeptides from Blighia sapida. Biochem. J. 60, 334–338.

Hassal, C.H., Reyle, K., Feng, P., 1954. Hypoglycin A, B, biologically active polypeptides from Blighia sapida. Nature 173, 356–357.

Henry, S.H., Page, S.W., Bolger, P.M., 1998. Hazard assessment of ackee fruit (Blighia sapida). Hum. Ecol. Risk Assess. 4, 1175–1187.

Hill, K.R., 1952. The vomiting sickness of Jamaica: a review. West Indian Med. J. 1, 243–264.

Höllerhage, M., Matusch, A., Champy, P., Lombès, A., Ruberg, M., Oertel, W.H., Höglinger, G.U., 2009. Natural lipophilic inhibitors of mitochondrial complex I are candidate toxins for sporadic neurodegenerative tau pathologies. Exp. Neurol. 220, 133–142.

Holson, D.A., 2015. Ackee Fruit Toxicity Medication. Medscape. https://emedicine.medscape.com/article/1008792-medication (accessed June 2017).

Hugon, J., Ludolph, A.C., Spencer, P.S., Gimenez Roldan, S., Dumas, J.L., 1993. Studies on the etiology and pathogenesis of motor neuron diseases. III. Magnetic cortical stimulation in patients with lathyrism. Acta Neurol. Scand. 88, 412–416.

Isenberg, S.L., Carter, M.D., Graham, L.A., Mathews, T.P., Johnson, D., Thomas, J.D., Pirkle, J.L., Johnson, R.C., 2015. Quantification of metabolites for assessing human exposure of soapberry toxins hypoglycin A and methylenecyclopropylglycine. Chem. Res. Toxicol. 28, 1753–1759.

Isenberg, S.L., Carter, M.D., Hayes, S.R., Graham, L.A., Johnson, D., Mathews, T.P., Harden, L.A., Takeoka, G.R., Thomas, J.D., Pirkle, J.L., Johnson, R.C., 2016. Quantification of toxins in soapberry (Sapindaceae) arils: hypoglycin A and methylenecyclopropylglycine. J. Agric. Food Chem. 64, 5607–5613.

Islam, S., 2012. Outbreak of illness and deaths among children living near lychee orchards in northern Bangladesh. Health Sci. Bull. 10, 15–21.

Islam, M.S., Sharif, A.R., Sazzad, H.M.S., Khan, A.K.M.D., Hasan, M., Akter, S., Rahman, M., Luby, S.P., Heffelfinger, J.D., Gurley, E.S., 2017. Outbreak of sudden death with acute encephalitis syndrome among children associated with exposure to lychee orchards in Northern Bangladesh, 2012. Am. J. Trop. Med. Hyg. http://dx.doi.org/10.4269/ajtmh.16-0856.

JECFA, Joint FAO/WHO Expert Committee on Food Additives, 2012. Cyanogenic glycosides. In: WHO Food Additives Series No. 65, Safety Evaluation of Certain Food Additives and Contaminants. WHO, Rome, pp. 171–323. https://whqlibdoc.who.int/publications/2012/9789241660655_eng.pdf(accessed June 2017).

Jiang, L., Ilag, L.L., 2014. Detection of endogenous BMAA in dinoflagellate (Heterocapsa triquetra) hints at evolutionary conservation and environmental concern. PubRaw Sci. 2, 1–8.

Jiang, L., Eriksson, J., Lage, S., Jonasson, S., Shams, S., Mehine, M., Ilag, L.L., Rasmussen, U., 2014. Diatoms: a novel source for the neurotoxin BMAA in aquatic environments. PLoS One 9, e84578.

John, T.J., Das, M., 2014. Acute encephalitis syndrome in children in Muzaffarpur: hypothesis of toxic origin. Curr. Sci. 106, 1184–1185.

Jones, D.A., 1998. Why are so many food plants cyanogenic? Phytochemistry 47, 155–162.

Joskow, R., Belson, M., Vesper, H., Backer, L., Rubin, C., 2006. Ackee fruit poisoning: an outbreak investigation in Haiti 2000–2001, and review of the literature. J. Clin. Toxicol. 44, 267–273.

Karlsson, O., Roman, E., Brittebo, E.B., 2009. Long-term cognitive impairments in adult rats treated neonatally with beta-N-methylamino-L-alanine. Toxicol. Sci. 112, 185–195.

Karlsson, O., Berg, A.L., Lindström, A.K., Hanrieder, J., Arnerup, G., Roman, E., Bergquist, J., Lindquist, N.G., Brittebo, E.B., Andersson, M., 2012. Neonatal exposure to the cyanobacterial toxin BMAA induces changes in protein expression and neurodegeneration in adult hippocampus. Toxicol. Sci. 130, 391–404.

Karlsson, O., Berg, A.L., Hanrieder, J., Arnerup, G., Lindström, A.K., Brittebo, E.B., 2015a. Intracellular fibril formation, calcification, and enrichment of chaperones, cytoskeletal, and intermediate filament proteins in the adult hippocampus CA1 following neonatal exposure to the nonprotein amino acid BMAA. Arch. Toxicol. 89, 423–436.

Karlsson, O., Jiang, L., Ersson, L., Malmstrom, T., Ilag, L.L., Brittebo, E.B., 2015b. Environmental neurotoxin interaction with proteins: dose-dependent increase of free and protein-associated BMAA (β-N-methylamino-L-alanine) in neonatal rat brain. Sci. Rep. 5, 15570. http://dx.doi.org/10.1038/srep15570.

Katibi, O.S., Olaosebikan, R., Abdulkadir, M.B., Ogunkunle, T.O., Ibraheem, R.M., Murtala, R., 2015. Ackee fruit poisoning in eight siblings: implications for public health awareness. Am. J. Trop. Med. Hyg. 93, 1122–1123.

Kedari, T.S., Khan, A.A., 2014. Guyabano (Annona muricata): a review of its traditional uses, phytochemistry and pharmacology. Am. J. Res. Commun. 210.

Khandare, A.L., Babu, J.J., Ankulu, M., Aparna, N., Shirfule, A., Rao, G.S., 2014. Grass pea consumption and present scenario of neurolathyrism in Maharashtra state of India. Indian J. Med. Res. 140, 96–101.

Kimani, S., Moterroso, V., Lasarev, M., Kipruto, S., Bukachi, F., Maitai, C., David, L., Tshala-Katumbay, D., 2013. Carbamoylation correlates of cyanate neuropathy and cyanide poisoning: relevance to the biomarkers of cassava cyanogenesis and motor system toxicity. Springerplus 2, 647.

Kimani, S., Moterroso, V., Morales, P., Wagner, J., Kipruto, S., Bukachi, F., Maitai, C., Tshala-Katumbay, D., 2014a. Cross-species and tissue variations in cyanide detoxification rates in rodents and non-human primates on protein-restricted diet. Food Chem. Toxicol. 66, 203–209.

Kimani, S., Sinei, K., Bukachi, F., Tshala-Katumbay, D., Maitai, C., 2014b. Memory deficits associated with sublethal cyanide poisoning relative to cyanate toxicity in rodents. Metab. Brain Dis. 29, 105–112.

Kisby, G.E., Ellison, M., Spencer, P.S., 1992. Content of the neurotoxins cycasin (methylazoxymethanol beta-D-glucoside) and BMAA (β-N-methylamino-L-alanine) in cycad flour prepared by Guam Chamorros. Neurology 42, 1336–1340.

Kisby, G.E., Fry, R.C., Lasarev, M.R., Bammler, T.K., Beyer, R.P., Churchwell, M., Doerge, D.R., Meira, L.B., Palmer, V.S., Ramos-Crawford, A.L., Ren, X., Sullivan, R.C., Kavanagh, T.J., Samson, L.D., Zarbl, H., Spencer, P.S., 2011a. The cycad genotoxin MAM modulates brain cellular pathways involved in neurodegenerative disease in a DNA-dependent manner. PLoS One 6, e20911.

Kisby, G., Palmer, V., Lasarev, M., Fry, R., Iordanov, M., Magun, E., Samson, L., Spencer, P., 2011b. Does the cycad genotoxin MAM implicated in Guam ALS-PDC induce disease-relevant changes in mouse brain that includes olfaction? Commun. Integr. Biol. 4, 731–734.

Krogsgaard-Larsen, P., Hansen, J.J., 1992. Naturally-occurring excitatory amino acids as neurotoxins and leads in drug design. Toxicol. Lett. 64, 409–416.

Kumar, S., Bejiga, G., Ahmed, S., Nakkoul, H., Sarker, A., 2011. Genetic improvement of grass pea for low neurotoxin (β-ODAP) content. Food Chem. Toxicol. 49, 589–600.

Kusama-Eguchi, K., Yamazaki, Y., Ueda, T., Suda, A., Hirayama, Y., Ikegami, F., Watanabe, K., May, M., Lambein, F., Kusama, T., 2010. Hind-limb paraparesis in a rat model for neurolathyrism associated with apoptosis and an impaired vascular endothelial growth factor system in the spinal cord. J. Comp. Neurol. 518, 928–942.

Kusama-Eguchi, K., Miyano, T., Yamamoto, M., Suda, A., Ito, Y., Ishige, K., Ishii, M., Ogawa, Y., Watanabe, K., Ikegami, F., Kusama, T., 2014. New insights into the mechanism of neurolathyrism: L-β-ODAP triggers [Ca2+]$_i$ accumulation and cell death in primary motor neurons through transient receptor potential channels and metabotropic glutamate receptors. Food Chem. Toxicol. 67, 113–122.

Lage, S., Costa, P.R., Moita, T., Eriksson, J., Rasmussen, U., Rydberg, S.J., 2014. BMAA in shellfish from two Portuguese transitional water bodies suggests the marine dinoflagellate *Gymnodinium catenatum* as a potential BMAA source. Aquat. Toxicol. 152, 131–138.

Langston, J.W., Irwin, I., Langston, E.B., Forno, L.S., 1984. 1-Methyl-4-phenylpyridinium ion (MPP+): identification of a metabolite of MPTP, a toxin selective to the substantia nigra. Neurosci. Lett. 48, 87–92.

Lannuzel, A., Michel, P.P., Caparros-Lefebvre, D., Abaul, J., Hocquemiller, R., Ruberg, M., 2002. Toxicity of Annonaceae for dopaminergic neurons: potential role in atypical parkinsonism in Guadeloupe. Mov. Disord. 17, 84–90.

Lannuzel, A., Michel, P.P., Höglinger, G.U., Champy, P., Jousset, A., Medja, F., Lombès, A., Darios, F., Gleye, C., Laurens, A., Hocquemiller, R., Hirsch, E.C., Ruberg, M., 2003. The mitochondrial complex I inhibitor annonacin is toxic to mesencephalic dopaminergic neurons by impairment of energy metabolism. Neuroscience 121, 287–296.

Lannuzel, A., Höglinger, G.U., Champy, P., Michel, P.P., Hirsch, E.C., Ruberg, M., 2006. Is atypical parkinsonism in the Caribbean caused by the consumption of Annonacae? J. Neural Transm. 70 (Suppl), 153–157.

Lannuzel, A., Höglinger, G.U., Verhaeghe, S., Gire, L., Belson, S., Escobar-Khondiker, M., Poullain, P., Oertel, W.H., Hirsch, E.C., Dubois, B., Riberg, M., 2007. Atypical parkinsonism in Guadeloupe: a common risk factor for two closely related phenotypes? Brain 130 (Pt. 3), 816–827.

Lannuzel, A., Ruberg, M., Michel, P.P., 2008. Atypical parkinsonism in the Caribbean island of Guadeloupe: etiological role of the mitochondrial complex I inhibitor annonacin. Mov. Disord. 23, 2122–2128.

Le Ven, J., Schmitz-Afonso, I., Lewin, G., Brunelle, A., Touboul, D., Champy, P., 2014. Identification of the environmental neurotoxins annonaceous acetogenins in an *Annona cherimolia* Mill. alcoholic beverage using HPLC-ESI-LTQ-Orbitrap. J. Agric. Food Chem. 62, 8696–8704.

Li, D., Agnihotri, G., Dakoji, S., Oh, E., Lantz, M., Liu, H.W., 1999. The toxicity of methylenecyclopropylglycine: studies of the inhibitory effects of (methylenecyclopropyl) formyl-CoA on enzymes involved in fatty acid metabolism and the molecular basis of its inactivation of enoyl-CoA hydratases. J. Am. Chem. Soc. 121, 9034–9042.

Lim, T.K., 2016. *Manihot esculenta*. In: Lim, T.K. (Ed.), *Edible Medicinal and Non Medicinal Plants 2016*, vol. 10: *Modified Stems, Roots, Bulbs*, Springer Verlag, Berlin, pp. 308–353.

Lobner, D., 2009. Mechanisms of β-N-methylamino-L-alanine induced neurotoxicity. Amyotroph. Lateral Scler. 10 (Suppl. 2), 56–60.

Ludolph, A.C., Hugon, J., Dwivedi, M.P., Schaumburg, H.H., Spencer, P.S., 1987. Studies on the aetiology and pathogenesis of motor neuron diseases. 1. Lathyrism: clinical findings in established cases. Brain 110 (Pt. 1), 149–165.

Madhusudanan, M., Mennon, A.K., Ummer, K., Radhakrishnanan, K., 2008. Clinical and etiological profile of tropical ataxic neuropathy in Kerala, South India. Eur. Neurol. 60, 21–26.

Masseret, E., Banack, S., Boumédiène, F., Abadie, E., Brient, L., Pernet, F., Juntas-Morales, R., Pageot, N., Metcalf, J., Cox, P., Camu, W., French Network on ALS Clusters Detection and Investigation, 2013. Dietary BMAA exposure in an amyotrophic lateral sclerosis cluster from southern France. PLoS One 8, e83406.

Matsuoka, Y., Rakonczay, Z., Giacobini, E., Naritoku, D., 1993. L-β-Methylamino-alanine-induced behavioral changes in rats. Pharmacol. Biochem. Behav. 44, 727–734.

Meda, H.A., Lison, D., Bucet, J.-P., Lison, D., Barennes, H., Ouangre, A., Sanou, M., Cousens, S., Tall, F., van de Perre, P., 1999. Epidemic of fatal encephalopathy in preschool children in Burkina Faso and consumption of unripe ackee (*Blighia sapida*) fruit. Lancet 353, 536–540.

Melde, K., Buettner, H., Boschert, W., Wolf, H.P.O., Ghisla, S., 1989. Mechanism of hypoglycemic action by methylenecyclopropylglycine. Biochem. J. 259, 921–924.

Melde, K., Jackson, S., Bartlett, H.S., Sherratt, A., Ghisla, S., 1991. Metabolic consequences of methylenecyclopropylglycine poisoning in rats. Biochem. J. 274, 395–400.

Meneely, J.P., Chevallier, O.P., Graham, S., Greer, B., Green, B.D., Elliott, C.T., 2016. β-Methylamino-L-alanine (BMAA) is not found in the brains of patients with confirmed Alzheimer's disease. Sci. Rep. 6, 36363. http://dx.doi.org/10.1038/srep36363.

Metcalf J.S., Lobner D., Banack S.A., Cox G.A., Nunn P.B., Wyatt P.B., Cox P.A., Analysis of BMAA enantiomers in cycads, cyanobacteria, and mammals: in vivo formation and toxicity of D-BMAA, *Amino Acids* 49, 1427–1439. http://dx.doi.org/10.1007/s00726-017-2445-y. Epub 2017 Jun 15.

Mikić, A., Mihailović, V., Ćupina, B., Durić, B., Krstić, D., Vasić, M., Vasiljević, S., Karagić, D., Dorđević, V., 2011. Towards the re-introduction of grass pea (*Lathyrus sativus*) in the West Balkan countries: the case of Serbia and Srpska (Bosnia and Herzegovina). Food Chem. Toxicol. 49, 650–654.

Miller, S., Nunn, P.B., Bridges, R.J., 1993. Induction of astrocyte glutamine synthetase activity by the *Lathyrus sativus* toxin β-N-oxalyl-L-a,β-diaminopropionic acid (β-L-ODAP). Glia 7, 329–336.

Mills, J., Melville, G.N., West, M., Castro, A., 1987. Effect of hypoglycin A on insulin release. Biochem. Pharmacol. 36, 495–497.

Mishra, V.N., Tripathi, C.B., Kumar, A., Nandmer, V., Ansari, A.Z., Kumar, B., Chaurasia, R.N., Joshi, D., 2014. Lathyrism: has the scenario changed in 2013? Neurol. Res. 36, 38–40.

Moghadamtousi, S.Z., Fadaeinasab, M., Nikzad, S., Mohan, G., Ali, H.M., Kadir, H.A., 2015. *Annona muricata* (Annonaceae): A review of Its traditional uses, Isolated cceto-genins and biological activities. Int. J. Mol. Sci. 16, 15625–15658.

Mohamed, S., Hamad, M.H., Hassan, H.H., Salih, M.A., 2015. Glutaric aciduria type 1 as a cause of dystonic cerebral palsy. Saudi Med. J. 36, 1354–1357.

Montine, T.J., Li, K., Perl, D.P., Galasko, D., 2005. Lack of β-methylamino-l-alanine in brain from controls, AD, or Chamorros with PDC. Neurology 65, 768–769.

Morton, J.F., 1987. *Fruits of Warm Climates*, J.F. Morton, Miami, Fl., pp 269-271. https://www.hort.purdue.edu/newcrop/morton/akee.html, https://www.hort.purdue.edu/newcrop/morton/akee.html, https://www.rt.purdue.edu/newcrop/morton/akee.html (accessed June 2017).

Moyal, P., 2014. *Blighia sapida*, the Ackee tree, still a sword of Damocles over children in Africa. HAL ID: <hal-01083624> https://horizon.documentation.ird.fr/exl-doc/pleins_textes/divers15-01/010063317.pdf (accessed June 2017).

Muñoz-Saez, E., de Munck, E., Arahuetes, R.M., Solas, M.T., Martínez, A.M., Miguel, B.G., 2013. β-*N*-Methylamino-L-alanine induces changes in both GSK3 and TDP-43 in human neuroblastoma. J. Toxicol. Sci. 38, 425–430.

Muñoz-Sáez, E., de Munck García, E., Arahuetes Portero, R.M., Martínez, A., Solas Alados, M.T., Miguel, B.G., 2015. Analysis of β-N-methylamino-L-alanine (L-BMAA) neuro-toxicity in rat cerebellum. Neurotoxicology 48, 192–205.

Murch, S.J., Cox, P.A., Banack, S.A., Steele, J.C., Sacks, O.W., 2004. Occurrence of β-methylamino-l-alanine (BMAA) in ALS/PDC patients from Guam. Acta Neurol. Scand. 110, 267–269.

Nambisan, B., 2011. Strategies for elimination of cyanogens from cassava for reducing tox-icity and improving food safety. Food Chem. Toxicol. 49, 690–693.

Netzahualcoyotzi, C., Tapia, R., 2015. Degeneration of spinal motor neurons by chronic AMPA-induced excitotoxicity in vivo and protection by energy substrates, Acta Neu-ropathol. Commun. 3, 27.

Nunn, P.B., Ponnusamy, M., 2009. β-*N*-Methylaminoalanine (BMAA): metabolism and metabolic effects in model systems and in neural and other tissues of the rat in vitro. Toxicon 54, 85–94.

Nunn, P.B., Seelig, M., Zagoren, J.C., Spencer, P.S., 1987. Stereospecific acute neuronotoxicity of 'uncommon' plant amino acids linked to human motor-system dis-eases. Brain Res. 410, 375–379.

Nunn, P.B., Bell, A., Watson, A.A., Nash, R.J., 2010. Toxicity of non-protein amino acids in humans and domestic animals. Nat. Prod. Commun. 5, 485–504.

Nzwalo, H., Cliff, J., 2011. Konzo: from poverty, cassava, and cyanogen intake to toxico-nutritional neurological disease. PLos Negl. Trop. Dis. 5, e1051.

Oloyede, O.B., Ajibove, T.O., Abdussalam, A.F., Adeleye, A.O., 2014. *Blighia sapida* leaves halt elevated blood glucose, dyslipidemia and oxidative stress in alloxan-induced diabetic rats. J. Ethnopharmacol. 157, 309–319.

Oluwole, O.S., 2015. Cyclical konzo epidemics and climate variability. Ann. Neurol. 77, 371–380.

Oluwole, O.S., Oludiran, A., 2013. Geospatial association of endemicity of ataxic poly-neuropathy and highly cyanogenic cassava cultivars. Int. J. Health Geogr. 12, 41.

Oluwole, O.S., Onabolu, A.O., Catgreave, I.A., Rosling, H., Persson, A., Link, H., 2003. Incidence of endemic ataxic polyneuropathy and its relation to exposure to cyanide in a Nigerian community. J. Neurol. Neurosurg. Psychiatry 74, 1417–1422.

Osmundsen, H., Billington, D., Taylor, J.R., Sherratt, H.A.S., 1978. The effects of hypo-glycin on glucose metabolism in the rat. Biochem. J. 170, 337–342.

Pablo, J., Banack, S.A., Cox, S.A., Johnson, T.E., Papapetropoulos, S., Bradley, W.G., Buck, A., Mash, D.C., 2009. Cyanobacterial neurotoxin BMAA in ALS and Alzheimer's disease. Acta Neurol. Scand. 120, 216–225.

Pai, K.S., Ravinbdranath, V., 1993. L-BOAA induces selective inhibition of brain mitochondrial enzyme, NADH-dehydrogenase. Brain Res. 621, 215–221.

Paireau, J., Tuan, N.H., Lefrançois, R., Buckwalter, M.R., Nghia, N.D., Hien, N.T., 2012. Litchi-associated acute encephalitis in children, northern Vietnam, 2004–2009. Emerg. Infect. Dis. 18, 1817–1824.

Pan, M., Mabry, T.J., Cao, P., Moini, M., 1997. Identification of nonprotein amino acids from cycad seeds as N-ethoxycarbonyl ethyl ester derivatives by positive chemical-ionization gas chromatography-mass spectrometry. J. Chromatogr. A 787, 288–294.

Pigeon, N., Campeau, P.M., Cyr, D., Lemieux, B., Clarke, J.T., 2009. Clinical heterogeneity in ethylmalonic encephalopathy. J. Child Neurol. 24, 991–996.

Potjewyd, G., Day, P.J., Shangula, S., Margison, G.P., Povey, A.C., 2017. L-β-N-Methylamino-l-alanine (BMAA) nitrosation generates a cytotoxic DNA damaging alkylating agent: an unexplored mechanism for neurodegenerative disease. Neurotoxicology 59, 105–109.

Potts, L.F., Luzzio, F.A., Smith, S.C., Hetman, M., Champy, P., Litvan, I., 2012. Annonacin in *Asimina triloba* fruit: implication for neurotoxicity. Neurotoxicology 33, 53–58.

Quere, M., Ogouassangni, A., Bokossa, A., Perra, A., Van Damme, W., 1999. Methylene blue and fatal encephalopathy from ackee fruit poisoning. Lancet 353, 1623.

Reis, J., Spencer, P.S., 2015. Lathyrism. In: Chopra, J., Sawhney, M.S., 2nd ed. *Neurology in the Tropics*. Elsevier, pp. 923–932.

Reis, J., Handschumacher, P., Palmer, V.S., Spencer, P.S., 2017. Climatic factors under the tropics. Chapter 3. in: Preux, P.-M., Dumas, M. (Eds.), *Neuroepidemiology in Tropical Health*, Elsevier, New York.

Reisman, N., 2015. Hypoglycemic plant potential. Medscape. https://emedicine.medscape.com/article/817325-overview (accessed June 2017).

Réveillon, D., Abadie, E., Séchet, V., Brient, L., Savar, V., Bardouil, M., Hess, P., Amzil, Z., 2015. Beta-N-methylamino-L-alanine: LC-MS/MS optimization, screening of cyanobacterial strains and occurrence in shellfish from Thau, a French Mediterranean lagoon. Mar. Drugs 12, 5441–5467.

Rocchiccioli, F., Aubourg, P., Bougnères, P.F., 1986. Medium- and long-chain dicarboxylic aciduria in patients with Zellweger syndrome and neonatal adrenoleukodystrophy. Pediatr. Res. 20, 62–66.

Rosén, J., Westerberg, E., Schmiedt, S., Hellenäs, K.E., 2016. BMAA detected as neither free nor protein bound amino acid in blue mussels. Toxicon 109, 45–50.

Ross, S.M., Seelig, M., Spencer, P.S., 1987. Specific antagonism of excitotoxic action of "uncommon" amino acids assayed in organotypic mouse cortical cultures. Brain Res. 425, 120–127.

Ross, S.M., Roy, D.N., Spencer, P.S., 1989. β-N-Oxalylamino-L-alanine action on glutamate receptors. J. Neurochem. 53, 710–715.

Rottscholl, R., Haegele, M., Jainsch, B., Xu, H., Respondek, G., Höllerhage, M., Rösler, T.W., Bony, E., Le Ven, J., Guérineau, V., Schmitz-Afonso, I., Champy, P., Oertel, W.H., Yamada, E.S., Höglinger, G.U., 2016. Chronic consumption of *Annona muricata* juice triggers and aggravates cerebral tau phosphorylation in wild-type and MAPT transgenic mice. J. Neurochem. 139, 624–639.

Sabri, M.I., Lystrup, B., Roy, D.N., Spencer, P.S., 1995. Action of β-N-oxalylamino-L-alanine on mouse brain NADH-dehydrogenase activity. J. Neurochem. 65, 1842–1848.

Sander, J., Terhardt, M., Sander, S., Janzen, N., 2017. Quantification of methylenecyclopropyl compounds and acyl conjugates by UPLC-MS/MS in the study of the biochemical effects of the ingestion of canned ackee (*Blighia sapida*) and lychee (*Litchi chinensis*). J. Agric. Food Chem. 65, 2603–2608.

Scott, H.H., 1916. On the "vomiting sickness of Jamaica" Ann. Trop. Med. Parasitol. 10, 1–63.

Seawright, A.A., Brown, A.W., Nolan, C.C., Cavanagh, J.B., 1990. Selective degeneration of cerebellar cortical neurons caused by cycad neurotoxin, L-β-methylaminoalanine (L-BMAA), in rats. Neuropathol. Appl. Neurobiol. 16, 153–169.

Shah, A., John, T.J., 2014. Recurrent outbreaks of hypoglycemic encephalopathy in Muzaffarpur, Bihar. Curr. Sci. 107, 570–571.

Shen, H., Kim, K., Oh, Y., Yoon, K.S., Baik, H.H., Kim, S.S., Ha, J., Kang, I., Choe, W., 2016. Neurotoxin β-N-methylamino-L-alanine induces endoplasmic reticulum stress-mediated neuronal apoptosis. Mol. Med. Rep. 14, 4873–4880.

Sherratt, H.S., Turnbull, D.M., 1999. Methylene blue and fatal encephalopathy from ackee fruit poisoning. Lancet 353, 1623–1624.

Shrivastava, A., Srikantiah, P., Kumar, A., Bhushan, G., Goel, K., Kumar, S., 2015. Outbreaks of unexplained neurologic illness—Muzaffarpur, India, 2013–2014. Morb. Mortal. Wkly. Rep. 64, 49–53.

Shrivastava, A., Kumar, A., Thomas, J.D., Laserson, K.F., Bhushan, G., Carter, M.D., Chhabra, M., Mittal, V., Khare, S., Sejvar, J.J., Dwivedi, M., Isenberg, S.L., Johnson, R., Pirkle, J.L., Sharer, J.D., Hall, P.L., Yadav, R., Velayudhan, A., Papanna, M., Singh, P., Somashekar, D., Pradhan, A., Goel, K., Pandey, R., Kumar, M., Kumar, S., Chakrabarti, A., Sivaperumal, P., Kumar, A.R., Schier, J.G., Chang, A., Graham, L.A., Mathews, T.P., Johnson, D., Valentin, L., Caldwell, K.L., Jarrett, J.M., Harden, L.A., Takeoka, G.R., Tong, S., Queen, K., Paden, C., Whitney, A., Haberling, D.L., Singh, R., Singh, R.S., Earhart, K.C., Dhariwal, A.C., Chauhan, L.S., Venkatesh, S., Srikantiah, P., 2017. Association of acute toxic encephalopathy with litchi consumption in an outbreak in Muzaffarpur, India, 2014: a case-control study. Lancet Glob. Health 5, e458–e466.

Singh, S.S., Rao, S.L., 2013. Lessons from neurolathyrism: a disease of the past and the future of Lathyrus sativus (Khesari dal). Indian J. Med. Res. 138, 32–37.

Smartt, J., Kaul, A., Araya, W.-A., Rahman, M.M., Kearnety, J., 1994. Grasspea (Lathyrus sativus L.) as a potentially safe legume food crop. In: Expanding the Production and Use of Cool Season Food Legumes. vol. 19 in Current Plant Science and Biotechnology in Agriculture, Springer, pp. 144–155. https://link.springer.com/chapter/10.1007/978-94-011-0798-3_7 (accessed June 2017).

Snyder, L.R., Cruz-Aguado, R., Sadilek, M., Galasko, D., Shaw, C.A., Mointine, T.L., 2009. Lack of cerebral BMAA in human cerebral cortex. Neurology 72, 1360–1361.

Spencer, P.S., 1987. Guam ALS/parkinsonism-dementia: a long-latency neurotoxic disorder caused by "slow toxin(s)" in food? The Grass Pea: Threat and Promise. Can. J. Neurol. Sci. 14 (Suppl. 3), 347–357.

Spencer, P.S., 1989. The Grass Pea: threat and promise. London. Third World Medical Research Foundation, New York, 244 p.

Spencer, P.S., 1999. Food toxins, AMPA receptors, and motor neuron diseases. Drug Metab. Rev. 31, 561–587.

Spencer, P.S., Palmer, V.S., 2017. The enigma of litchi toxicity: an emerging health concern in southern Asia. Lancet Glob. Health 5, e383–e384.

Spencer, P.S., Hugon, J., Ludolph, A., Nunn, P.B., Ross, S.M., Roy, D.N., Schaumburg, H.H., 1987a. Discovery and partial characterization of primate motor-system toxins. In: Bock, G., O'Connor, M. (Eds.), Selective Neuronal Death. John Wiley & Sons, Chichester, pp. 221–238.

Spencer, P.S., Nunn, P.B., Hugon, J., Ludolph, A.C., Ross, S.M., Roy, D.N., Robertson, R.C., 1987b. Guam amyotrophic lateral sclerosis-parkinsonism-dementia linked to a plant excitant neurotoxin. Science 237, 517–522.

Spencer, P., Fry, R.C., Kisby, G.E., 2012a. Unraveling 50-year-old clues linking neurodegeneration and cancer to cycad toxins: are microRNAs common mediators? Front. Genet. 3, 192.

Spencer, P.S., Fry, R.C., Palmer, V., Kisby, G.E., 2012b. Western Pacific ALS-PDC: a prototypical neurodegenerative disorder linked to DNA damage and aberrant proteogenesis? Front. Neurol. 3, 180.

Spencer, P.S., Gardner, E., Palmer, V.S., Kisby, G.E., 2015a. Environmental neurotoxins linked to a prototypical neurodegenerative disease. In: Aschner, M., Costa, L. (Eds.), Environmental Factors in Neurodevelopment and Neurodegenerative Disorders. Elsevier, New York, pp. 212–237.

Spencer, P.S., Palmer, V.S., Mazumder, R., 2015b. Probable toxic cause for suspected lychee-linked viral encephalitis. Emerg. Infect. Dis. 21, 904–905.

Spencer, P.S., Gardner, C.E., Palmer, V.S., Kisby, G.E., 2016a. Vervets and macaques: similarities and differences in their responses to L-BMAA. Neurotoxicology 56, 284–286.

Spencer, P.S., Palmer, V.S., Kisby, G.E., 2016b. Seeking environmental causes of neurodegenerative disease and envisioning primary prevention. Neurotoxicology 56, 269–283.

Tan, R.-Y., Xing, G.-Y., Zhou, G.-M., Li, F.-M., Hu, W.-T., Lambein, F., Xiong, J.-L., Zhang, S.-X., Kong, H.-Y., Li, Z.-X., Xiong, Y.-C., 2017. Plant toxin β-ODAP activates integrin β1 and focal adhesion: a critical pathway to cause neurolathyrism. Sci. Rep. 7, 40677.

Tanaka, K., 1972. On the mode of action of hypoglycin A. III. Isolation and identification of cis-4-decene-1,10-dioic, cis, cis-4,7-decadiene-1,10-dioic, cis-4-octene-1,8-dioic, glutaric, and adipic acids, N-(methylenecyclopropyl)acetylglycine, and N-isovalerylglycine from urine of hypoglycin A-treated rats. J. Biol. Chem. 247, 7465–7478.

Tanaka, K., Ikeda, Y., 1990. Hypoglycin and Jamaica vomiting sickness. Prog. Clin. Biol. Res, 327, 167–184.

Tanaka, K., Ramsdell, H.S., Baretz, B.H., Keefe, M.B., Kean, E.A., Johnson, B., 1976. Identification of ethylmalonic acid in urine of two patients with the vomiting sickness of Jamaica. Clin. Chim. Acta 69, 105–112.

Tor-Agbidye, J., Sabri, M.I., Spencer, P.S., Palmer, V.S., Lasarev, M.R., Craig, M.A., Blythe, L.L., 1999. Bioactivation of cyanide to cyanate in sulfur amino acid deficiency: relevance to neurological diseases in humans subsisting on cassava. Toxicol. Sci. 50, 228–235.

Torbick, N., Ziniti, B., Stommel, E., Linder, E., Andrew, A., Caller, T., Haney, J., Bradley, W., Henegan, P.L., Shi, X., 2017. Assessing cyanobacterial harmful algal blooms as risk factors for amyotrophic lateral sclerosis. Neurotox. Res. http://dx.doi.org/10.1007/s12640-017-9740-y.

Tshala-Katumbay, D., Mumba, N., Okitundu, L., Kazadi, K., Banea, M., Tylleskär, T., Boivin, M., Muyembe-Tamfum, J.J., 2013. Cassava food toxins, konzo disease, and neurodegeneration in sub-Sahara Africans. Neurology 80, (10), 949–951.

U.S. Centers for Disease Control and Prevention, 1992. Toxic hypoglycemic syndrome—Jamaica, 1989–1991. Morb. Mortal. Wkly. Rep. 41, 53–55.

U.S. Food and Drug Administration, 2014. FDA Issues Final Guidance on Ackee Products. CFSAN Constituent Update. https://www.fda.gov/Food/NewsEvents/Constituent Updates/ucm393073.htm (accessed June 2017).

Vanderpas, J., Bourdoux, P., Lagasse, R., Rivera, M., Dramaix, M., Lody, D., Nelson, G., Delange, F., Ermans, A.M., Thilly, C.H., 1984. Endemic infantile hypothyroidism in a severe endemic goitre area of central Africa. Clin. Endocrinol. (Oxf) 20, 327–340.

van Moorhem, M., Decrock, E., Coussee, E., Faes, L., De Vuyst, E., Vranckx, K., De Bock, M., Wang, N., D'Herde, K., Lambein, F., Callewaert, G., Leybaert, L., 2010. L-β-ODAP alters mitochondrial Ca2+ handling as an early event in excitotoxicity. Cell Calcium 47, 287–296.

van Moorhem, M., Decrock, E., De Vuyst, E., De Bock, M., Wang, N., Lambein, F., van Den Bosch, L., Leybaert, L., 2011. L-β-N-oxalyl-α,β-diaminopropionic acid toxicity in motor neurons. Neuroreport 22, 131–135.

van Onselen, R., Cook, N.A., Phelan, R.R., Downing, T.G., 2015. Bacteria do not incorporate β-N-methylamino-L-alanine into their proteins. Toxicon 102, 55–61, 2015.

van Onselen, R., Venables, L., van de Venter, M., Downing, T.G., 2017. β-N-Methylamino-L-alanine toxicity in PC12: excitotoxicity vs. misincorporation. Neurotox. Res. http://dx.doi.org/10.1007/s12640-017-9743-9748.

Vega, A., Bell, E.A., 1967. α-Amino-β-methylaminopropionic acid, a new amino acid from seeds of Cycas circinalis. Phytochemistry 6, 759–762.

Vetter, J., 2000. Plant cyanogenic glycosides. Toxicon 38, 11–36.

Vega, A., Bell, E.A., Nunn, P.B., 1968. The preparation of L- and D-a-amino-b-methylaminopropionic acids and the identification of the compound isolated from Cycascircinalis as the L-isomer. Phytochemistry 7, 1885–1887.

von Holt, C., Chang, J., von Holt, M., Bohm, J., 1964. Metabolism and metabolic effects of hypoglycin. Biochim. Biophys. Acta 90, 611–613.

von Holt, C., von Holt, M., Bohm, H., 1966. Metabolic effects of hypoglycin and methylenecyclopropaneacetic acid. Biochim. Biophys. Acta 125, 11–21.

WHO, World Health Organization, 2004. Hydrogen Cyanide and Cyanides: Human Health Aspects. Concise International Chemical Assessment Document 61. WHO, Geneva. https://www.who.int/ipcs/publications/cicad/en/cicad61.pdf (accessed June 2017).

Woldemanuel, Y.H., Hassan, A., Zenenbve, G., 2012. Neurolathyrism: two Ethiopian case reports and review of the literature. J. Neurol. 259, 1263–1268.

Xie, X., Basile, M., Mash, D.C., 2013. Cerebral uptake and protein incorporation of cyanobacterial toxin β-N-methylamino-L-alanine. Neuroreport 24, 779–784.

Yamada, E.S., Respondek, G., Müssner, S., de Andrade, A., Höllerhage, M., Depienne, C., Rastetter, A., Tarze, A., Friguet, B., Salama, M., Champy, P., Oertel, W.H., Höglinger, G.U., 2014. Annonacin, a natural lipophilic mitochondrial complex I inhibitor, increases phosphorylation of tau in the brain of FTDP-17 transgenic mice. Exp. Neurol. 253, 113–125.

Yan, Z.Y., Spencer, P.S., Li, Z.X., Liang, Y.M., Wang, Y.F., Wang, C.Y., Li, F.M., 2006. Lathyrus sativus (grass pea) and its neurotoxin ODAP. Phytochemistry 67, 107–121.

Yin, H.Z., Yu, S., Hsu, C.I., Liu, J., Acab, A., Wu, R., Tao, A., Chiang, B.J., Weiss, J.H., 2014. Intrathecal infusion of BMAA induces selective motor neuron damage and astrogliosis in the ventral horn of the spinal cord. Exp. Neurol. 2261, 1–9.

Yuan, S.S., Chang, H.L., Chen, H.W., Yeh, Y.T., Kao, Y.H., Kao, Y.H., Lin, K.H., Wu, Y.C., Su, J.H., 2003. Annonacin, a mono-tetrahydrofuran acetogenin, arrests cancer cells at the G1 phase and causes cytotoxicity in a Bax- and caspase-3-related pathway. Life Sci. 72, 2853–2861.

Zhang, Z.X., Anderson, D.W., Mantel, N., Román, G.C., 1996. Motor neuron disease on Guam: geographic and familial occurrence, 1956–85. Acta Neurol. Scand. 94, 51–59.